Optimal Control for Chemical Engineers

Optimal Control for Chemical Engineers

Simant Ranjan Upreti

CRC Press
Taylor & Francis Group
Boca Raton London New York

CRC Press is an imprint of the
Taylor & Francis Group, an **informa** business

CRC Press
Taylor & Francis Group
6000 Broken Sound Parkway NW, Suite 300
Boca Raton, FL 33487-2742

First issued in paperback 2017

Version Date: 2012912

ISBN 13: 978-1-4398-3894-5 (hbk)
ISBN 13: 978-1-138-07483-5 (pbk)

Library of Congress Cataloging-in-Publication Data

Upreti, Simant Ranjan.
    Optimal control for chemical engineers / Simant Ranjan Upreti.
        p. cm.
    Includes bibliographical references and index.
    ISBN 978-1-4398-3894-5 (hardback)
    1. Chemical process control. 2. Mathematical optimization. I. Title.

TP155.75.U67 2012
519.6--dc23   2012030747

Visit the Taylor & Francis Web site at
http://www.taylorandfrancis.com

and the CRC Press Web site at
http://www.crcpress.com

This book is dedicated to

my parents,
Drs. Hema and Murari Lal Upreti,

Grand Master Choa Kok Sui,
Shaykh Khwaja Shamsuddin Azeemi,

and
Lord Gaṇapati

Contents

Preface

It is my pleasure to present this book on optimal control geared toward chemical engineers. The application of optimal control is a logical step when it comes to pushing the envelopes of unit operations and processes undergoing changes with time and space.

This book is essentially a summary of important concepts I have learned in the last 16 years from the classroom, self-study, and research along with interaction with some great individuals including teachers, authors, peers, and students. The goal of this book is to provide a sufficiently detailed treatment of optimal control that will enable readers to formulate optimal control problems and solve them. With this emphasis, the book provides necessary mathematical analyses and derivations of important results. It is assumed that the reader is at the level of a graduate student.

Chapter 1 stimulates interest in optimal control by describing various processes and introducing the mathematical description of optimal control problems. Against this backdrop, readers are introduced to the basic concepts of optimal control in Chapter 2. The notion of optimality is presented and analyzed in Chapter 3. The ubiquitous Lagrange multipliers are introduced in this chapter. They are elaborated later in Chapter 4 along with important theorems and rules of application. Chapter 5 presents the celebrated Pontryagin's principle of optimal control. With this background, Chapter 6 puts together different types of optimal control problems and the necessary conditions for optimality. Chapter 7 describes important numerical methods and computational algorithms in a lucid manner to solve a wide range of optimal control problems. Chapter 8 introduces the optimal control of processes that are periodic and provides relevant numerical methods and algorithms. A brief review of mathematical concepts is provided in Chapter 9. Chapter-end bibliographies contain the cited references as well as important sources on which I have relied.

For an introductory one-semester course, instructors can consider Chapters 1–3, the main results from Chapters 4 and 5, and Chapters 6 and 7. An advanced course may include all chapters with obviously less time devoted to the first three. Chapters 7 and 8 may form a part of an advanced optimization course.

Containing all relevant mathematical results and their derivations, the book encourages self-study. During initial readings, some readers might want to skip a derivation, accept the result temporarily, and focus more on the applications. A working knowledge of computer programming is highly recommended to solve optimal control problems — whether one intends to write one's own programs or use software and programs developed by others.

Optimal control is the result of tremendous contributions of wonderful mathematicians, scientists, and engineers. To list their achievements is a formidable task. What I have presented in this book is what I could understand and have first-hand experience with. I hope the savants will help me in improving this book and the students will find the book useful.

I am profoundly grateful to Dr. Anil Mehrotra, Dr. Ayodeji Jeje, and Dr. Robert Heidemann for their assiduous mentoring and training during my doctorate and postdoctoral fellowship at the University of Calgary. I am thankful to my graduate students, especially Amir Sani, Hameed Muhamad, Dinesh Kumar Patel, and Vishalkumar Patel for helping with the proofreading. I acknowledge Allison Shatkin, Karen Simon, and Marsha Pronin at CRC Press who have offered superb assistance in the writing of this book.

I am very appreciative of the outstanding contributions of the developers of TEX, LATEX, MiKTEX, Xfig, Asymptote, AUCTEX, and GNU Emacs — the primary applications I have used to prepare the book.

Finally, my wife Deepa provided unstinting support and encouragement to follow through with this project. My children Jahnavi and Pranav were very patient with me all along. To the three of them I am deeply indebted.

Toronto   *Simant R. Upreti*

Notation

Vectors

We will use lower case bold face letters for vectors. For example,

$$\mathbf{y} \equiv \begin{bmatrix} y_1 \\ y_2 \\ \vdots \\ y_n \end{bmatrix}$$

is a column vector. It has n components. The transpose of $\mathbf{y}$ is

$$\mathbf{y}^\top \equiv \begin{bmatrix} y_1 & y_2 & \cdots & y_n \end{bmatrix}$$

where $\mathbf{y}^\top$ is a row vector. Also, $\mathbf{y}(t)$ means that each component of $\mathbf{y}$ is time dependent.

Function Vectors

A function vector is a vector of functions. For example, the n-component function vectors $\mathbf{f}(t)$ and $\mathbf{g}(\mathbf{y}, \mathbf{u})$ are, respectively,

$$\begin{bmatrix} f_1(t) \\ f_2(t) \\ \vdots \\ f_n(t) \end{bmatrix} \quad \text{and} \quad \begin{bmatrix} g_1(\mathbf{y}, \mathbf{u}) \\ g_2(\mathbf{y}, \mathbf{u}) \\ \vdots \\ g_n(\mathbf{y}, \mathbf{u}) \end{bmatrix} \quad \text{or} \quad \begin{bmatrix} g_1(y_1, y_2, \ldots, y_n, u_1, u_2, \ldots, u_m) \\ g_2(y_1, y_2, \ldots, y_n, u_1, u_2, \ldots, u_m) \\ \vdots \\ g_n(y_1, y_2, \ldots, y_n, u_1, u_2, \ldots, u_m) \end{bmatrix}$$

where $\mathbf{u}$ is an m-component vector. The function arguments can be functions themselves, as in $\mathbf{g}[\mathbf{y}(t), \mathbf{u}(t)]$.

Matrices

We will most often use upper case bold face for matrices, e.g.,

$$
\mathbf{A} \equiv
\begin{bmatrix}
a_{11} & a_{12} & \cdots & a_{1n} \\
a_{21} & a_{22} & \cdots & a_{2n} \\
\vdots & \vdots & \cdots & \vdots \\
a_{m1} & a_{m2} & \cdots & a_{mn}
\end{bmatrix}
$$

which is an $m \times n$ matrix. The matrix components can be functions.

Derivatives

We will use an over dot $\dot{}$ to denote the derivative with respect to time t. A prime $'$ will denote the derivative with respect to an independent variable other than time such as x according to the context. Thus,

$$
\dot{y} \equiv \frac{\mathrm{d}y}{\mathrm{d}t}, \quad y' \equiv \frac{\mathrm{d}y}{\mathrm{d}x}, \quad \text{and} \quad y'' \equiv \frac{\mathrm{d}y'}{\mathrm{d}x} \equiv \frac{\mathrm{d}^2 y'}{\mathrm{d}x^2}
$$

Partial Derivatives

Often, we will use the subscript notation for partial derivatives. For example,

$$
H_u \equiv \frac{\partial H}{\partial u} \quad \text{and} \quad H_{uy} \equiv \frac{\partial H_y}{\partial u} = \frac{\partial}{\partial u}\left(\frac{\partial H}{\partial y}\right)
$$

Derivatives Involving Vectors

The derivative of a scalar f with respect to vector $\mathbf{y}$ is a vector $f_{\mathbf{y}}$ made of components that are partial derivatives. Thus,

$$
f_{\mathbf{y}} \equiv
\begin{bmatrix}
\dfrac{\partial f}{\partial y_1} & \dfrac{\partial f}{\partial y_2} & \cdots & \dfrac{\partial f}{\partial y_n}
\end{bmatrix}^{\mathsf{T}}
$$

The derivative of a vector $\mathbf{f}$ with respect to a scalar, say, t, is again a vector of partial derivatives. For example,

$$\mathbf{f}_t \equiv \left[\frac{\partial f_1}{\partial t} \quad \frac{\partial f_2}{\partial t} \quad \cdots \quad \frac{\partial f_n}{\partial t} \right]^\top$$

The derivative of a vector $\mathbf{f}$ with respect to another vector $\mathbf{u}$ is a Jacobian matrix of partial derivatives. Thus,

$$\mathbf{f}_\mathbf{u} \equiv \begin{bmatrix} \dfrac{\partial f_1}{\partial u_1} & \dfrac{\partial f_1}{\partial u_2} & \cdots & \dfrac{\partial f_1}{\partial u_m} \\[2ex] \dfrac{\partial f_2}{\partial u_1} & \dfrac{\partial f_2}{\partial u_2} & \cdots & \dfrac{\partial f_2}{\partial u_m} \\[2ex] \vdots & \vdots & \cdots & \vdots \\[2ex] \dfrac{\partial f_n}{\partial u_1} & \dfrac{\partial f_n}{\partial u_2} & \cdots & \dfrac{\partial f_n}{\partial u_m} \end{bmatrix}$$

States, Costates, and Controls

In the optimal control problems, we will most often use
- $\mathbf{y}$ to denote the vector of state variables
- $\mathbf{u}$ to denote the vector of controls
- $\boldsymbol{\lambda}$ to denote the vector of costate variables

The above vectors will depend upon an independent variable, which will usually be time.

Miscellaneous Symbols

A few miscellaneous symbols are as follows.

H	Hamiltonian
J, L, M	augmented functionals
t	time
t_f	final time
σ	transformed time in the range 0–1
τ	time period

Chapter 1

Introduction

This chapter introduces optimal control with the help of several examples taken from chemical engineering applications. The examples elucidate the use of control functions to achieve what is desired in those applications. The mathematical underpinnings illustrate the formulation of optimal control problems. The examples help build up the notion of objective functionals to be optimized using control functions.

1.1   Definition

An **optimal control** is a function that optimizes the performance of a system changing with time, space, or any other independent variable. That function is a relation between a selected system input or property and an independent variable. The appellation "control" signifies the use of a function to control the state of the system and obtain some desired performance. As a subject, **optimal control** is the embodiment of principles that characterize optimal controls, and help determine them in what we call optimal control problems.

Consider a well-mixed batch reactor, shown in Figure 1.1, with chemical species A and B reacting to form a product C. The reactivities are dependent on the reactor temperature, T, which can be changed with time, t. At

Figure 1.1 A batch reactor operating with temperature as a function of time over a certain time duration

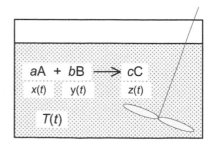

any time, however, the temperature is the same or uniform throughout the reactor because of perfect mixing. Such a system is described by the mass balances of the involved species or the equations of change. They are differential equations, which have time as the independent variable in the present case.

An optimal control problem for the batch reactor is to find the temperature versus time function, the application of which maximizes the product concentration at the final time t_f. That function is the optimal control $\hat{T}(t)$ among all possible control functions, such as those shown in Figure 1.2.

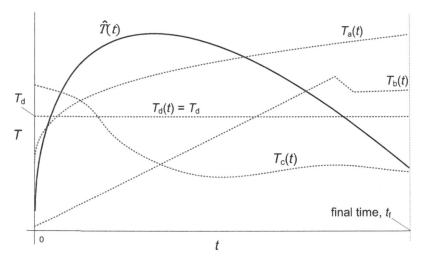

Figure 1.2 Optimal control $\hat{T}(t)$ and other possible control functions $T_a(t)$–$T_d(t)$

Let us formulate the above problem for the elementary reaction

$$aA + bB \longrightarrow cC$$

where a, b, and c are the stoichiometric coefficients of the species A, B, and C. Denoting their respective concentrations by x, y, and z at any time t, the batch reaction process may be described by the following equations of change:

$$\frac{dx}{dt} = -k_0 \exp\left[-\frac{E}{RT}\right] ax^a(t)y^b(t) \tag{1.1}$$

$$\frac{dy}{dt} = -k_0 \exp\left[-\frac{E}{RT}\right] bx^a(t)y^b(t) \tag{1.2}$$

$$\frac{dz}{dt} = k_0 \exp\left[-\frac{E}{RT}\right] cx^a(t)y^b(t) \tag{1.3}$$

with the initial conditions

$$x(0) = x_0, \quad y(0) = y_0, \quad \text{and} \quad z(0) = 0.$$

In the above equations, k_0 is the Arrhenius constant, E is the constant activation energy of the reaction, R is the universal gas constant, and T is the absolute temperature dependent on t. The temperature $T(t)$ is a control function, which is undetermined. It can be suitably changed to affect the product concentration z at the final time t_f. The objective is to find the optimal control function $T(t)$ that maximizes $z(t_f)$, i. e.,

$$I = z(t_f) = \int_{z(0)}^{z(t_f)} dz = \int_0^{t_f} \frac{dz}{dt} \, dt$$

$$= ck_0 \int_0^{t_f} \exp\left[-\frac{E}{RT(t)}\right] x^a(t) y^b(t) \, dt \tag{1.4}$$

subject to the satisfaction of Equations (1.1) and (1.2) with the specified initial concentrations x_0 and y_0 for A and B, respectively.

There could be other constraints as well. For example, $T(t)$ should never exceed a maximum temperature T_{max}, and the concentration of A should never fall below a threshold level x_{min}. In this case, the inequalities

$$T(t) \le T_{max} \tag{1.5}$$

$$x(t) \ge x_{min} \tag{1.6}$$

also need to be satisfied in the time interval $[0, t_f]$.

Objective Functional

Observe that I defined by Equation (1.4) is a function of x, y, and T, which in turn are functions of t. Thus, symbolically,

$$I = I[x(t), y(t), T(t)]$$

A function such as I depending on one or more functions is known as a **functional**. It will be explained fully in Chapter 2. In the present problem, I being the objective to be optimized is an **objective functional** with the function $T(t)$ as the optimization parameter.

Note that T is an undetermined function of t. The functions $x(t)$ and $y(t)$ depend implicitly on $T(t)$ through Equations (1.1) and (1.2) for specified constants and initial conditions. The evaluation of I for any particular form of $T(t)$ [say $T_a(t)$ in Figure 1.2] requires all the function $[T_a(t)]$ values over the specified time interval $[0, t_f]$. The objective is to find an optimal form of $T(t)$ or $\hat{T}(t)$, which yields the maximum value of I simultaneously satisfying Equations (1.1) and (1.2) and any other constraints such as Equations (1.5) and (1.6).

1.2   Optimal Control versus Optimization

It is easy to perceive from the above example that optimal control involves optimization of an objective functional subject to the equations of change in a system and additional constraints, if any. Because of this fact, optimal control is also known as *dynamic* or *trajectory optimization.*

The salient feature of optimal control is that it uses functions as optimization parameters. These functions are called control functions or simply controls. The routine, static optimization is a special case of optimal control using uniform or single-valued controls such as $T_{\mathrm{d}}(t)$ or T_{d} in Figure 1.2. Had we prescribed the invariance of temperature with respect to time in the previous example, the problem would have been that of the routine optimization with the goal to find the optimal time invariant temperature from all possible choices restricted to be time invariant like T_{d}. Because this restriction (or invariance with respect to independent variable) does not exist in optimal control, it has a significant advantage over routine optimization. Let us get more details.

Infinite Optimization Parameters

A control function used in optimal control comprises a number of values, one for each value of the independent variable. That number is infinity if at least a part of the function is continuous. Thus, in the previous example of the batch reactor, the control $T(t)$ is a set of optimization parameters

$$T_1, \; T_2, \; \ldots, \; T_i, \; \ldots, \; T_n$$

where

$$T_i = T(t_i), \qquad t_1 = 0, \qquad t_i = (t_{i-1} + \Delta t_i) \quad \text{for } 1 < i < n, \qquad t_n = t_{\mathrm{f}}$$

and Δt_i tends to zero as n tends to infinity. Hence, from the standpoint of the routine optimization, optimal control is equivalent to multi-parameter optimization. With a significantly greater number of parameters available to optimize in general, optimal control unlocks a considerably extensive region to search for optimal solutions otherwise unobtainable from the routine optimization. This striking feature, along with remarkable progress in high-speed computing, has made optimal control increasingly relevant today to processes and products facing tougher market competitions, stricter regulations, and thinner profit margins.

1.3   Examples of Optimal Control Problems

To gain further understanding of the applications of optimal control, let us study some examples of optimal control problems. In each problem, there is a system changing with time or some other independent variable. The system is mathematically described or modeled with the help of differential equations. At least one such differential equation is needed in order to have an optimal control problem. Appearing in the model is a set of undetermined control functions, which determines the dependent variables. The set of controls and dependent variables in turn determine the objective functional of the optimal control problem.

1.3.1   Batch Distillation

Figure 1.3 shows a schematic of the batch distillation process for the separation of a volatile compound from a binary liquid mixture. It is heated in the bottom still to generate vapors, which condense at the top to yield distillate having a higher concentration of a the volatile compound. A part of the distillate is withdrawn as product while the rest is recycled to the still. An optimal control problem is to maximize the production of distillate of a desired purity over a fixed time duration by controlling the distillate production rate with time (Converse and Gross, 1963).

   Assuming a constant boil-up rate in the still and no liquid hold-up in the column, the process can be modeled as

$$\frac{\mathrm{d}m}{\mathrm{d}t} = -u(t), \qquad\qquad m(0) = m_0 \qquad\qquad (1.7)$$

$$\frac{\mathrm{d}x}{\mathrm{d}t} = \frac{u(t)}{m}(x - y), \qquad\qquad x(0) = x_0 \qquad\qquad (1.8)$$

where m is the mass of mixture in the still, t is time, u is the mass of distillate product withdrawn per unit time, and x and y are the mass fractions of the volatile compound in the still and distillate, respectively. The initial mass m_0 and the mass fraction x_0 in the still are known. The distillate mass fraction is at any time is known from a relation specified as

$$y = y(m, x) \qquad\qquad (1.9)$$

Thus, u, m, x, and y are time dependent. The optimal control problem is to find the control function $u(t)$ that maximizes the objective functional

$$I = \int_0^{t_f} u(t)\, \mathrm{d}t$$

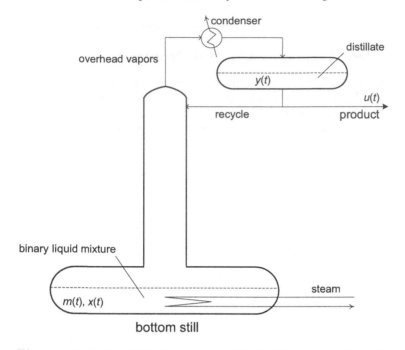

Figure 1.3 Batch distillation process with distillate production rate versus time or $u(t)$ as the control function

subject to Equations (1.7)–(1.9) as well as the purity specification y^* for the distillate product given by

$$y^* = \frac{\int\limits_0^{t_f} yu(t)\,\mathrm{d}t}{\int\limits_0^{t_f} u(t)\,\mathrm{d}t} \qquad \text{or} \qquad \int\limits_0^{t_f} u(y - y^*)\,\mathrm{d}t = 0 \qquad (1.10)$$

over a certain time duration t_f.

1.3.2 Plug Flow Reactor

Pressure plays an important role in reversible gas phase reactions such as

$$\text{A} \underset{k_2}{\overset{k_1}{\rightleftharpoons}} 2\text{B}$$

where the number of moles of species changes along a reaction path. In the above reaction, while lower pressure favors the forward path, higher pressure does the opposite. For that reaction carried out isothermally in a plug flow reactor (Figure 1.4), it is desired to obtain maximum product (or equivalently minimum reactant) concentration by controlling pressure along the reactor length.

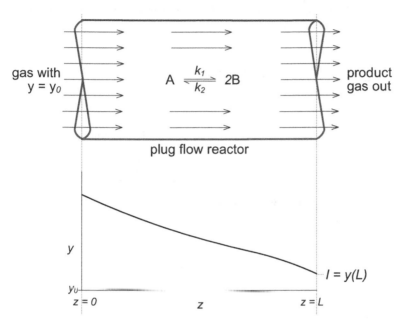

Figure 1.4 Gas-phase reaction in a plug flow reactor

Assuming the ideal gas law to hold, the model for the reactor of uniform cross-section is given by

$$\frac{dy}{dz} = \tau S \left[\frac{-k_1 y P(z)}{2y_0 - y} + \frac{4k_2 (y_0 - y)^2 P(z)^2}{(2y_0 - y)^2} \right], \quad y(0) = y_0 \qquad (1.11)$$

where S is the cross-section area, τ is the residence time of species inside the reactor, y is the concentration of the reactant A, z is the independent variable denoting the reactor length, P is pressure, and k_1 and k_2 are the forward and backward reaction rate coefficients (van de Vusse and Voetter, 1961).

The optimal control problem is to find the control function $P(z)$ that minimizes y at the reactor end $z = L$ subject to Equation (1.11). The minimum y is the objective functional given by

$$I = y(L) = y(0) + \int\limits_{y(0)}^{y(L)} dz = y_0 + \int\limits_0^L \frac{dy}{dz} \, dz \qquad (1.12)$$

1.3.3   Heat Exchanger

Figure 1.5 shows a single-tube heat exchanger used to heat (or cool) the fluid flowing inside the tube by controlling its wall temperature T_w as a function of time t (Huang et al., 1969). At any time, T_w is uniform along the z-direction, i. e., the length of the heat exchanger.

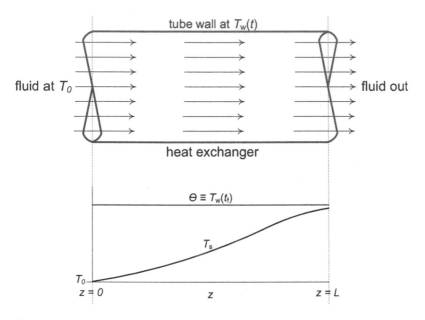

Figure 1.5 A single-tube heat exchanger

Corresponding to a given wall temperature θ, it is desired to attain a steady state temperature distribution $T_s(z)$ of the fluid in a specified time interval $[0, t_f]$. For the control to be meaningful, t_f is less than the time spent by the fluid inside the heat exchanger. A simple heat transfer model of the heat exchanger is given by

$$\frac{\partial T}{\partial t} = -v\frac{\mathrm{d}T}{\mathrm{d}z} + \frac{h}{\rho C_\mathrm{p}}[T_\mathrm{w}(t) - T] \tag{1.13}$$

where t and z are the independent variables denoting time and heat exchanger length, T and T_w are the temperatures of the fluid and the wall, v is the average fluid velocity, h is the wall-to-fluid heat transfer coefficient, and C_p is the specific heat capacity of the fluid of density ρ. While T depends on z and t, T_w is a function of t only. The initial and boundary conditions are

$$T(z, 0) = T(0, t) = T_0 \tag{1.14}$$

where T_0 is the fluid temperature at the inlet of the heat exchanger.

The steady state temperature $T_s(z)$ is the temperature defined by Equation (1.13) with the time derivative set to zero and $T_w(t)$ replaced with θ. Thus,

$$\frac{dT_s}{dz} = \frac{h}{v\rho C_p}[\theta - T] \tag{1.15}$$

Subject to the satisfaction of Equations (1.13)–(1.15), the optimal control problem is to find the control function $T_w(t)$ that brings in time t_f, the final unsteady state fluid temperature closest to the steady state wall temperature. Hence it is desired to minimize the objective functional

$$I = \int\limits_0^L [T(z, t_f) - T_s(z)]^2 \, dz$$

where L is the length of the heat exchanger. It may also be required that $T_w(t)$ does not exceed some maximum value, i. e.,

$$T_w(t) \leq T_{max}$$

throughout the time interval $[0, t_f]$.

1.3.4 Gas Diffusion in a Non-Volatile Liquid

This example shows the application of optimal control to determine a system property (diffusivity) as a function of another system property (concentration).

Consider the diffusion of a gas into an underlying layer of a non-volatile liquid such as heavy oil or polymer (Upreti and Mehrotra, 2000; Tendulkar et al., 2009) inside a closed vessel of uniform cross-section area A (Figure 1.6). As the gas penetrates the liquid layer, the pressure inside the vessel goes down. The system is at constant temperature throughout the duration t_f of this process with negligible change in the thickness L of the polymer layer. The mass concentration c of gas in the layer at any time t and depth z is given by

$$\frac{\partial c}{\partial t} = D(c)\left(1 + \frac{c}{\rho}\right)\frac{\partial^2 c}{\partial z^2} + \left[\left(1 + \frac{c}{\rho}\right)\frac{dD(c)}{dc} + \frac{D(c)}{\rho}\right]\left(\frac{\partial c}{\partial z}\right)^2 \tag{1.16}$$

where D is the concentration-dependent diffusivity of the gas in the liquid, and ρ is its density. Equation (1.16) has the following initial conditions:

$$c(0,0) = c_{sat}(t = 0) \qquad \text{at the gas–liquid interface} \tag{1.17}$$
$$c(z,0) = 0 \qquad \text{for } 0 < z \leq L \tag{1.18}$$

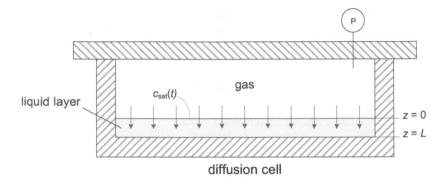

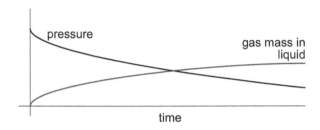

Figure 1.6 Diffusion of gas into the liquid layer

and boundary conditions:

$$c(0, t) = c_{\text{sat}}(t) \qquad\qquad \text{for } 0 < t \leq t_f \qquad (1.19)$$

$$\left.\frac{\partial c}{\partial z}\right|_{z=L} = 0 \qquad\qquad \text{for } 0 \leq t \leq t_f \qquad (1.20)$$

where c_{sat} is the equilibrium saturation concentration of the gas at the interface.

The objective in this problem is to determine the unknown concentration-dependent gas diffusivity such that its use in Equations (1.16)–(1.20) yields the calculated mass of gas in the layer

$$m_c(t) = \int_0^L c(z, t) A \, dz \qquad (1.21)$$

in agreement with the experimental gas mass $m_e(t)$. The latter mass is already known using system pressure recorded with time and experimental pressure-volume-temperature data for the gas. Thus, the optimal control problem is to find the control function $D(c)$ that minimizes the objective functional

$$I = \int_0^{t_f} [m_c(t) - m_e(t)]^2 \, dt$$

subject to Equations (1.16)–(1.21).

1.3.5 Periodic Reactor

This reactor poses a optimal periodic control problem, which involves periodic control functions. Their application can result in better performance relative to steady state operation and help achieve difficult performance criteria such as those involving molecular weight distribution (MWD).

Figure 1.7 shows one such application to polymerization carried out in a constant volume stirred tank reactor or CSTR fed by liquid streams of initiator and monomer. The feed streams react to form polymer, which is discharged in the output stream. It is desired to find periodic feed flow rates versus time that would produce a polymer of specified polydispersity index (PI), a measure of polymer MWD.

A model for the above process is given by

$$\frac{\mathrm{d}x}{\mathrm{d}t} = \frac{F_x x_f}{V} - \frac{(F_x + F_y)x}{V} - k_x xy \tag{1.22}$$

$$\frac{\mathrm{d}y}{\mathrm{d}t} = \frac{F_y y_f}{V} - \frac{(F_x + F_y)y}{V} - k_x xy - k_p y \mu_0 \tag{1.23}$$

$$\frac{\mathrm{d}\mu_0}{\mathrm{d}t} = k_x xy - \frac{(F_x + F_y)\mu_0}{V} \tag{1.24}$$

$$\frac{\mathrm{d}\mu_1}{\mathrm{d}t} = k_x xy - \frac{(F_x + F_y)\mu_1}{V} + k_p y \mu_0 \tag{1.25}$$

$$\frac{\mathrm{d}\mu_2}{\mathrm{d}t} = k_x xy - \frac{(F_x + F_y)\mu_2}{V} + k_p y(\mu_0 + 2\mu_1) \tag{1.26}$$

where x and y are the initiator and monomer concentrations in the reactor, $F_x(t)$ and $F_y(t)$ are the undetermined initiator and monomer feed flow rates of time period τ, k_x and k_p are the reaction rate coefficients for initiation and polymerization, V is the reactor volume, and μ_0, μ_1, and μ_2 are the first three moments of number distribution of polymer chains (Frontini et al., 1986). The subscript f indicates the feed stream property.

Subject to Equations (1.22)–(1.26), the optimal control problem is to find the control functions $F_x(t)$ and $F_y(t)$ that repeat over a given time period τ to produce polymer of a specific PI, i.e., minimize the objective functional

$$I = \left[\frac{1}{\tau} \int_0^\tau \frac{\mu_0(t)\mu_2(t)}{\mu_1^2(t)} \, \mathrm{d}t - D^* \right]^2$$

where the integrand and D^* are the instantaneous and specified PIs, respectively. The periodicity conditions

$$s(t) = s(t + \tau), \quad s = \{F_x, F_y, x, y, \mu_0, \mu_1, \mu_2\}$$

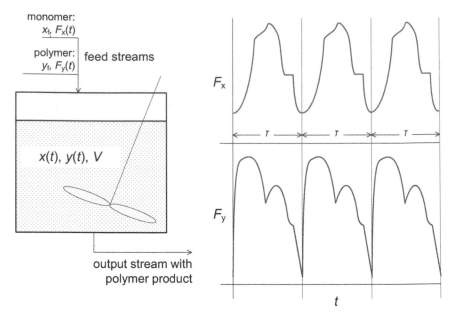

Figure 1.7 A constant volume stirred tank reactor with periodic feed flow rates like those on the right

must be satisfied simultaneously.

1.3.6 Nuclear Reactor

Consider nuclear fission in a reactor where neutrons react with large fissile nuclei to produce more neutrons and smaller fissile nuclei called precursors. The latter subsequently absorb more neutrons to produce "delayed" neutrons. The kinetic energy of these products is converted into thermal energy when they collide with neighboring atoms. Thus, the power output of the reactor depends on the concentration of neutrons available to carry out nuclear fission.

The power output can be changed according to the demand by inserting (retracting) a neutron-absorbing control rod into (from) the reactor, as shown in Figure 1.8. The control rod upon insertion absorbs neutrons, thereby reducing the heat flux and consequently the power output. The opposite happens when the rod is retracted.

The reaction kinetics is given by

$$\frac{\mathrm{d}x}{\mathrm{d}t} = \frac{rx - \alpha x^2 - \beta x}{\tau} + \mu y, \quad x(0) = x_0 \tag{1.27}$$

$$\frac{\mathrm{d}y}{\mathrm{d}t} = \frac{\beta x}{\tau} - \mu y, \quad\quad y(0) = y_0 \tag{1.28}$$

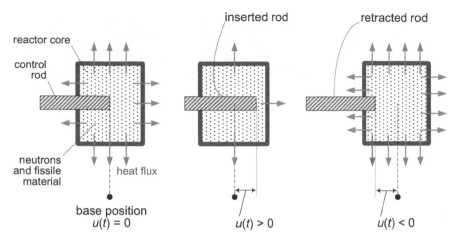

Figure 1.8 Control rod positions in a nuclear reactor

where x and y are the concentrations of neutrons and precursors, t is time,

$$r = r[u(t)] \tag{1.29}$$

is the degree of change in neutron multiplication as a known function of control rod displacement $u(t)$, α is the reactivity coefficient, β is the fraction of delayed neutrons, μ is the decay constant for precursors, and τ is the average time taken by a neutron to produce a neutron or precursor (Fan, 1966).

It is desired to change x from x_0 to a stable value of x_f at time t_f with minimum displacement of the control rod. Thus, the optimal control problem is to find the control function $u(t)$ that minimizes the objective functional

$$I = \int_0^{t_f} u^2(t)\,\mathrm{d}t \tag{1.30}$$

with the final conditions

$$x(t_f) = x_f$$

$$\left.\frac{\mathrm{d}x}{\mathrm{d}t}\right|_{t=t_f} = 0$$

subject to Equations (1.27)–(1.29) as well as the constraint

$$|u(t)| \le u_{\max}$$

1.3.7 Vapor Extraction of Heavy Oil

Vapor extraction or Vapex involves the extraction of heavy oil from a porous reservoir (Figure 1.9) using a vaporized solvent close to the dew point. Upon being injected from the top, the solvent drastically reduces the viscosity of oil, causing it to drain under gravity and get produced at the bottom. We would like to maximize the oil production in Vapex by considering solvent pressure versus time as a control function to influence solvent concentration at the top.

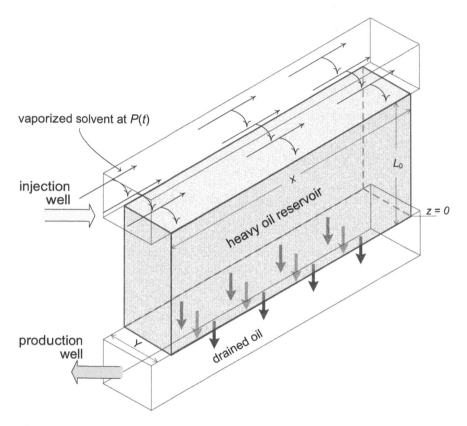

vaporized solvent at $P(t)$

injection well

L_0

x

z = 0

heavy oil reservoir

production well

drained oil

Figure 1.9 Vapor extraction of heavy oil from a reservoir

Assuming constant temperature and oil density, uniform reservoir porosity and permeability, and changes only along the vertical z direction, the solvent mass balance is given by

$$\frac{\partial \omega}{\partial t} = -\frac{\partial}{\partial z}\left[D(\omega)\frac{\partial \omega}{\partial z}\right] + \frac{1}{\phi}\frac{\partial}{\partial z}(v\omega) \qquad (1.31)$$

where t is time, ω is the solvent mass fraction in the oil, D is the dispersion coefficient of the solvent into the reservoir medium of porosity ϕ, and v is the downward velocity of the oil. Typically, D is a known function of ω. The initial and boundary conditions are

$$w(z, 0) = 0, \quad 0 \le z < L_0 \tag{1.32}$$

$$w[L(t), t] = w_i[P(t)] \tag{1.33}$$

where L is the height of oil in the reservoir with the initial value L_0, w_i is the solvent mass fraction in the solvent–oil interface at the top, and P is the pressure of the solvent. The relationship between w_i and P is known a priori from experiments. The change in L due to oil drainage is given by

$$\frac{\mathrm{d}L}{\mathrm{d}t} = -v[w(0, t)], \quad z(0) = L_0 \tag{1.34}$$

Finally, assuming that Darcy's law holds, we have

$$v = \frac{K_{\mathrm{r}} K \rho g}{\mu(w)} \tag{1.35}$$

where K_{r} is relative permeability, g is gravity, ρ is the oil density, and μ is oil viscosity, which is a known function of w.

Given the reservoir of length X and thickness Y, the optimal control problem is to determine the control function $P(t)$ that maximizes the mass m of oil produced at the bottom, i.e., the objective functional

$$m = \rho XY \int_0^{L(t_{\mathrm{f}})} \mathrm{d}z$$

by satisfying Equations (1.31)–(1.35) as well as any other constraint such as

$$P \le P_{\max}$$

1.3.8 Chemotherapy

The drug concentration y_1 and the number of immune, healthy, and cancer cells (y_2, y_3, and y_4) in an organ at any time t during chemotherapy can be expressed as

$$\frac{\mathrm{d}y_1}{\mathrm{d}t} = u(t) - \gamma_6 y_1 \tag{1.36}$$

$$\frac{\mathrm{d}y_2}{\mathrm{d}t} = \dot{y}_{2,\mathrm{in}} + r_2 \frac{y_2 y_4}{\beta_2 + y_4} - \gamma_3 y_2 y_4 - \gamma_4 y_2 - \alpha_2 y_2 \left(1 - e^{-y_1 \lambda_2}\right) \tag{1.37}$$

$$\frac{\mathrm{d}y_3}{\mathrm{d}t} = r_3 y_3 (1 - \beta_3 y_3) - \gamma_5 y_3 y_4 - \alpha_3 y_3 \left(1 - e^{-y_1 \lambda_3}\right) \tag{1.38}$$

$$\frac{\mathrm{d}y_4}{\mathrm{d}t} = r_1 y_4 (1 - \beta_1 y_4) - \gamma_1 y_3 y_4 - \gamma_2 y_2 y_4 - \alpha_1 y_4 \left(1 - e^{-y_1 \lambda_1}\right) \tag{1.39}$$

where $\dot{y}_{2,\text{in}}$ is the constant rate of immune cells that enter the organ to fight cancer cells, and $u(t)$ is the rate of drug injection into the organ (de Pillis and Radunskaya, 2003). The r_is and β_is are constants in the growth terms, while α_is and λ_is are the constants in the decay terms arising due to the action of the drug. The γ_is are the constants in the remaining decay terms. Note that the drug has toxic side effects since it kills the immune and healthy cells as well.

To treat cancer using chemotherapy, an objective could be to minimize the number of cancer cells in a specified time t_f using minimum drug to reduce its toxic effects. The optimal control problem in this case is to find the control function $u(t)$ that minimizes the objective functional

$$I = y_4(t_f) + \int_0^{t_f} u(t)\,\mathrm{d}t \tag{1.40}$$

subject to Equations (1.36)–(1.39), and the initial values of drug concentration and cell numbers. It is expected that $u(t)$ is never less than zero.

Other constraints may be present. For example, the number of healthy cells during treatment should not fall below a certain minimum, i.e.,

$$y_3(t) \geq y_{3,\text{min}}$$

Also, there could be an upper limit to drug dosage, i.e.,

$$u(t) \leq u_{\text{max}}$$

1.3.9 Medicinal Drug Delivery

A polymer loaded with medicinal drug provides a means to administer it to specific body parts. The initial drug distribution in the polymer matrix could be controlled to obtain a desirable pattern of drug release with time (Lu et al., 1998).

The one-dimensional release of the drug from the matrix to a tissue in contact (Figure 1.10) can be modeled as

$$\frac{\partial c}{\partial t} = \frac{\partial}{\partial x}\left(D\frac{\partial c}{\partial x}\right) \tag{1.41}$$

where c is the drug concentration, t is the time variable, x is the variable along the thickness L of the matrix, and D is drug diffusivity.

The initial condition is

$$c(x,0) = u(x) \tag{1.42}$$

where $u(x)$ is the undetermined drug distribution. The boundary conditions for $t > 0$ are

$$\left.\frac{\partial c}{\partial x}\right|_{x=0} = 0 \quad \text{and} \quad c(L,t) = 0 \tag{1.43}$$

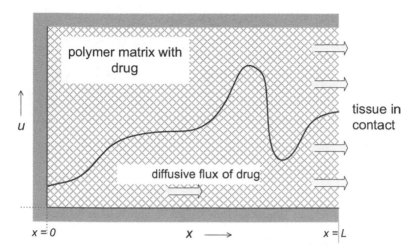

Figure 1.10 Drug release from a polymer matrix to a tissue in contact

At any time, the flux of the drug into the tissue at the point of contact is given by

$$J(t) = -D\left[\frac{\partial c}{\partial x}\right]_{x=L} \tag{1.44}$$

It is desired to match $J(t)$ with a specified drug release versus time relation, $J^*(t)$ over a certain time duration t_f.

Hence, the optimal control problem is to find the control function $u(x)$ that minimizes the objective functional

$$I = \int\limits_0^{t_f} [J(t) - J^*(t)]^2 \, dt$$

subject to Equations (1.41)–(1.44).

1.3.10 Blood Flow and Metabolism

The peripheral blood flow and metabolism rates versus time in a tissue could be non-invasively determined through skin surface temperature, which follows the circadian rhythm of sunrise and sunset under resting conditions (Upreti and Jeje, 2004).

Figure 1.11 shows the cross-section of the skin layer around a cylindrical limb such as a big toe.

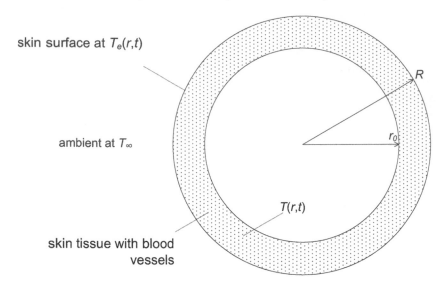

skin surface at $T_e(r,t)$

ambient at T_∞

R

r_0

$T(r,t)$

skin tissue with blood
vessels

Figure 1.11 Cross-section of the skin layer surrounding a cylindrical limb

The heat transfer model for the skin layer is given by

$$\frac{\partial T}{\partial t} = \frac{k}{r\rho C_p}\frac{\partial}{\partial r}\left(r\frac{\partial T}{\partial r}\right) + \frac{F(t)C_b(T_b - T)}{C_p} + \frac{\Delta H(t)}{\rho C_p} \qquad (1.45)$$

where T is temperature, t and r are time and radial variables, k and C_p are the thermal conductivity and specific heat capacity of the skin tissue, C_b and T_b are the specific heat capacity and temperature of the blood, and $F(t)$ and $\Delta H(t)$ are the undetermined rates of blood flow and metabolism in the tissue. The boundary conditions are given by

$$-k\left[\frac{dT}{dr}\right]_{r=R} = h(T - T_\infty) \qquad (1.46)$$

$$T(r,0) = T(r,\tau), \quad r_0 \le r \le R \qquad (1.47)$$

where h is the convective heat transfer coefficient, T_∞ is the ambient temperature, r_0 and R are the inner and outer radii of the skin tissue, and τ is the time period of the circadian rhythm. This rhythm enforces Equation (1.47) and similar periodicity of $F(t)$ and $\Delta H(t)$, i.e., the equations

$$F(0) = F(\tau) \quad \text{and} \quad \Delta H(0) = \Delta H(\tau)$$

Thus, the optimal control problem is to find the two periodic control functions, $F(t)$ and $\Delta H(t)$, the incorporation of which in Equations (1.45)–(1.47) minimizes the difference between the model-predicted $T(R,t)$ and its experimental counterpart $T_e(R,t)$, i.e., the objective functional

$$I = \int\limits_0^\tau \left[1 - \frac{T(R,t)}{T_e(R,t)}\right]^2 dt$$

1.4   Structure of Optimal Control Problems

The above examples help us identify the structure of optimal control problems. As shown in Figure 1.12, an optimal control problem involves one or more controls and other inputs to a system under change. The optimal controls provide the desired output from the system.

A set of controls in a system is a combination of

1. system inputs such as temperature versus time relation in a batch reaction,
2. system properties such as diffusivity versus concentration relation in a gas–polymer system, and
3. entities generated in the system, e.g., the rate of heat generation versus time relation in a skin tissue.

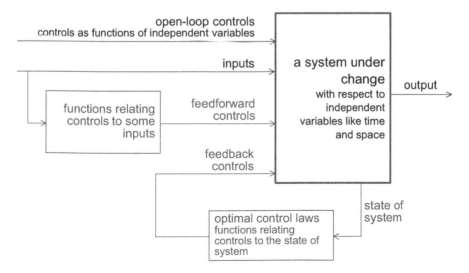

Figure 1.12 Structure of optimal control problems

Open-Loop Control

In most optimal control problems, it is not possible to obtain *optimal control laws*, i. e., optimal controls as explicit functions of system state. Note that system state is the set of system properties such as temperature, pressure, and concentration. They are subject to change with independent variables like time and space. In the absence of an optimal control law, the optimal control needs to be determined all over again if the initial system state changes.

The controls that are not given by optimal control laws are often called *open-loop controls*. They simply are functions of independent variables and specific to the initial system state. The application of open-loop controls is termed *open-loop control*, which is the subject matter of this book.

Closed-Loop Control

Sometimes, it is possible to derive optimal control laws when the underlying mathematical models describing the system are simple enough. In many problems though, obtaining optimal control laws mostly requires drastic simplifications of the underlying mathematical models, thereby compromising on the accuracy of control.

Nonetheless, when determined, optimal control laws are easier to implement for the control of continuous processes where inputs are susceptible to change during long operation periods. Optimal controls are then readily obtained from the system state and applied or fed back to the system, as shown in Figure 1.12. These controls are called *feedback controls* and the control strategy is termed *feedback control*, which is a type of *closed-loop control*.

A different type of closed-loop control is *feedforward control*, in which optimal controls are explicitly obtained in advance from the inputs in conjunction with the mathematical model of a system. As shown in the above figure, *feedforward controls* are applied to the system without having to wait for the system state the inputs and controls would later generate.

Bibliography

A.O. Converse and G.D. Gross. Optimal distillate-rate policy in batch distillation. *Ind. Eng. Chem. Fundamen.*, 2(3):217–221, 1963.

L.G. de Pillis and A. Radunskaya. The dynamics of an optimally controlled tumor model: A case study. *Math. Comput. Model.*, 37:1221–1244, 2003.

L.T. Fan. *The Continuous Maximum Principle*, Chapter 4, pages 113–115. John Wiley & Sons, New York, 1966.

G.L. Frontini, G.E. Eliçabe, D.A. Couso, and G.R. Meira. Optimal periodic control of a continuous "living" anionic polymerization. I. Theoretical study. *J. Appl. Poly. Sci.*, 31:1019–1039, 1986.

H.-S. Huang, L.T. Fan, and C.L. Hwang. Optimal wall temperature control of a heat exchanger. Technical Report 15, Institute for Systems Design and Optimization, Kansas State University, Manhattan, 1969.

S. Lu, W.F. Ramirez, and K.S. Anseth. Modeling and optimization of drug release from laminated polymer matrix devices. *AIChE J.*, 44(7):1689–1696, 1998.

J. Tendulkar, S.R. Upreti, and A. Lohi. Experimental determination of concentration-dependent carbon dioxide diffusivity in LDPE. *J. App. Sci.*, 111:380–387, 2009.

S.R. Upreti and A.A. Jeje. A noninvasive technique to determine peripheral blood flow and heat generation in a human limb. *Chem. Eng. Sci.*, 59: 4415–4423, 2004.

S.R. Upreti and A.K. Mehrotra. Experimental measurement of gas diffusivity in bitumen. *Ind. Eng. Chem. Res.*, 39:1080–1087, 2000.

J.G. van de Vusse and H. Voetter. Optimum pressure and concentration gradients in tubular reactors. *Chem. Eng. Sci.*, 14(1):90–98, 1961.

Exercises

1.1 Revise the batch distillation problem in Section 1.3.1 (p. 5) in order to minimize the mass fraction of the volatile compound in the bottom still in a certain time duration t_f.

1.2 Using temperature $T(z)$ as a control function at constant pressure, formulate the problem of the plug flow reactor in Section 1.3.2 (p. 6) to maximize the concentration of the intermediate product B in the following sequence of elementary first order reactions

$$A \xrightarrow{\;k_1(T)\;} B \xrightarrow{\;k_2(T)\;} C$$

where k_1 and k_2 are the reaction rate coefficients dependent on the temperature T.

1.3 Modify the objective functional of the nuclear reactor example in Section 1.3.6 (p. 12) to additionally enable the minimization of oscillations in the neutron concentration.

1.4 Revise the drug delivery example in Section 1.3.9 (p. 16) to minimize the consumption of drug, simultaneously ensuring that its flux does not fall below a threshold value over a certain time duration t_f.

1.5 What new conditions would be required if the optimal periodic control problem of Section 1.3.5 (p. 11) is changed to a non-periodic one?

1.6 Compare the use of $u(t)$ in the objective functionals of Equations (1.30) and (1.40).

Chapter 2

Fundamental Concepts

This chapter introduces the fundamental concepts of optimal control. Beginning with a functional and its domain of associated functions, we learn about the need for them to be in linear or vector spaces and be quantified based on size measures or norms. With this background, we establish the differential of a functional and relax its definition to variation in order to include a broad spectrum of functionals. A number of examples are presented to illustrate how to obtain the variation of an objective functional in an optimal control problem.

2.1 From Function to Functional

Let us understand the concept of a functional in the light of what we already know about a function. A **function** associates a function value with a set of variables, each of which assumes a *single* value for function evaluation. For example, in the equation

$$f = x^2, \quad a \le x \le b$$

f is a function associating a function value $f(x)$ with the variable x in its domain $[a, b]$. Thus, $f(x)$ is equal to the square of x. This variable assumes a single value each time $f(x)$ is evaluated, as shown in Figure 2.1.

A **functional**, on the other hand, associates a functional value with a set of functions, each of which assumes in its respective domain *a set of values* for functional evaluation. For example, in the equation

$$I = \int_a^b f(x)\, \mathrm{d}x \tag{2.1}$$

where $f(x)$ is a continuous function. I is a functional, which associates a functional value $I(f)$ with the function f in its domain $[a, b]$. We know that value to be the area between the x-axis from a to b and the curve f. Note that

Figure 2.1 Evaluation of
a function f for $x = a$ and
$x = b$

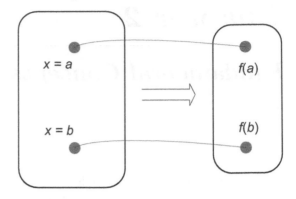

for the evaluation of $I(f)$, the function f assumes *not a single value* but the set of all values from $f(a)$ to $f(b)$. This fact becomes explicit in the discrete equivalent of Equation (2.1), i.e.,

$$I = \sum_{i=1}^{n \to \infty} f_i \cdot \Delta x_i \atop \lim \Delta x_i \to 0 \qquad (2.2)$$

where

$$f_i = f(x_i); \quad x_1 = a; \quad x_i = x_{i-1} + \Delta x_i, \ i = 2, 3, \dots, (n-1); \quad \text{and} \quad x_n = b.$$

Note that as the Δx_i values tend to zero, n tends to infinity. Not a single function value but *a set of* all function values

$$\{f_1, f_2, \dots, f_n | \ n \to \infty\}$$

is needed to evaluate the functional value $I(f)$. Figure 2.2 shows the evaluation of I for the functions g and h, which are two different forms of the function f.

To sum up, a function depends on a set of variables, each of which assumes *a single value* for function evaluation. On the other hand, a functional depends

Figure 2.2 Evaluation of a
functional I for $f = g$ and
$f = h$

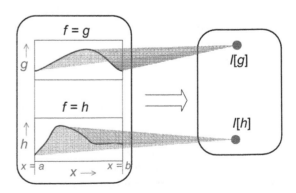

on a set of functions, each of which assumes *a set of values* for functional evaluation.

2.1.1 Functional as a Multivariable Function

A functional dependent on a function is analogous to a multivariable function dependent on a vector of variables. For example, consider the functional

$$K(f) = \frac{f_1^{\,2} f_2}{f_3^{\,2}}$$

dependent on a function f comprising three components f_1, f_2, and f_3. For an evaluation of $K(f)$, f assumes *a set of values*, i. e., a value for each of its components. We can say that the components behave as variables. From this viewpoint, the functional is equivalent to a multivariable function dependent on the variable vector

$$\mathbf{f} = \begin{bmatrix} f_1 & f_2 & f_3 \end{bmatrix}^{\mathsf{T}}$$

To extend the analogy further, consider a functional dependent on an integral of a continuous function. The latter has an infinite number of components over a non-zero range of integration. Thus, the functional is equivalent to a multivariable function dependent on the variable vector comprising those components. For example, the functional I in Equation (2.1) is equivalent to a multivariable function dependent on a vector of infinite components of f.

An interesting upshot of the above analogy is that a continuous function is equivalent to a vector of infinite components. Thus, $f(x)$ in Equation (2.1) is equivalent to the vector

$$\mathbf{f} = \begin{bmatrix} f_1 & f_2 & \cdots & f_n \end{bmatrix}^{\mathsf{T}}, \quad \lim n \to \infty$$

whose infinitely many components are shown in Equation (2.2).

2.2 Domain of a Functional

The domain of a functional is a space that holds all possible forms of the associated functions or vectors with some common specification. For example, the domain of the functional I in Equation (2.1) is the space holding all forms of $f(x)$ with the common specification that each form be continuous for $a \leq x \leq b$.

We require domains that are **linear** or **vector spaces**. Such a space contains all linear combinations of its elements. The details and the rationale of linear spaces are as follows.

2.2.1 Linear or Vector Spaces

In optimal control, we desire to find the minimum or maximum value of a functional defined over a specified domain. The analytical procedure is to continuously change the associated function from some reference form and examine the corresponding change in the functional. The new form of the function is, in fact, the result of a linear combination of the reference form and some other form of the function in the same domain. This examination can continue only if the new form of the function lies within the specified domain each time the function is changed. Otherwise, the corresponding new functional may not exist or be valid. The validity of the functional is ensured by having the specified domain be a **linear** or **vector space**. This space holds within itself all linear combinations of its elements (functions), which are called vectors. A precise definition of linear space is provided in Section 9.19 (p. 278).

In this book, we will deal with functions or vectors that belong to linear or vector spaces. Examples of these spaces include a space of vectors of specified n components, a space of continuous functions dependent on an independent variable varying in a specified interval, etc.

2.2.2 Norm of a Function

The aforementioned analytical procedure furthermore requires a single-valued measure for the size of the function or vector in the specified domain of a functional. That measure is termed the **norm**. We use it to quantify the difference between two functions, or equivalently, the change in the function from some reference form.

The norm of a vector or function y is denoted by $\|y\|$. A norm has the following properties:

1. It is zero for a zero vector or function* and non-zero otherwise.
2. For any real number α,
$$\|\alpha y\| = |\alpha| \|y\|.$$
3. The norm obeys the **triangle inequality**, i.e.,
$$\|y + z\| \le \|y\| + \|z\|$$
   where z is another vector or function.

Examples

In the two-dimensional Cartesian coordinate system, the length of a vector $\mathbf{y} = \begin{bmatrix} y_1 & y_2 \end{bmatrix}^\top$ is the norm given by

$$\|\mathbf{y}\| = \sqrt{y_1^2 + y_2^2} \tag{2.3}$$

---

* It has all components zero or is zero throughout.

For a continuous function $f(x)$ in the x-interval $[a, b]$, a definition of the norm could be

$$\|f\| = \sqrt{\int_a^b [f(x)]^2 \mathrm{d}x} \tag{2.4}$$

We can easily verify that Equations (2.3) and (2.4) satisfy the properties of a norm.

Thus, given a norm, we can compare the sizes of any two vectors. Considering a change in a vector or a function, which is reflected in its norm, we can study the optimality of the dependent functional by examining how it changes. The approach is similar to that in routine optimization where we consider a change in a variable and examine the corresponding change in the dependent function.

2.3 Properties of Functionals

We need three important properties for functionals — continuity, linearity, and homogeneity. These properties are required to develop the concept of the differential of a functional.

A functional I is continuous in its domain at a particular function z if for every other function y in the neighborhood

$$\lim_{y \to z} I(y) = I(z)$$

A linear functional I over a specified domain satisfies the relation

$$I(\alpha y + \beta z) = \alpha I(y) + \beta I(z) \tag{2.5}$$

where y and z are any two functions in the domain, and α and β are any two real numbers. A homogeneous functional I of the first degree, on the other hand, satisfies the relation

$$I(\alpha y) = \alpha I(y)$$

Note that a linear functional is inherently homogeneous but a homogeneous functional is not necessarily linear. The following example illustrates this important point.

Example 2.1

Consider the following two functionals:

$$J = \int_a^b y \, \mathrm{d}x \qquad \text{and} \qquad K = \int_a^b \frac{y^2}{y - y'} \, \mathrm{d}x$$

where y' denotes the derivative $\mathrm{d}y/\mathrm{d}x$. The functional J is linear and therefore homogeneous. However, the functional K is homogeneous but not linear. If $z = \alpha y$, then $z' = \alpha y'$ and

$$K(\alpha y) = K(z) = \int_a^b \frac{\alpha^2 y^2}{\alpha y - \alpha y'}\,\mathrm{d}x = \alpha \int_a^b \frac{y^2}{y - y'}\,\mathrm{d}x = \alpha K(y)$$

showing that K is a homogeneous functional. But K does not satisfy the linearity relation, i. e., Equation (2.5), since

$$\underbrace{\int_a^b \frac{(\alpha y + \beta z)^2}{(\alpha y + \beta z) - (\alpha y' + \beta z')}\,\mathrm{d}x}_{K(\alpha y + \beta z)} \neq \underbrace{\alpha \int_a^b \frac{y^2}{y - y'}\,\mathrm{d}x}_{\alpha K(y)} + \underbrace{\beta \int_a^b \frac{z^2}{z - z'}\,\mathrm{d}x}_{\beta K(z)}$$

☐

2.4 Differential of a Functional

The concept of a differential enables the study of the change in a functional by rendering it linear and continuous over a sufficiently small change in the associated function. The notion follows the definition of the differential of a function explained in Section 9.8 (p. 271).

Consider a change in a functional due to a change in the associated function. The **differential** of the functional is defined as the functional change that is a linear and continuous functional depending on the function change that is sufficiently small. Thus, given a function change h of size greater than zero but less than some positive real number δ, the differential $\mathrm{d}I(y_0; h)$ of a functional $I(y)$ at $y = y_0$ is defined by

$$I(y_0 + h) - I(y_0) = \mathrm{d}I(y_0; h) + \epsilon(h) \qquad (2.6)$$

where the error $\epsilon(h)$ vanishes *faster* than the size of h. In other words, the ratio of $\epsilon(h)$ to some size measure of h vanishes with the measure. When that happens, the error is bound to become zero at some non-zero size of h. Then for h of that size or smaller, the change in functional can be conveniently represented by the simple functional $\mathrm{d}I(y_0; h)$ called the differential, which is a linear and continuous functional of the function h.

2.4.1 Fréchet Differential

The *Fréchet differential* is defined by Equation (2.6) when the norm of h is used for its size. In this case, therefore, the ratio $\epsilon(h)/\|h\|$ is required to vanish with $\|h\|$.

Dividing both sides of Equation (2.6) by $\|h\|$ and taking its limit to zero, we get the following equivalent definition for the Fréchet differential:

$$\lim_{\|h\|\to 0} \frac{I(y_0 + h) - I(y_0) - dI(y_0; h)}{\|h\|} = \lim_{\|h\|\to 0} \frac{\epsilon(h)}{\|h\|} = 0 \qquad (2.7)$$

Placing no restriction on the form of h, i.e., the shape of the curve $h(x)$, this requirement implies that the error should decrease uniformly with h, as explained below.

Let I be a functional of a function y having only two components. Figure 2.3 graphs the following changes in y from a reference form y_0:

$$h_{k,1}, \ h_{k,2}, \ h_{l,1}, \ h_{l,2}, \ h_{m,1}, \ \text{and} \ h_{m,2}$$

These changes are different forms of h and are shown as arrow vectors. Thus, $h_{k,1}$ is the first change having the norm (or radius) k with a component along each of the two mutually perpendicular axes. The norm is given by

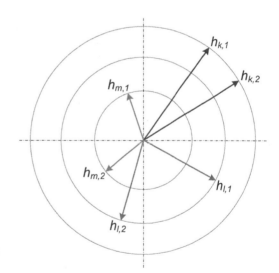

Figure 2.3 Changes in a two-component function for a Fréchet differential

Equation (2.3). Having the same norm k, $h_{k,2}$ is the second change in a different direction. The existence of the Fréchet differential at y_0 requires that the ratio $\epsilon(h)/\|h\|$ should decrease as *any* h having some norm k or less is downsized to *any* h having a smaller norm.

For simplicity, let us adhere to two arbitrary directions of h for the same norm (i.e., radius), as shown in the figure. Thus, for $dI(y_0; h)$ to exist, the ratio $\epsilon(h)/\|h\|$ should decrease as an h having the norm k is downsized in *all* possible ways — from (i) $h_{k,1}$ to $h_{l,1}$, (ii) $h_{k,1}$ to $h_{l,2}$, (iii) $h_{k,2}$ to $h_{l,1}$, and (iv) $h_{k,2}$ to $h_{l,1}$.

Figure 2.4 illustrates an analogous example with y as a continuous function and the norm defined by Equation (2.4). In this case, $h_{k,1}(x)$ and $h_{k,2}(x)$ are two different forms of $h(x)$ of norm k with different directions (Section 9.20, p. 279), which are characterized by different shapes of the plotted curves.

In general, we have the following condition for a Fréchet differential to exist. As long as $\|h\|$ decreases, the ratio $\epsilon(h)/\|h\|$, and consequently the error $\epsilon(h)$, should decrease uniformly with h, i.e., regardless of its direction. If this condition is satisfied, then at some sufficiently small and non-zero norm m of h, the error itself would become zero. Then for any h having the norm m or less, the functional change is representable by the Fréchet differential, which is a linear and continuous functional of h.

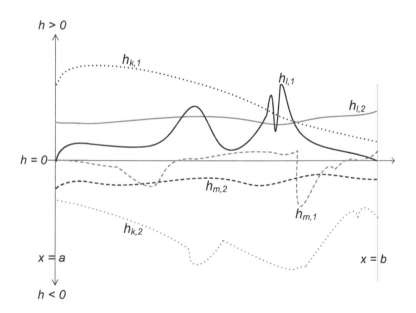

Figure 2.4 Changes in a continuous function for the Fréchet differential

2.4.2 Gâteaux Differential

The Fréchet differential is based on the aforementioned condition of uniform error disappearance. This condition is too stringent to be satisfied by a large class of functionals. If we loosen that condition by preserving the direction of h during its downsizing, the resulting differential is known as the *Gâteaux differential*. Denoted by $dI(y_0; h)$, it is a linear and continuous functional of h defined by

$$I(y_0 + \alpha h) - I(y_0) = dI(y_0; \alpha h) + \epsilon_1(\alpha h) \tag{2.8}$$

where the ratio of $\epsilon_1(\alpha h)$ to a scalar multiplier α vanishes with α itself for all h in the domain of I with $0 < \|\alpha h\| < \delta$. Note that $\epsilon_1(\alpha h)/\alpha$ is the ratio $\epsilon(\alpha h)/\|\alpha h\|$ with the introduction of $\epsilon_1(\alpha h) \equiv \epsilon(\alpha h)/\|h\|$.

Similar to Equation (2.7), the equivalent definition of the Gâteaux differential is

$$\lim_{\alpha \to 0} \frac{I(y_0 + h) - I(y_0) - dI(y_0; h)}{\alpha} = 0$$

which is obtained by dividing Equation (2.8) by α, taking its limit to zero, and replacing αh by h in the end.

Figures 2.5 and 2.6 show the downsizing of h when it is, respectively, a two-component vector and a continuous function. Observe the preservation of the direction of a vector or the shape of a function during the downsizing process.

Figure 2.5 Changes in a two-component function for the Gâteaux differential

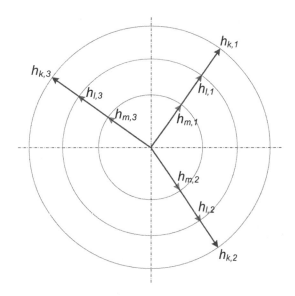

When $\epsilon_1(\alpha h)$ becomes zero for a sufficiently small α, then from Equation (2.8)

$$I(y_0 + \alpha h) - I(y_0) = dI(y_0; \alpha h) = \alpha dI(y_0; h)$$

since $dI(y_0; \alpha h)$ is linear. The above equation can be written in terms of a scalar variable β as

$$dI(y_0; h) = \lim_{\alpha \to 0} \frac{\{I[y_0 + (\alpha + \beta)h] - I(y_0 + \beta h)\}_{\beta=0}}{\alpha}$$

Introducing $\gamma \equiv \beta$ as the reference value undergoing the change ($\Delta\gamma \equiv \alpha$)

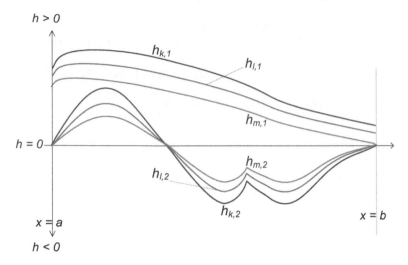

Figure 2.6 Changes in a continuous function for the Gâteaux differential

from β to $(\beta + \alpha)$, we get

$$dI(y_0; h) = \lim_{\Delta\gamma \to 0} \frac{\{I[y_0 + (\gamma + \Delta\gamma)h] - I(y_0 + \gamma h)\}_{\gamma=0}}{\Delta\gamma}$$

$$= \frac{\mathrm{d}}{\mathrm{d}\gamma} I(y_0 + \gamma h)_{\gamma=0}$$

Replacing γ by α, the Gâteaux differential can be written as

$$dI(y_0; h) = \frac{\mathrm{d}}{\mathrm{d}\alpha} I(y_0 + \alpha h)_{\alpha=0} \tag{2.9}$$

Equation (2.9) delivers a very important result. It shows that the Gâteaux differential of a functional is equal to its directional derivative (Section 9.9.1, p. 272) along h, i.e., the change in the associated function y at y_0. Thus, the value of the differential is the derivative of the functional with respect to the scalar multiplier to the function change, when evaluated at the zero value of the multiplier.

Example 2.2
Find the Gâteaux differential of the functional

$$I(y) = \int_0^1 (y^2 - 2y + 1) \, \mathrm{d}x$$

corresponding to the reference y-function y_0, and the variation h given, respectively, by

$$y_0 = 2x \quad \text{and} \quad h = x + 1$$

Let $y \equiv (y_0 + \alpha h)$ be a function in the vicinity of y_0. Then Equation (2.9), along with the above definition of y and specifications of y_0 and h, provides

$$dI(y_0; h) = \frac{d}{d\alpha} I(\underbrace{y_0 + \alpha h}_{y})_{\alpha=0} = \frac{d}{d\alpha} \int_0^1 [y^2 - 2y + 1]_{\alpha=0} dx$$

$$= \int_0^1 \frac{d}{d\alpha} [y^2 - 2y + 1]_{\alpha=0} dx = \int_0^1 \left[\frac{d}{dy}(y^2 - 2y + 1) \frac{dy}{d\alpha} \right]_{\alpha=0} dx$$

$$= \int_0^1 [2\underbrace{(y_0 + \alpha h)}_{y} - 2]_{\alpha=0} h\, dx = 2 \int_0^1 (y_0 - 1)h\, dx$$

$$= 2 \int_0^1 (\underbrace{2x}_{y_0} - 1)(\underbrace{x+1}_{h})\, dx = \frac{1}{3}$$

Observe that the Gâteaux differential of I is the functional

$$dI(y_0; h) = 2 \int_0^1 (y_0 - 1)h\, dx$$

depending on $y_0(x)$ and $h(x)$. As we show below, $dI(y_0; h)$ is linear and continuous with respect to h, as required by the definition.

Linearity of the Gâteaux Differential

The linearity can be easily verified through

$$dI(y_0; \alpha h + \beta k) = \alpha dI(y_0; h) + \beta dI(y_0; k)$$

Continuity of the Gâteaux Differential

To establish the continuity, we need to show that

$$\lim_{k \to h} dI(y_0; k) = dI(y_0; h)$$

Equivalently, for any function h and a real number $\epsilon > 0$ there exists another real number $\delta > 0$ such that

$$\left| dI(y_0; k) - dI(y_0; h) \right| < \epsilon$$

for any function k satisfying

$$\|h - k\| < \delta$$

Let γ be the maximum absolute value of $(h - k)$ in the x-interval $[0, 1]$. Then

$$\left| dI(y_0; h) - dI(y_0; k) \right| = \left| 2 \int_0^1 (y_0 - 1)(h - k) \, dx \right| \leq \left| 2 \int_0^1 (y_0 - 1)\gamma \, dx \right|$$

With γ related to $\|h - k\|$ as

$$\gamma = \kappa \|h - k\|$$

where κ is some positive real number, we have

$$\left| dI(y_0; k) - dI(y_0; h) \right| \leq 2\kappa \|h - k\| \underbrace{\left| \int_0^1 (y_0 - 1) \, dx \right|}_{I_0}$$

We can always find a positive real number ϵ greater than the right-hand side of the above inequality, i.e.,

$$2\kappa \|h - k\| I_0 < \epsilon$$

Thus, we can write

$$\left| dI(y_0; h) - dI(y_0; k) \right| < \epsilon$$

for which the second-last inequality provides

$$\|h - k\| < \underbrace{\epsilon/(2\kappa I_0)}_{\delta}$$

with the right-hand side standing for a positive real number δ. ▯

The above example introduces an important result. The Gâteaux differential of the functional

$$I = \int_a^b F(y) \, dx$$

is given by

$$dI(y_0; h) = \frac{d}{d\alpha} \left[\int_a^b F \, dx \right]_{\alpha=0} = \int_a^b \left[\frac{dF}{dy} \right]_{\alpha=0} h \, dx \qquad (2.10)$$

provided that dF/dy exists and is continuous with respect to y, which is defined as $y \equiv y_0 + \alpha h$.

Example 2.3

Find the Gâteaux differential of the functional

$$I(y) = \int_0^1 |y| \, dx$$

corresponding to the reference y-function y_0 and the variation h given by $y_0 = 0$ and $h = 2x$, respectively.

Let $y \equiv (y_0 + \alpha h)$ be a function in the vicinity of y_0. Applying Equation (2.10),

$$dI(y_0; h) = \int_0^1 \left[\frac{d}{dy}|y|\right]_{\alpha=0} h \, dx = 2\int_0^1 \left[\frac{d}{dy}|y|\right]_{\alpha=0} x \, dx$$

Now at $\alpha = 0$, we have $y = y_0 = 0$, for which the derivative $d|y|/dy$ is not defined. The reason is that the right and left-hand derivatives corresponding, respectively, to the change in y (i.e., Δy) greater and less than zero are not equal as follows:

$$\left[\frac{d}{dy}|y|\right]_{\alpha=0} = \lim_{\Delta y \to 0}\left[\frac{|y_0 + \Delta y| - |y_0|}{\Delta y}\right]_{y_0=0}$$

$$= \lim_{\Delta y \to 0}\frac{|\Delta y|}{\Delta y} = \begin{cases} 1 & \text{if} \quad \Delta y > 0 \\ -1 & \text{if} \quad \Delta y < 0 \end{cases}$$

As a consequence, the Gâteaux differential of I does not exist or is not defined at $y_0 = 0$. ▯

Example 2.4

In the previous example, if we specify the reference function as

$$y_0 = x + 1$$

then $y_0 > 0$ in the range of the integration. For this specification, both the right and left-hand derivatives of $|y|$ with respect to y have $\Delta y > 0$ as Δy tends to zero. Hence

$$\left[\frac{d}{dy}|y|\right]_{\alpha=0} = 1$$

and

$$dI(y_0; h) = 2\int_0^1 \left[\frac{d}{dy}|y|\right]_{\alpha=0} x \, dx = 1$$

▯

In fact, the Gâteaux differential of I in Example 2.3 is (i) 1 if $y_0 > 0$, and (ii) -1 if $y_0 < 0$ in the range of the integration.

2.4.3 Variation

To deal with an even wider class of functionals, the concept of the Gâteaux differential is further relaxed to the *Gâteaux variation*, or simply the variation. Similar to the Gâteaux differential, the variation $\delta I(y_0; h)$ of a functional I is

1. defined by
$$I(y_0 + \alpha h) - I(y_0) = \delta I(y_0; \alpha h) + \epsilon_1(\alpha h) \qquad (2.11)$$

   where $\epsilon_1(\alpha h)$ becomes zero for a sufficiently small α; and

2. is equal to the directional derivative along h, i.e.,

$$\boxed{\delta I(y_0; h) = \frac{\mathrm{d}}{\mathrm{d}\alpha} I(y_0 + \alpha h)_{\alpha=0}} \qquad (2.12)$$

However, $\delta I(y_0; h)$ does not have to be linear and continuous with respect to h as opposed to the Gâteaux differential.

2.4.3.1 Homogeneity of Variation

Note that $\delta I(y_0; h)$ is inherently a homogeneous functional of degree one. It means that
$$\delta I(y_0; \beta h) = \beta \delta I(y_0; h)$$

for any real number β. Recall from Example 2.1 (p. 27) that a homogeneous functional is not necessarily linear. The homogeneity of the variation arises from Equation (2.9) as follows. Let $\gamma \equiv \alpha\beta$. Then, from Equation (2.12),

$$\begin{aligned}
\delta I(y_0; \beta h) &= \frac{\mathrm{d}}{\mathrm{d}\alpha} I[y_0 + \alpha(\beta h)]_{\alpha=0} = \frac{\mathrm{d}}{\mathrm{d}\alpha} I(y_0 + \gamma h)_{\alpha=0} \\
&= \frac{\mathrm{d}\gamma}{\mathrm{d}\alpha} \frac{\mathrm{d}}{\mathrm{d}\gamma} I(y_0 + \gamma h)_{\gamma=0} = \beta \delta I(y_0; h)
\end{aligned}$$

Example 2.5

Find the variation of $I(y)$ in Example 2.2 (p. 32).

Similar to the Gâteaux differential, the variation is equal to the directional derivative
$$\frac{\mathrm{d}}{\mathrm{d}\alpha} I(y_0 + \alpha h)_{\alpha=0}$$

Thus, $\delta I(y_0; h)$ is the same as the $\mathrm{d}I(y_0; h)$ in Example 2.2 and is given by

$$\delta I(y_0; h) = 2 \int_0^1 (y_0 - 1) h \, \mathrm{d}x = \frac{1}{3}$$

The above variation is of course homogeneous with respect to h since

$$\delta I(y_0; \alpha h) \;=\; 2 \int_0^1 (y_0 - 1)\alpha h\, \mathrm{d}x \;=\; 2\alpha \int_0^1 (y_0 - 1)h\, \mathrm{d}x \;=\; \alpha \delta I(y_0; h)$$

☐

Generalization to Several Functions

For a functional dependent on a vector **y** of several functions in general, the above definitions for differentials apply with the vectors $\mathbf{y}_0$ and **h** replacing y_0 and h, respectively. For example, the general form of Equation (2.12) is

$$\delta I(\mathbf{y}_0; \mathbf{h}) = \frac{\mathrm{d}}{\mathrm{d}\alpha} I(\mathbf{y}_0 + \alpha \mathbf{h})_{\alpha=0} \tag{2.13}$$

Example 2.6
Find the variation of the functional

$$I(\mathbf{y}) = \int_0^1 (y_1^2 - 2y_2^2 + 1)\, \mathrm{d}x, \quad \mathbf{y} = \begin{bmatrix} y_1(x) & y_2(x) \end{bmatrix}^{\mathsf{T}}$$

corresponding to the reference **y**-function $\mathbf{y}_0$, and the variation **h** given, respectively, by

$$\begin{bmatrix} y_{0,1} \\ y_{0,2} \end{bmatrix} = \begin{bmatrix} 2x \\ 3x \end{bmatrix} \quad \text{and} \quad \begin{bmatrix} h_1 \\ h_2 \end{bmatrix} = \begin{bmatrix} x+1 \\ x \end{bmatrix}$$

Let $\mathbf{y} \equiv (\mathbf{y}_0 + \alpha \mathbf{h})$ be a function vector in the vicinity of $\mathbf{y}_0$. Then Equation (2.13), along with the above definition of **y** and specifications of $\mathbf{y}_0$ and **h**, provides

$$\delta I(\mathbf{y}_0; \mathbf{h}) \;=\; \frac{\mathrm{d}}{\mathrm{d}\alpha} I(\underbrace{\mathbf{y}_0 + \alpha \mathbf{h}}_{\mathbf{y}})_{\alpha=0} \;=\; \frac{\mathrm{d}}{\mathrm{d}\alpha} \int_0^1 [y_1^2 - 2y_2^2 + 1]_{\alpha=0}\, \mathrm{d}x$$

$$= \int_0^1 \frac{\mathrm{d}}{\mathrm{d}\alpha} [y_1^2 - 2y_2^2 + 1]_{\alpha=0}\, \mathrm{d}x \;=\; \int_0^1 \left[2y_1 \frac{\mathrm{d}y_1}{\mathrm{d}\alpha} - 4y_2 \frac{\mathrm{d}y_2}{\mathrm{d}\alpha} \right]_{\alpha=0} \mathrm{d}x$$

$$= \int_0^1 [2\underbrace{(y_{0,1} + \alpha h_1)}_{y_1} h_1 - 4\underbrace{(y_{0,2} + \alpha h_2)}_{y_2} h_2]_{\alpha=0}\, \mathrm{d}x$$

$$= 2 \int_0^1 (y_{0,1} h_1 - 2y_{0,2} h_2)\, \mathrm{d}x \;=\; -\frac{2}{3}$$

It can be easily verified that the above variation is homogeneous, i. e.,

$$\delta I(\mathbf{y}_0; \alpha \mathbf{h}) = \alpha \delta I(\mathbf{y}_0; \mathbf{h})$$

□

2.4.4　Summary of Differentials

In summary, we have three types of differentials of a functional:

1. Fréchet differential,

2. Gâteaux differential, and

3. Gâteaux variation, or simply variation

in the order of decreasing strictness for their existence. The Fréchet differential requires the error to decrease with function change regardless of its direction (or shape). On the other hand, the Gâteaux differential as well as the variation require the error to decrease with function change along its direction for all possible changes. While both Fréchet and Gâteaux differentials should be linear and continuous with respect to a change in function, a variation does not need to be so. It is inherently homogeneous with respect to the function change.

In other words, functionals are more likely to have variations than differentials. Thus, most of the time, we will use the variation of functionals in optimal control analysis. Fréchet and Gâteaux differentials will be invoked only when their typical properties are needed.

2.4.5　Relations between Differentials

The Fréchet differential of a functional is also the Gâteaux differential. In turn, the Gâteaux differential of a functional is also the variation. Thus, for a functional, the existence of the Fréchet differential implies the existence of the Gâteaux differential. In turn, the existence of the Gâteaux differential implies the existence of the variation. However, there is no guarantee that the reverse relations hold. For example, a functional may have the variation but not the Gâteaux differential. Using conditional statements, these relations are

$$\boxed{\begin{array}{c}\text{Fréchet}\\\text{differential}\end{array}} \longrightarrow \boxed{\begin{array}{c}\text{Gâteaux}\\\text{differential}\end{array}} \longrightarrow \boxed{\text{Variation}}$$

The following two examples illustrate these relations.

Example 2.7

Consider the functional

$$
I(y) = \begin{cases} \dfrac{y_1 y_2}{y_1 + y_2} & \text{if } y \neq 0 \\ 0 & \text{if } y = 0 \end{cases}
$$

where the function (or equivalently the vector) y has two real components, y_1 and y_2. Find the variation of I at $y_0 = 0$, i. e., for $y_{0,1} = 0$ and $y_{0,2} = 0$.

In the vicinity of y_0, we have

$$
\underbrace{\begin{bmatrix} y_1 \\ y_2 \end{bmatrix}}_{y} = \underbrace{\begin{bmatrix} 0 \\ 0 \end{bmatrix}}_{y_0} + \alpha \underbrace{\begin{bmatrix} h_1 \\ h_2 \end{bmatrix}}_{h} = \underbrace{\begin{bmatrix} \alpha h_1 \\ \alpha h_2 \end{bmatrix}}_{\alpha h}
$$

corresponding to which

$$
I(y_0 + h) = I(y) = \frac{\alpha h_1 h_2}{h_1 + h_2}
$$

From Equation (2.12), the variation of I is given by

$$
\delta I(y_0; h) = \frac{\mathrm{d}}{\mathrm{d}\alpha} I(y_0 + h)_{\alpha=0} = \frac{h_1 h_2}{h_1 + h_2}
$$

which fulfills the requirement of being first-degree homogeneous, i. e.,

$$
\delta I(y_0; \gamma h) = \gamma \delta I(h)
$$

where γ is any real number. This result can be easily verified by replacing h by γh in the expression for $\delta I(y_0; h)$. ▯

Example 2.8

In the previous example, does the Gâteaux differential of I exist at $y_0 = 0$?

The Gâteaux differential of I does not exist at $y_0 = 0$ since

$$
J(h) \equiv \frac{\mathrm{d}}{\mathrm{d}\alpha} I(y)_{\alpha=0} = \frac{h_1 h_2}{h_1 + h_2}
$$

is not linear with respect to h. Observe that

$$
J(h) + J(k) \neq J(h + k)
$$

where $k = \begin{bmatrix} k_1 & k_2 \end{bmatrix}^\top$ is some vector different from $h = \begin{bmatrix} h_1 & h_2 \end{bmatrix}^\top$. ▯

2.5   Variation of an Integral Objective Functional

Let us derive the variation of the integral objective functional

$$I(y) = \int_a^b F(y, y') \, \mathrm{d}x$$

where y and y' (or $\mathrm{d}y/\mathrm{d}x$) are functions of the independent variable x. The variation has to be obtained in the neighborhood of a reference function $y_0(x)$ having its derivative $y_0'(x)$ with respect to x. If the change from $y_0(x)$ to $y(x)$ is $\alpha h(x)$, then

$$y(x) = y_0(x) + \alpha h(x) \tag{2.14}$$
$$y'(x) = y_0'(x) + \alpha h'(x) \tag{2.15}$$

From Equation (2.12), the variation of I is given by

$$\delta I(y_0; h) = \frac{\mathrm{d}}{\mathrm{d}\alpha} I(y_0 + \alpha h)_{\alpha=0} = \frac{\mathrm{d}}{\mathrm{d}\alpha} \left[\int_a^b F(y_0 + \alpha h, y_0' + \alpha h') \, \mathrm{d}x \right]_{\alpha=0}$$

$$= \left[\int_a^b \frac{\mathrm{d}}{\mathrm{d}\alpha} F(y_0 + \alpha h, y_0' + \alpha h') \, \mathrm{d}x \right]_{\alpha=0}$$

Using Equations (2.14) and (2.15) in the above equation,

$$\delta I(y_0; h) = \left[\int_a^b \frac{\mathrm{d}}{\mathrm{d}\alpha} F(y, y') \, \mathrm{d}x \right]_{\alpha=0} = \left[\int_a^b \left(\frac{\partial F}{\partial y} \frac{\mathrm{d}y}{\mathrm{d}\alpha} + \frac{\partial F}{\partial y'} \frac{\mathrm{d}y'}{\mathrm{d}\alpha} \right) \mathrm{d}x \right]_{\alpha=0}$$

$$= \left[\int_a^b [F_y(y, y')h + F_{y'}(y, y')h'] \, \mathrm{d}x \right]_{\alpha=0}$$

where F_y and $F_{y'}$ are partial derivatives of the integrand F with respect to y and y', respectively. Expanding y and y' in the above equation using Equations (2.14) and (2.15), and substituting $\alpha = 0$, we get

$$\delta I(y_0; h) = \int_a^b [F_y(y_0, y_0')h + F_{y'}(y_0, y_0')h'] \, \mathrm{d}x$$

It is convenient to express the above variation simply as

$$\delta I = \int_a^b [F_y(y, y')\delta y + F_{y'}(y, y')\delta y'] \, \mathrm{d}x \tag{2.16}$$

where it is understood that

- y is the reference function $y_0(x)$, and y' is the corresponding derivative function $y_0'(x)$; and

- δy and δy are the functions $h(x)$ and $h'(x)$, respectively.

In the above equation, the coefficient of δy (i.e., F_y) is called the **variational derivative** of the functional I with respect to y. Similarly, $F_{y'}$ is called the variational derivative of I with respect to y'. If $F = F(y')$, then obviously

$$\delta I = \int_a^b F_y(y)\delta y \, dx \tag{2.17}$$

Example 2.9
Find the variation of the functional

$$I = \int_0^1 [y^3(x) - y'^2(x)] \, dx$$

at $y = x^3 - 1$ for $\delta y = -x$.

Using Equation (2.16), the variation of I is given by

$$\delta I = \int_0^1 [3y^2\delta y - 2y'\delta y'] \, dx$$

When $y = x^3 - 1$, $y' = 3x^2$ so that

$$\delta I = \int_0^1 [3(x^3 - 1)^2\delta y - 6x^2\delta y'] \, dx$$

Upon substituting $\delta y = x$ and the corresponding $\delta y' = -1$ in the above equation and integrating, we obtain

$$\delta I = 1\frac{13}{40}$$

□

2.5.1 Equivalence to Other Differentials

Consider the integral functional

$$I = \int_a^b F(y) \, dx \tag{2.18}$$

where y is a function of x. If dF/dy exists and is continuous with respect to x, then the variation of I is also the Fréchet differential as well as the Gâteaux differential.

2.5.1.1 Equivalence to the Fréchet Differential

The Fréchet differential $dI(y; h)$ is defined by

$$\lim_{h \to 0} \frac{I(y+h) - I(y) - dI(y; h)}{\|h\|} = 0 \tag{2.7}$$

The above expression is equivalent to the following statement:

Given an $\epsilon > 0$ there is a $\delta > 0$ such that for $\|h\| < \delta$,

$$\left| \frac{I(y+h) - I(y) - dI(y; h)}{\|h\|} \right| < \epsilon$$

If the variation $\delta I(y; h)$ is equivalent to the Fréchet differential, then the above statement must hold for the variation $\delta I(y; h)$ substituted for $dI(y; h)$.

Let us define

$$p \equiv \left| I(y+h) - I(y) - \delta I(y; h) \right|$$

Then we need to show that $p < \epsilon \|h\|$.

From the definition of I in Equation (2.18) and Equation (2.17), we have

$$p = \left| \int_a^b \left[F(y+h) - F(y) - F_y(y)h(x) \right] dx \right|$$

Applying the Mean Value Theorem for derivatives (Section 9.14.1, p. 276) at any x in $[a, b]$, we obtain

$$\frac{F(y+h) - F(y)}{h} = \frac{dF}{dy} \bigg|_{\bar{y}} = F_y(\bar{y})$$

for some $\bar{y}$ in the interval $[y - h, y + h]$. Substituting this result in the expression for p,

$$p = \left| \int_a^b \left[F_y(\bar{y}, x) - F_y(y, x) \right] h(x) \, dx \right|$$

The function $F_y(y)$, being a continuous function of y and x in the closed interval $[a, b]$, is uniformly continuous therein. Thus, for an $\epsilon_1 > 0$ there exists a $\delta > 0$ such that for $\|\bar{y} - y\| < \delta$,

$$|F_y(\bar{y}, x) - F_y(y, x)| < \epsilon_1$$

Now in terms of $\left| \bar{h} \right|$, the maximum absolute value of h in the interval $[a, b]$, we obviously have

$$|h(x)| \leq \left| \bar{h} \right| \equiv \beta \|h\|$$

where $|\bar{h}|$ is defined in terms of a positive real number β and the norm $\|h\|$. Multiplying the last two inequalities and integrating with respect to x over the interval $[a, b]$, we get

$$\int_a^b \left| F_y\left[(\bar{y}, x) - F_y(y, x)\right] h(x) \right| \, dx \; < \; \underbrace{\beta(b-a)\epsilon_1 \|h\|}_{\epsilon}$$

where we introduce $\epsilon \equiv \beta(b-a)\epsilon_1$. From the triangle inequality for integrals (Section 9.22, p. 280),

$$\underbrace{\left| \int_a^b [F_y(\bar{y}, x) - F_y(y, x)] h(x) \, dx \right|}_{p} \; \leq \; \int_a^b \left| [F_y(\bar{y}, x) - F_y(y, x)] h(x) \right| \, dx$$

The last two inequalities yield

$$p = \left| \int_a^b [F_y(\bar{y}, x) - F_y(y, x)] h(x) \, dx \right| \; < \; \epsilon \|h\|$$

which was to be proved.

2.5.1.2 Equivalence to the Gâteaux Differential

To show this equivalence, we first show that the Fréchet differential of a functional is also its Gâteaux differential.

The Fréchet differential, $dI(y_0; h)$, is defined by Equation (2.6) if the ratio $\epsilon(h)/\|h\|$ tends to zero with $\|h\|$. Let $h = \alpha k$ where α is a real number and k is a function. Then the definition can be written as

$$I(y_0 + \alpha k) - I(y_0) = dI(y_0; \alpha k) + \epsilon(\alpha k), \qquad \lim_{\|\alpha k\| \to 0} \frac{\epsilon(\alpha k)}{\|\alpha k\|} = 0$$

Since $\lim \|\alpha k\| \to 0$ is equivalent to $\lim \alpha \to 0$ for a given k, we obtain

$$I(y_0 + \alpha k) - I(y_0) = dI(y_0; \alpha k) + \epsilon(\alpha k), \qquad \lim_{\alpha \to 0} \frac{\epsilon_1(\alpha k)}{\alpha} = 0$$

which is the definition of the Gâteaux differential [see Equation (2.8), p. 30]. Hence, the Fréchet differential is inherently the Gâteaux differential of a functional. This result plus the equivalence of the variation to the Fréchet differential of an integral functional, as shown in the last section, proves the following:

The variation of an integral functional is equivalent to the Fréchet as well as the Gâteaux differential provided that the derivative of the integrand exists and is continuous with respect to the variable of integration.

2.5.2   Application to Optimal Control Problems

In an optimal control problem, we arrive at an integral objective functional having the general form

$$J = \int_0^{t_f} f(\mathbf{y}, \dot{\mathbf{y}}, \mathbf{u}, \boldsymbol{\lambda}) \, dt \tag{2.19}$$

where the integrand depends on

1. the vector of state variables:

$$\mathbf{y} = \begin{bmatrix} y_1(t) & y_2(t) & \cdots & y_n(t) \end{bmatrix}^\top,$$

2. the derivative of $\mathbf{y}$ with respect to time t:

$$\dot{\mathbf{y}} = \begin{bmatrix} \dot{y}_1(t) & \dot{y}_2(t) & \cdots & \dot{y}_n(t) \end{bmatrix}^\top,$$

3. the vector of controls:

$$\mathbf{u} = \begin{bmatrix} u_1(t) & u_2(t) & \cdots & u_m(t) \end{bmatrix}^\top, \text{ and}$$

4. the vector of certain undetermined multiplier functions or costates:

$$\boldsymbol{\lambda} = \begin{bmatrix} \lambda_1(t) & \lambda_2(t) & \cdots & \lambda_n(t) \end{bmatrix}^\top$$

It is desired to determine the optimal control $\mathbf{u}$ that optimizes (i. e., either minimizes or maximizes) J. To that end, we need to obtain the variation of J to help determine the optimum. The variation of J is a straightforward generalization of Equation (2.16) and is given by

$$\boxed{\delta J = \int_0^{t_f} \left\{ \sum_{i=1}^n f_{y_i} \delta y_i + \sum_{i=1}^n f_{\dot{y}_i} \delta \dot{y}_i + \sum_{i=1}^m f_{u_i} \delta u_i + \sum_{i=1}^n f_{\lambda_i} \delta \lambda_i \right\} dt} \tag{2.20}$$

where the partial derivatives — f_{y_i}, $f_{\dot{y}_i}$, f_{u_i}, and f_{λ_i} — are the variational derivatives of the functional J with respect to y_i, $\dot{y}_i$, u_i, and λ_i, respectively. Note that each of δy_i, $\delta \dot{y}_i$, δu_i, and $\delta \lambda_i$ is a function of t.

In compact vector notation,

$$\delta J = \int_0^{t_f} \left\{ f_{\mathbf{y}}^\top \delta \mathbf{y} + f_{\dot{\mathbf{y}}}^\top \delta \dot{\mathbf{y}} + f_{\mathbf{u}}^\top \delta \mathbf{u} + f_{\boldsymbol{\lambda}}^\top \delta \boldsymbol{\lambda} \right\} dt$$

where $\delta\mathbf{y}$, $\delta\dot{\mathbf{y}}$, $\delta\mathbf{u}$, and $\delta\boldsymbol{\lambda}$ are the variation vectors corresponding to $\mathbf{y}$, $\dot{\mathbf{y}}$, $\mathbf{u}$, and $\boldsymbol{\lambda}$, respectively. Thus, for example,

$$\delta\mathbf{y} = \begin{bmatrix} \delta y_1(t) & \delta y_2(t) & \cdots & \delta y_n(t) \end{bmatrix}^{\top}$$

Example 2.10

A simpler version of the batch reactor problem (p. 2) for the reaction

$$a\mathrm{A} \longrightarrow c\mathrm{C}$$

without any inequality constraints is as follows.

Find the $T(t)$ that maximizes

$$I = ck_0 \int_0^{t_{\mathrm{f}}} \exp\left[-\frac{E}{RT(t)}\right] x^a(t)\, dt \qquad (2.21)$$

subject to the satisfaction of the differential equation constraint

$$\dot{x} = -ak_0 \exp\left[-\frac{E}{RT(t)}\right] x^a(t), \quad x(0) = x_0 \qquad (2.22)$$

where $\dot{x}$ is the derivative of x with respect to t, and x_0 is the initial concentration of A.

To proceed, we need to apply the Lagrange Multiplier Rule, the details of which will be provided later in Chapter 4. According to this rule, the above constrained problem is equivalent to the problem of finding the control $T(t)$ that maximizes the following augmented functional:

$$J = ck_0 \int_0^{t_{\mathrm{f}}} \exp\left(-\frac{E}{RT}\right) x^a\, dt + \int_0^{t_{\mathrm{f}}} \lambda\left[-\dot{x} - ak_0 \exp\left(-\frac{E}{RT}\right) x^a\right] dt$$

$$= \int_0^{t_{\mathrm{f}}} \underbrace{\left\{ ck_0 \exp\left(-\frac{E}{RT}\right) x^a - \lambda\left[\dot{x} + ak_0 \exp\left(-\frac{E}{RT}\right) x^a\right]\right\}}_{f} dt \qquad (2.23)$$

subject to $x(0) = x_0$ with λ as an undetermined multiplier function of t. To avoid clutter, we will from now on show the bracketed independent variables of functions only if needed for clarity.

At this point it is sufficient to learn how the augmented functional J is formed by adjoining the constraint [Equation (2.22)] to the original functional I after bringing all constraint terms to the right-hand side and multiplying them by an undetermined multiplier λ.

In the present example, we will find the variation of J. Denoting the integrand in Equation (2.23) by f and considering the dependence

$$f = f(x, \dot{x}, T, \lambda)$$

the variation of J is given by

$$\delta J = \int_0^{t_f} (f_x \delta x + f_{\dot{x}} \delta \dot{x} + f_T \delta T + f_\lambda \delta \lambda)\, dt$$

where f_x, $f_{\dot{x}}$, f_T, and f_λ are partial derivatives of f with respect to x, $\dot{x}$, T, and λ, respectively. Therefore,

$$\delta J = \int_0^{t_f} \left\{ \left[k_0 \exp\left(-\frac{E}{RT} \right)(c - a\lambda)ax^{a-1} \right] \delta x - [\lambda]\delta \dot{x} + \left[\frac{k_0 E}{RT^2} \exp\left(-\frac{E}{RT} \right) \right. \right.$$

$$\left. (c - a\lambda)x^a \right] \delta T - \left[\dot{x} + ak_0 \exp\left(-\frac{E}{RT} \right) x^a \right] \delta \lambda \right\} dt$$

where δx, $\delta \dot{x}$, δT, and $\delta \lambda$ are functions of t. The coefficient of $\delta \lambda$ in the above equation is zero because of the differential equation constraint, Equation (2.22), which must be satisfied at each t in the interval $[0, t_f]$. Hence, the final expression for δJ is

$$\delta J = \int_0^{t_f} \left\{ \left[k_0 \exp\left(-\frac{E}{RT} \right)(c - a\lambda)ax^{a-1} \right] \delta x - [\lambda]\delta \dot{x} \right.$$

$$\left. + \left[\frac{k_0 E}{RT^2} \exp\left(-\frac{E}{RT} \right)(c - a\lambda)x^a \right] \delta T \right\} dt$$

$$\square$$

Example 2.11
Consider the simplified heat exchanger problem (Section 1.3.3, p. 8) in which it is desired to find the $T_w(t)$ that minimizes

$$I = \int_0^L [T(x, t_f) - T^*(x)]^2\, dx$$

subject to the satisfaction of the differential equation constraint

$$\dot{T} = -vT' + \frac{h}{\rho C_p}[T_w(t) - T] \tag{2.24}$$

and $T(z, 0) = T(0, t) = T_0$ with $\dot{T}$ and T' as partial derivatives of T with respect to t and x, respectively. The temperature profile $T^*(x)$ is specified. With the application of the Lagrange Multiplier Rule, this constrained problem is equivalent to the minimization of the following augmented objective functional:

$$J = \int_0^L \underbrace{\left[T(x,t_{\mathrm f}) - T^*(x)\right]^2}_{F} \mathrm{d}x + \int_0^L \int_0^{t_{\mathrm f}} \underbrace{\lambda\left[-\dot{T} - vT' + \frac{h}{\rho C_p}(T_{\mathrm w} - T)\right]}_{G} \mathrm{d}t\,\mathrm{d}x$$

$$(2.25)$$

subject to $T(z,0) = T(0,t) = T_0$ where the undetermined multiplier λ is a function of the two independent variables x and t. Note the presence of the double integral corresponding to these variables. Let us now find the variation of J.

Denoting the first and second integrand of Equation (2.25) by F and G, respectively, and considering the dependencies

$$F = F[T(x,t_{\mathrm f})] \quad \text{and} \quad G = G(\dot{T}, T', T, T_{\mathrm w}, \lambda)$$

the variation of J is given by

$$\delta J = \int_0^L F_{T(x,t_{\mathrm f})} \delta T(x,t_{\mathrm f})\,\mathrm{d}x + \int_0^L \int_0^{t_{\mathrm f}} \Big\{ G_{\dot{T}}\delta\dot{T} + G_{T'}\delta T' + G_T\delta T + G_{T_{\mathrm w}}\delta T_{\mathrm w}$$

$$+ G_\lambda \delta\lambda \Big\}\mathrm{d}t\,\mathrm{d}x$$

where $F_{T(x,t_{\mathrm f})}$ is the partial derivative of F with respect to $T(x,t_{\mathrm f})$, $G_{\dot{T}}$ is the partial derivative of G with respect to $\dot{T}$, and so on. We have

$$G_\lambda = -\dot{T} - vT' + \frac{h}{\rho C_p}(T_{\mathrm w} - T) = 0$$

because of the differential equation constraint, Equation (2.24). Thus, the variation of J is given by

$$\delta J = 2 \int_0^L \left[T(x,t_{\mathrm f}) - T^*(x)\right]\delta T(x,t_{\mathrm f})\,\mathrm{d}x - \int_0^L \int_0^{t_{\mathrm f}} \Big\{\lambda\Big[\delta\dot{T} + v\delta T' + \Big(\frac{h}{\rho C_p}\Big)\delta T$$

$$- \Big(\frac{h}{\rho C_p}\Big)\delta T_{\mathrm w}\Big]\Big\}\mathrm{d}t\,\mathrm{d}x$$

$$\square$$

Example 2.12
The batch distillation problem of Section 1.3.1 (p. 5) is equivalent to the minimization of

$$I = \int_0^{t_{\mathrm f}} \{u + \mu u(y - y^*)\}\,\mathrm{d}t \qquad (2.26)$$

subject to

$$\dot{m} = -u, \tag{2.27}$$

$$\dot{x} = \frac{u}{m}(x - y), \tag{2.28}$$

$m(0) = m_0$, and $x(0) = x_0$ with μ as a time invariant multiplier used to incorporate the purity specification, Equation (1.10), in Equation (2.26).

The problem at hand has only two constraints, namely, Equations (2.27) and (2.28). With the application of the Lagrange Multiplier Rule, this problem is equivalent to the minimization of the following augmented objective functional:

$$J = \int_0^{t_f} \underbrace{\left\{ u + \mu u(y - y^*) + \lambda_1(-\dot{m} - u) + \lambda_2\left[-\dot{x} + \frac{u}{m}(x - y)\right] \right\}}_{f} dt$$

subject to $m(0) = m_0$ and $x(0) = x_0$. Let us now find the variation of J.

We denote the integrand of the above equation by f and observe that

$$f = f(m, \dot{m}, x, \dot{x}, y, u, \mu, \lambda_1, \lambda_2)$$

where all arguments of f except μ are functions of t. Then the variation of J is given by

$$\delta J = \int_0^{t_f} \Big\{ f_m \delta m + f_{\dot{m}} \delta\dot{m} + f_x \delta x + f_{\dot{x}} \delta\dot{x} + f_y \delta y + f_u \delta u + f_\mu \delta\mu + f_{\lambda_1} \delta\lambda_1$$
$$+ f_{\lambda_2} \delta\lambda_2 \Big\} dt$$

The last three terms of the above equation are zero since

$$\int_0^{t_f} f_\mu \delta\mu \, dt = \delta\mu \int_0^{t_f} u(y - y^*) \, dt = 0, \qquad f_{\lambda_1} = -\dot{m} - u = 0, \qquad \text{and}$$

$$f_{\lambda_2} = -\dot{x} + \frac{u}{m}(x - y) = 0$$

respectively, due to the constraints — Equations (1.10), (2.27), and (2.28) — which have to be satisfied by necessity. In general, the coefficient of the variation of a multiplier will always be zero due to the presence of the corresponding constraint. Hence, the variation of J is given by

$$\delta J = \int_0^{t_f} \Bigg\{ \left[\frac{\lambda_2 u}{m^2}(y - x)\right]\delta m - [\lambda_1]\delta\dot{m} + \left[\frac{\lambda_2 u}{m}\right]\delta x - [\lambda_2]\delta\dot{x} + \left[u\left(\mu - \frac{\lambda_2}{m}\right)\right]\delta y$$

$$+ \left[1 + \mu(y - y^*) - \lambda_1 + \frac{\lambda_2}{m}(x - y)\right]\delta u \Bigg\} dt \tag{2.29}$$

$\square$

Further Simplification

The variations of integral objective functionals can be further simplified at their optima by integrating by parts the terms involving derivative functions such as $\delta\dot{x}$ and $\delta\dot{m}$ in Equation (2.29). As a matter of fact, this simplification is an important step in deriving the optimality conditions. Therefore, it is very important to familiarize oneself with the following formula of **integration by parts**:

$$
\int_a^b yz\,\mathrm{d}x = \left[y \int_a^x z\,\mathrm{d}x \right]_a^b - \int_a^b \left\{ \frac{\mathrm{d}y}{\mathrm{d}x} \int_a^x z\,\mathrm{d}x \right\} \mathrm{d}x \tag{2.30}
$$

where y and z are continuous functions of x, with y being additionally differentiable. Note that the right-hand side of the above equation carries only the integral of the second function z — a simplification in case z is a derivative function like $\delta\dot{x}$.

Example 2.13

For the batch distillation problem, let us simplify δJ given by Equation (2.29) when J is optimal. We can get rid of $\delta\dot{m}$ and $\delta\dot{x}$ in the following two terms of δJ:

$$
- \int_0^{t_f} [\lambda_1]\delta\dot{m}\,\mathrm{d}t \quad \text{and} \quad - \int_0^{t_f} [\lambda_2]\delta\dot{x}\,\mathrm{d}t
$$

using integration by parts. It is possible since the multiplier functions — λ_1 and λ_2 — are differentiable whenever J is optimal. We will prove the differentiability of multiplier functions later in Chapter 3 on p. 78. Note that the derivative function has to be the second function z when utilizing Equation (2.30). Its application yields

$$
- \int_0^{t_f} [\lambda_1]\delta\dot{m}\,\mathrm{d}t = - \left[\lambda_1 \int_0^t \delta\dot{m}\,\mathrm{d}t \right]_0^{t_f} + \int_0^{t_f} \left\{ \frac{\mathrm{d}\lambda_1}{\mathrm{d}t} \int_0^t \delta\dot{m}\,\mathrm{d}t \right\} \mathrm{d}t
$$

$$
= - \left[\lambda_1 \delta m \right]_0^{t_f} + \int_0^{t_f} \dot{\lambda}_1 \delta m\,\mathrm{d}t
$$

and similarly

$$
- \int_0^{t_f} [\lambda_2]\delta\dot{x}\,\mathrm{d}t = - \left[\lambda_2 \delta x \right]_0^{t_f} + \int_0^{t_f} \dot{\lambda}_2 \delta x\,\mathrm{d}t
$$

Upon substituting the last two equations into Equation (2.29), we get

$$
\delta J = \int_0^{t_f} \left\{ \left[\dot{\lambda}_1 - \frac{\lambda_2 u}{m^2}(x - y) \right] \delta m + \left[\dot{\lambda}_2 + \frac{\lambda_2 u}{m} \right] \delta x + \left[u \left(\mu - \frac{\lambda_2}{m} \right) \right] \delta y \right.
$$

$$
\left. + \left[1 + \mu(y - y^*) - \lambda_1 + \frac{\lambda_2}{m}(x - y) \right] \delta u \right\} dt
$$

$$
- \left[\lambda_1 \delta m \right]_0^{t_f} - \left[\lambda_2 \delta x \right]_0^{t_f}
$$

The variation of J thus obtained is readily amenable to further analysis for the determination of optimal control. ⬚

2.6 Second Variation

Let $\delta I(y_0; h)$ exist at y_0 as well as $(y_0 + \alpha h)$ for sufficiently small α. Then the variation of $\delta I(y_0; h)$ at y_0 is the second variation of the functional I and is denoted by $\delta^2 I(y_0; h)$. Using Equation (2.12), we obtain

$$
\delta^2 I(y_0; h) = \frac{d}{d\alpha} \delta I(y_0; \alpha h)_{\alpha=0} = \lim_{\alpha \to 0} \frac{\delta I(y_0 + \alpha h; \alpha h) - \delta I(y_0; \alpha h)}{\alpha}
$$

$$
= \lim_{\alpha \to 0} \frac{\frac{d}{d\beta} I(y_0 + \alpha h + \beta h)_{\beta=0} - \frac{d}{d\beta} I(y_0 + \beta h)_{\beta=0}}{\alpha}
$$

With $\gamma \equiv \beta$, and $\Delta\gamma \equiv \alpha$, we get

$$
\delta^2 I(y_0; h) = \lim_{\Delta\gamma \to 0} \frac{\frac{d}{d\gamma} I[y_0 + (\gamma + \Delta\gamma)h]_{\gamma=0} - \frac{d}{d\gamma} I(y_0 + \gamma h)_{\gamma=0}}{\Delta\gamma}
$$

$$
= \frac{d}{d\gamma} \left[\frac{d}{d\gamma} I(y_0 + \gamma h)_{\gamma=0} \right] = \frac{d^2}{d\gamma^2} I(y_0 + \gamma h)_{\gamma=0}
$$

Replacing γ by α, we obtain the second variation of I at y_0 as

$$
\boxed{\delta^2 I(y_0; h) = \frac{d^2}{d\alpha^2} I(y_0 + \alpha h)_{\alpha=0}} \tag{2.31}
$$

2.6.1 Second Degree Homogeneity

We now show that the second variation is homogeneous of the second degree in h, i.e.,

$$
\delta^2 I(y_0; \beta h) = \beta^2 \delta^2 I(y_0; h)
$$

Replacing h by βh in Equation (2.31), we get

$$\delta^2 I(y_0; \beta h) = \frac{d^2}{d\alpha^2} I(y_0 + \alpha\beta h)_{\alpha=0}$$

Let $\gamma \equiv \alpha\beta$. Then using the chain rule of differentiation,

$$\delta^2 I(y_0; \beta h) = \frac{d}{d\gamma}\left[\frac{d}{d\gamma} I(y_0 + \alpha\beta h)_{\gamma=0}\frac{d\gamma}{d\alpha}\right]\frac{d\gamma}{d\alpha}$$

$$= \beta^2 \frac{d^2}{d\gamma^2} I(y_0 + \gamma h)_{\gamma=0} = \beta^2 \delta^2 I(y_0; h)$$

2.6.2   Contribution to Functional Change

The functional change, $I(y_0 + h) - I(y_0)$, can be expressed in terms of the second variation provided $(d^2/d\alpha^2)I(y_0+\alpha h)$ exists and is continuous at $\alpha = 0$ for all h. Using the second order Taylor expansion (Appendix 2.A, p. 52) along a function g,

$$I(y_0 + \alpha g) - I(y_0) = \delta I(y_0; g)\alpha + \left[\frac{d^2}{d\beta^2} I(y_0 + \beta g)_{\beta=\gamma\alpha}\right]\frac{\alpha^2}{2}, \quad |\gamma| < 1$$

$$= \delta I(y_0; g)\alpha + \left[\underbrace{\frac{d^2}{d\beta^2} I(y_0 + \beta g)_{\beta=\gamma\alpha} - \frac{d^2}{d\alpha^2} I(y_0 + \alpha g)_{\alpha=0}}_{\varepsilon_2(\alpha g)}\right.$$

$$\left. + \delta^2 I(y_0; g)\right]\frac{\alpha^2}{2}, \quad |\gamma| < 1$$

where we have subtracted and added $\delta^2 I(y_0; g)$, and used Equation (2.31) for the subtracted term. Note that the term $\varepsilon_2(\alpha g)$ shown above tends to zero with α due to the continuity of $(d^2/d\alpha^2)I(y_0+\alpha g)$ with respect to α at $\alpha = 0$.

Since $\delta I(y_0; g)$ and $\delta^2 I(y_0; g)$ are homogeneous of the respective first and second degrees,

$$I(y_0 + \alpha g) - I(y_0) = \delta I(y_0; \alpha g) + \frac{1}{2}\delta^2 I(y_0; \alpha g) + \varepsilon_2(\alpha g)\frac{\alpha^2}{2}$$

With $h \equiv \alpha g$, and $\epsilon_2 \equiv \varepsilon_2\alpha^2/2$, we obtain

$$\boxed{I(y_0 + h) - I(y_0) = \delta I(y_0; h) + \frac{1}{2}\delta^2 I(y_0; h) + \epsilon_2(h)} \qquad (2.32)$$

where the error ϵ_2/α^2 tends to zero with α^2. The error is zero for sufficiently small α, or equivalently the change along a given h.

Generalization to Several Functions

The above result can be generalized to a functional dependent on several functions. Thus,

$$I(\mathbf{y}_0 + \mathbf{h}) - I(\mathbf{y}_0) = \delta I(\mathbf{y}_0; \mathbf{h}) + \frac{1}{2}\delta^2 I(\mathbf{y}_0; \mathbf{h}) + \epsilon_2(\mathbf{h}) \qquad (2.33)$$

where I depends on the function vector $\mathbf{y}$ given by

$$\begin{bmatrix} y_1(x) \\ y_2(x) \\ \vdots \\ y_n(x) \end{bmatrix} = \underbrace{\begin{bmatrix} y_{0,1}(x) \\ y_{0,2}(x) \\ \vdots \\ y_{0,n}(x) \end{bmatrix}}_{\mathbf{y}_0} + \alpha \underbrace{\begin{bmatrix} h_1(x) \\ h_2(x) \\ \vdots \\ h_n(x) \end{bmatrix}}_{\mathbf{h}}$$

and ϵ/α^2 tends to zero with α^2. The error is zero for sufficiently small α, or equivalently the change along $\mathbf{h}$.

We end this chapter by pointing out that the variation of an objective functional will provide us with important clues about its optimum, similar to what a differential does for an objective function. The second variation will provide some auxiliary conditions and help in the search for optimal solutions. The necessary and sufficient conditions for the optimum of an objective functional will be the topic of the next chapter. As expected, those conditions will use the concepts we have developed here.

2.A   Second-Order Taylor Expansion

Considering $I(y_0 + \alpha h)$ to be a function of α, the Fundamental Theorem of Calculus (Section 9.13, p. 275) gives

$$I(y_0 + \alpha h) - I(y_0) = \int_0^\alpha I'(y_0 + \beta h)\,\mathrm{d}\beta$$

where $'$ denotes the derivative with respect to β. In terms of

$$v \equiv I'(y_0 + \beta h) \quad \text{and} \quad w \equiv (\alpha - \beta)$$

we can write

$$I(y_0 + \alpha h) - I(y_0) = \int_0^\alpha \underbrace{I'(y_0 + \beta h)}_{v} \times \underbrace{1}_{w'} \, d\beta = \int_0^\alpha vw' \, d\beta$$

$$= \Big[vw \Big]_{\beta=0}^{\beta=\alpha} - \int_0^\alpha v'w \, d\beta \quad \text{(upon integration by parts)}$$

$$= \underbrace{\alpha \, I'(y_0 + \beta h)_{\beta=0}}_{\delta I(y_0;h)} + \underbrace{\int_0^\alpha [I''(y_0 + \beta h)](\alpha - \beta) \, d\beta}_{R_2} \quad (2.34)$$

where $I''(y_0 + \beta h)$ is the second derivative of $I(y_0 + \beta h)$ with respect to β.

If m and M are the respective minimum and maximum values of $I''(y_0 + \beta h)$ in the interval $0 \leq \beta \leq \alpha$, then

$$m \int_0^\alpha (\alpha - \beta) \, d\beta \leq R_2 \leq M \int_0^\alpha (\alpha - \beta) \, d\beta$$

or
$$m \frac{\alpha^2}{2} \leq R_2 \leq M \frac{\alpha^2}{2}$$

or
$$m \leq \frac{R_2}{\alpha^2/2} \leq M$$

Hence $R_2/(\alpha^2/2)$ is bounded by the minimum and maximum values of the second derivative, $I''(y_0 + \beta h)$, which is continuous. Therefore, by the Intermediate Value Theorem (Section 9.15, p. 276), $R_2/(\alpha^2/2)$ must be equal to $I''(y_0 + \beta h)$ for *some* β in the interval $0 < \beta < \alpha$.

The interval is $-\alpha < \beta < 0$ if α is negative. Hence the combined result for a non-zero α is

$$R_2 = \frac{\alpha^2}{2} I''(y_0 + \beta h), \quad \beta < |\alpha|$$

Let $\gamma \equiv \beta/\alpha$. Then

$$\beta = \alpha\gamma < |\alpha| \quad \text{or} \quad |\gamma| < 1$$

and Equation (2.34) can be finally written as [compare with Equation (9.1) in Section 9.6 (p. 270)]

$$\boxed{I(y_0 + \alpha h) - I(y_0) = \alpha \delta I(y_0; h) + \frac{\alpha^2}{2} \frac{d^2}{d\beta^2} I(y_0 + \beta h)_{\beta=\gamma\alpha}, \quad |\gamma| < 1}$$

Bibliography

D.G. Luenberger. *Optimization by Vector Space Methods*, Chapter 7, pages 169–175. John Wiley & Sons Inc., New York, 1969.

M.Z. Nashed. Some remarks on variations and differentials. *Am. Math. Mon.*, 73(4):63–76, 1966.

H. Sagan. *Introduction to the Calculus of Variations*, Chapter 1, pages 9–29. Dover Publications Inc., New York, 1969.

D.R. Smith. *Variational Methods in Optimization*, Chapter 1, pages 9–30. Dover Publications Inc., New York, 1998.

Exercises

2.1 Given two functionals $I[y(x)]$ and $J[y(x)]$, find the variation of

a. $I[y(x)]J[y(x)]$, and

b. $\dfrac{I[y(x)]}{J[y(x)]}$ where $J[y(x)] \neq 0$

2.2 In the plug flow reactor problem of Section 1.3.2 (p. 6), substitute Equation (1.11) into Equation (1.12) and find the variation of the resulting objective functional.

2.3 Derive the augmented functional for the optimal periodic control problem in Section 1.3.5 (p. 11). Find the variation of that functional.

2.4 Repeat the previous exercise for the nuclear reactor problem in Section 1.3.6 (p. 12) ignoring the inequality constraints.

2.5 Simplify the variations obtained in the last two exercises using integration by parts.

2.6 Follow the approach of Example 2.11 (p. 46) to derive the augmented functional for the blood flow and metabolism problem in Section 1.3.10 (p. 17). Find the variation of that functional.

2.7 Repeat the previous exercise for the medicinal drug delivery problem in Section 1.3.9 (p. 16).

2.8 The objective functional of an optimal control problem depends on functions as well as their derivatives. Under what circumstances would the variation of such a functional become the differential?

2.9 Show that the Gâteaux differential of the functional in Example 2.3 (p. 35) is (i) -1 if $y_0 < 0$, and (ii) 1 if $y_0 > 1$ in the range of the integration.

Chapter 3

Optimality in Optimal Control Problems

This chapter presents the conditions related to the optimality of a functional. We derive the necessary conditions for optimal control to exist and apply them to optimal control problems. We also present sufficient conditions assuring the optimum under certain conditions. Readers are encouraged to review the logic of the conditional statement in Section 9.25 (p. 282).

3.1   Necessary Condition for Optimality

Let I be a functional of a function y. If I is minimum at $\hat{y}$, then the values of I for all other admissible functions in the vicinity of $\hat{y}$ cannot be lower than the value of I at $\hat{y}$. Precisely,

$$I(y) - I(\hat{y}) \geq 0$$

is the necessary condition for the minimum of I where y is any admissible function in the vicinity and the norm of the function change $(y - \hat{y})$ is less than some positive number. Denoting $(y - \hat{y})$ by δy in the above inequality, we have

$$I(\underbrace{\hat{y} + \alpha\delta y}_{y}) - I(\hat{y}) \geq 0 \tag{3.1}$$

where α is a non-zero scalar variable with absolute value less than or equal to unity. In other words, $-1 \leq \alpha < 0$ and $0 < \alpha \leq 1$ are the two intervals of α.

Now if I has a variation at $\hat{y}$, then from the definition of variation [Equation (2.11), p. 36] and its homogeneity property

$$I(\hat{y} + \alpha\delta y) - I(\hat{y}) = \delta I(\hat{y}; \alpha\delta y) = \alpha\delta I(\hat{y}; \delta y) \equiv \alpha\delta I \tag{3.2}$$

for sufficiently small α in its prescribed intervals. Hence, from Inequality (3.1) and Equation (3.2)

$$\alpha\delta I \geq 0$$

According to the above inequality, δI should be either

1. greater than or equal to zero when α is greater than zero, or

2. less than or equal to zero when α is less than zero

Thus, $\delta I = 0$ is the only non-contradicting condition that is applicable for the minimum of I.

In fact, we obtain the same condition if I happens to be maximum at $\hat{y}$. In this case, we proceed with the observation that $I(y) - I(\hat{y}) \leq 0$ and do the same analysis as above involving α.

As a consequence,

$$\boxed{\delta I = 0} \tag{3.3}$$

is the **necessary** condition for the optimum (either minimum or maximum) of the functional I. Not guaranteeing an optimum of I, Equation (3.3) is just a logical consequence of I being an optimum.

We will now apply the above necessary condition to optimal control problems and derive equations to help identify optimal solutions.

3.2   Application to Simplest Optimal Control Problem

Consider the simplest optimal control problem in which it is desired to find a continuous control function $u(t)$ that optimizes the objective functional

$$I = \int_{0}^{t_f} F(y, u)\,\mathrm{d}t \tag{3.4}$$

subject to the constraint of the differential equation

$$\dot{y} = g(y, u) \quad \text{or} \quad G(y, \dot{y}, u) \equiv -\dot{y} + g(y, u) = 0 \tag{3.5}$$

with the initial condition

$$y(0) = y_0 \tag{3.6}$$

Equation (3.5) is called the **state equation** since it describes the state of the system through the **state variable** y as a function of the **independent variable** t. It is assumed that F and g have continuous partial derivatives with respect to y and u. Note that this problem is autonomous in the sense that the independent variable does not appear explicitly in F or g. When it does, the problem is easily convertible to the autonomous form (see Exercise 3.4).

Effect of Control

Observe that the objective functional I in Equation (3.4) is influenced by the control u directly as well as indirectly. While the direct influence is through the integrand F, the indirect influence stems from y being affected by u through the state equation constraint, i.e., Equation (3.5).

Hence, to solve the problem, we need to first obtain an explicit solution $y = y(u)$ and then substitute it in the expression of F. However, such solutions do not exist for most optimal control problems, which are typically constrained by highly non-linear state equations.

The recourse is to adjoin the constraints of an optimal control problem to the objective functional using new variables called Lagrange multipliers. Their introduction obviates the need to obtain explicit solutions of state variables in terms of controls. The optimization of the resulting augmented functional is then equivalent to the constrained optimization of the original functional. This outcome — which we shall call the Lagrange Multiplier Rule — springs from the Lagrange Multiplier Theorem, the details of which will be presented in the next chapter. For now we accept this result and work with the augmented functional.

3.2.1 Augmented Functional

In the present problem defined by Equation (3.4)–(3.6), we adjoin the state equation constraint to I using a Lagrange multiplier λ and obtain the augmented functional

$$J = \int_0^{t_f} \underbrace{[F(y, u) + \lambda G(y, \dot{y}, u)]}_{f}\, \mathrm{d}t \qquad (3.7)$$

For the given initial condition, i.e., Equation (3.6), the optimization of J is then equivalent to the optimization of I constrained by the state equation. Observe how the integrand f in Equation (3.7) is formed by multiplying λ by G, which consists of all terms of the state equation constraint moved to the right-hand side. We will follow this approach, which will later on enable us to introduce a useful mnemonic function called the Hamiltonian.

The Lagrange multiplier λ is also known as the **adjoint** or **costate variable**. It is an undetermined function of an independent variable, which is t in the present problem. Both λ and the optimal u are determined by the necessary condition for the optimum of J. The subsequent analysis expands the necessary condition, which is terse as such, into a set of workable equations or necessary conditions to be satisfied at the optimum.

3.2.2 Optimal Control Analysis

From Equation (3.3), if J is optimum, then $\delta J = 0$. This result, on the basis of Section 2.5.2 (p. 44), can be expressed as

$$\delta J = \int_0^{t_f} (\delta F + \lambda \delta G + G \delta \lambda)\, \mathrm{d}t = 0 \qquad (3.8)$$

where

$$\delta F = F_y \delta y + F_u \delta u \qquad (3.9)$$

and

$$\delta G = G_y \delta y + G_{\dot{y}} \delta \dot{y} + G_u \delta u \qquad (3.10)$$

We rewrite Equation (3.8) as

$$\delta J = \underbrace{\int_0^{t_f} (\delta F + \lambda \delta G)\, dt}_{J_1(\delta y, \delta u)} + \underbrace{\int_0^{t_f} G \delta \lambda\, dt}_{J_2(\delta \lambda)} = 0$$

where the first integral (J_1) depends on arbitrary functions δy and δu, and the second integral (J_2) depends on another arbitrary function $\delta \lambda$. The function $\delta \dot{y}$ is not arbitrary but depends on δy.

Note that J_1 is independent of $\delta \lambda$ while J_2 is independent of δy and δu. Because of this fact, both J_1 and J_2 must be individually zero. Otherwise, $J_1(\delta y, \delta u) = -J_2(\delta \lambda)$, implying the contradiction that J_1 depends on $\delta \lambda$ and J_2 depends on δy and δu. Therefore, $J_1 = 0$ and $J_2 = 0$ must be satisfied in order to ensure $\delta J = 0$.

Satisfaction of $J_2 = 0$

We consider the equation

$$J_2 = \int_0^{t_f} G \delta \lambda\, dt = 0$$

Since $\delta \lambda$ is arbitrary in the above equation, $J_2 = 0$ is satisfied by having

$$\boxed{G = 0} \qquad (3.11)$$

throughout the interval $[0, t_f]$. Keep in mind that we have considered the optimum of J without supposing that G is zero, as given by Equation (3.5). The above equation, $G = 0$, comes forth as a necessary condition for the optimum.

Satisfaction of $J_1 = 0$

Next, we consider

$$\boxed{J_1 = \int_0^{t_f} (\delta F + \lambda \delta G)\, dt = 0} \qquad (3.12)$$

Arising from $\delta J = 0$, both Equations (3.11) and (3.12) are the necessary conditions for the optimum of J. On the basis of the Lagrange Multiplier Rule, these two equations are also the necessary conditions for the constrained optimum of I. Both optima are still subject to the given initial condition, i. e., Equation (3.6). The equivalence between the two optima will be shown later in Section 4.3.3.

We now proceed to obtain workable equations that will ensure the satisfaction of Equation (3.12), and consequently, be necessary along with Equation (3.11) for the optimum of I.

With the help of Equations (3.9) and (3.10), Equation (3.12) can be expressed as

$$
J_1 = \underbrace{\int_0^{t_f} \left[(F_y + \lambda G_y)\delta y + \lambda G_{\dot{y}}\delta\dot{y} \right] dt}_{J_3(\delta y)} + \underbrace{\int_0^{t_f} (F_u + \lambda G_u)\delta u\, dt}_{J_4(\delta u)} = 0
$$

where the first integral (J_3) depends on the arbitrary function δy, while the second integral (J_4) depends on another arbitrary function δu.

Note that J_3 does not depend on δu, and J_4 does not depend on δy. Therefore, for the above equation to hold, both integrals must be individually zero. Otherwise, $J_3(\delta y) = -J_4(\delta u)$, implying that J_3 depends on δu and J_4 depends on δy, which is contradictory. Thus,

$$
J_3 = \int_0^{t_f} \left[(F_y + \lambda G_y)\delta y + \lambda G_{\dot{y}}\delta\dot{y} \right] dt = 0 \tag{3.13}
$$

and

$$
J_4 = \int_0^{t_f} (F_u + \lambda G_u)\delta u\, dt = 0 \tag{3.14}
$$

must be satisfied in order to ensure $J_1 = 0$.

Satisfaction of $J_4 = 0$

For Equation (3.14) to hold for any arbitrary δu,

$$
\boxed{F_u + \lambda G_u = 0, \quad 0 \le t \le t_f} \tag{3.15}
$$

The above equation is known as the **stationarity condition**.

Satisfaction of $J_3 = 0$

Upon substituting $G_{\dot{y}} = -1$, which is obtained from the definition of G in Equation (3.5), Equation (3.13) becomes

$$J_3 = \int_0^{t_f} (F_y + \lambda G_y)\delta y \, dt - \int_0^{t_f} \lambda \delta \dot{y} \, dt = 0 \tag{3.16}$$

Before simplifying the above equation further, we need to express the second integral in terms of δy. Applying integration by parts to the second integral, we get

$$\int_0^{t_f} (F_y + \lambda G_y)\delta y \, dt - \Big[\lambda \delta y\Big]_{t=t_f} + \underbrace{\Big[\lambda \delta y\Big]_{t=0}}_{\text{is zero}} + \int_0^{t_f} \dot{\lambda}\delta y \, dt = 0 \tag{3.17}$$

The above step assumes that we can differentiate λ with respect to t in Equation (3.16). It is indeed true, as shown in Appendix 3.A (p. 78).

Because of the initial condition, i.e., Equation (3.6), there cannot be any variation in y at $t = 0$ where y is fixed. Thus, the third term in Equation (3.17) is zero. Rearranging the remaining terms of the equation, we get

$$\int_0^{t_f} \left(F_y + \lambda G_y + \dot{\lambda}\right)\delta y \, dt - \Big[\lambda \delta y\Big]_{t=t_f} = 0 \tag{3.18}$$

From the Mean Value Theorem for integrals (Section 9.14.2, p. 276),

$$\int_0^{t_f} \underbrace{\left(F_y + \lambda G_y + \dot{\lambda}\right)}_{p(t)} \delta y \, dt = p(t_1)\delta y(t_1)(t_f - 0)$$

where $p(t)$ is the coefficient of δy and $0 \leq t_1 \leq t_f$. The above equation shows that the integral depends on an arbitrary function value $\delta y(t_1)$. We contend that the above integral does not depend on $\delta y(t_f)$ though. In other words, the above equation is valid for t_1 *less* than t_f. To prove it, let t_1 be equal to t_f in the equation and $p(t)$ be given over the interval $[0, t_f]$. Then, from the above equation,

$$\int_0^{t_f} p\delta y \, dt = p(t_f)\delta y(t_f)t_f \tag{3.19}$$

and the integrand $p\delta y$

- either stays constant so that it has the same value $p(t_f)\delta y(t_f)$ at any t_1 *less* than t_f (**Case 1**)

- or changes with t, in which case $p\delta y$ attains *at least once* the same value $p(t_f)\delta y(t_f)$ at t_1 *less* than t_f (**Case 2**).

  Suppose this conclusion is not true, so that

$$p(t_1)\delta y(t_1) > p(t_f)\delta y(t_f), \quad 0 \le t_1 < t_f$$

Then $p(t_f)\delta y(t_f)$ is the minimum integrand value. Thus,

$$\int_0^{t_f} p\delta y \, dt > p(t_f)\delta y(t_f)t_f$$

which contradicts Equation (3.19). The other situation when

$$p(t_1)\delta y(t_1) < p(t_f)\delta y(t_f), \quad 0 \le t_1 < t_f$$

is similarly contradictory. Hence, $p\delta y = p(t_f)\delta y(t_f)$ at least once for $t_1 < t_f$.

In either case,

$$\int_0^{t_f} p\delta y \, dt = p(t_1)\delta y(t_1)(t_f - 0)$$

where t_1 is less than t_f. Thus, the integral is relieved from dependence on $\delta y(t_f)$, which therefore affects only the last term of Equation (3.18). As reasoned in similar instances previously, the two terms of Equation (3.18) must therefore be individually zero. Hence

$$\int_0^{t_f} \left(F_y + \lambda G_y + \dot{\lambda}\right)\delta y \, dt = 0 \tag{3.20}$$

and

$$\lambda(t_f)\delta y(t_f) = 0 \tag{3.21}$$

Now Equation (3.20) is certainly satisfied if the coefficient of δy is zero, or equivalently

$$\boxed{\dot{\lambda} = -(F_y + \lambda G_y), \quad 0 \le t \le t_f} \tag{3.22}$$

The above equation is known as the **Euler–Lagrange equation** in honor of Swiss mathematician *Leonard Euler* (1707–1783) and French mathematician *Joseph Luis de Lagrange* (1736–1813). The Euler–Lagrange equation is also called the **adjoint** or **costate equation** since it defines the adjoint or costate variable λ.

A question arises whether the coefficient of δy in Equation (3.20) really needs to vanish at $t = 0$ where δy is already zero. The answer is yes and the proof is provided in Appendix 3.B (p. 81).

Table 3.1 Necessary conditions for the optimum of I in the problem defined by Equations (3.4)–(3.6)

1. The state equation	$\dot{y} = g(y, u),$	$0 \le t \le t_f$	(3.5)
with initial condition	$y(0) = y_0$		(3.6)
2. The costate equation	$\dot{\lambda} = -(F_y + \lambda G_y),$	$0 \le t \le t_f$	(3.22)
with final condition	$\lambda(t_f) = 0$		(3.23)
3. The stationarity condition	$F_u + \lambda G_u = 0,$	$0 \le t \le t_f$	(3.15)

Finally, we are left with Equation (3.21), which is satisfied if the coefficient of the arbitrary $\delta y(t_f)$ is zero, i.e.,

$$\boxed{\lambda = 0 \quad \text{at} \quad t = t_f} \qquad (3.23)$$

The above equation is in fact the *initial* condition for the *backward* integration (from $t = t_f$ to 0) of the costate equation, i.e., Equation (3.22).

Summary of Necessary Conditions

Table 3.1 collects the equations we have used to satisfy $\delta J = 0$. These equations are the necessary conditions for the optimum of J or, equivalently, of I subject to the state equation constraint. The solution of the necessary conditions provides the u and the corresponding y as well as the λ at *an* optimum of I. Put differently, for I to be optimal, u must be such that these conditions are satisfied. Otherwise I cannot be optimal.

Keep in mind that the satisfaction of these equations does not guarantee the optimum but provides a candidate for the optimum. For example, several optima may exist and the optimum provided by the necessary conditions may not be the best among all optima. (We will explore this issue further in Section 3.3.1 later on.)

Most important, if the equations are not satisfied, then the optimum cannot exist. The above facts conform to the logic of a conditional statement explained in Section 9.25 (p. 282). For now, let us apply these conditions to optimal control problems.

Example 3.1

Find the necessary conditions for the minimum of

$$I = \int_0^{t_f} (u^2 - yu)\, dt$$

subject to the satisfaction of the differential equation constraint

$$\dot{y} = -\sqrt{y} + u \qquad (3.24)$$

where $u(t)$ is the control and $y(t)$ is the state variable with the initial condition $y(0) = 0$.

The augmented functional for this problem is

$$J = \int_0^{t_f} \Big\{ \underbrace{u^2 - yu}_{F} + \underbrace{\lambda \big[-\dot{y} \underbrace{-\sqrt{y} + u}_{g} \big]}_{G} \Big\}\, dt$$

with the functions F, g, and G as indicated in the above equation. From Table 3.1, the necessary conditions for the minimum of I are

1. the state equation, Equation (3.24), with the initial condition $y(0) = 0$;

2. the costate equation, $\dot{\lambda} = -(F_y + \lambda G_y)$, i.e.,

$$\dot{\lambda} = u + \frac{\lambda}{2\sqrt{y}}$$

with the final condition $\lambda(t_f) = 0$; and

3. the stationarity condition $F_u + \lambda G_u = 0$, i.e.,

$$2u - y + \lambda = 0$$

throughout the time interval $[0, t_f]$. □

Example 3.2

Find the necessary conditions for the maximum of

$$I = \int_0^{t_f} ckx^a\, dt, \qquad k = k_o \exp\left(-\frac{E}{RT}\right)$$

subject to the satisfaction of the differential equation constraint

$$\dot{x} = -akx^a \qquad (3.25)$$

with $T(t)$ as the control and $x(0) = x_0$ as the initial condition.

This is the simple batch reactor control problem (see Example 2.10, p. 45) for which the augmented functional is

$$J = \int_0^{t_f} \left\{ \underbrace{ckx^a}_{F} + \lambda \underbrace{\left[-\dot{x} - akx^a \right]}_{g} \right\} dt$$

$$\underbrace{\hspace{6cm}}_{G}$$

with the functions F, g, and G as indicated above. Referring to Table 3.1, T is the control u as a function of the independent variable t and x is the state variable y. Using the table, the necessary conditions for the maximum of I are

1. the state equation, Equation (3.25), with the initial condition $x(0) = x_0$;

2. the costate equation, $\dot{\lambda} = -(F_x + \lambda G_x)$, i.e.,

$$\dot{\lambda} = -akx^{a-1}(c - a\lambda) \tag{3.26}$$

   with the final condition $\lambda(t_f) = 0$; and

3. the stationarity condition $F_T + \lambda G_T = 0$, i.e.,

$$\frac{kE}{RT^2} x^a (c - a\lambda) = 0 \tag{3.27}$$

over the time interval $[0, t_f]$.

$\square$

Hamiltonian

For the optimal control problem defined by Equations (3.4)–(3.6), a mnemonic to remember the necessary conditions for the minimum is the Hamiltonian function defined as

$$H \equiv F + \lambda g$$

It is easy to verify that

$$\boxed{\frac{dH}{d\lambda} \equiv H_\lambda = \dot{y}} \tag{3.28}$$

provides $\quad \dot{y} = g(y, u) \tag{3.5}$

state equation

$$\boxed{\frac{dH}{dy} \equiv H_y = -\dot{\lambda}} \tag{3.29}$$

provides $\quad \dot{\lambda} = -(F_y + \lambda G_y) \tag{3.22}$

costate equation

and

$$\boxed{\frac{dH}{du} \equiv H_u = 0} \tag{3.30}$$

provides $\quad F_u + \lambda G_u = 0 \tag{3.15}$

stationarity condition

Example 3.3

Repeat the previous example utilizing the Hamiltonian.

The Hamiltonian is given by

$$H = ckx^a + \lambda(-akx^a)$$

Therefore, the necessary conditions for the maximum of I are

1. the state equation, Equation (6.10), obtained from $\dot{x} = H_\lambda$ and the initial condition $x(0) = 0$;

2. the costate equation, Equation (3.26), obtained from $\dot{\lambda} = -H_x$ and the final condition $\lambda(t_f) = 0$; and

3. the stationarity condition, Equation (3.27), obtained from $H_T = 0$

over the time interval $[0, t_f]$. ⬜

3.2.3 Generalization

For the simplest optimal control problem with m control functions and n state variables in general, the objective functional to be optimized is

$$I = \int_0^{t_f} F(\underbrace{y_1, y_2, \ldots, y_n}_{\mathbf{y}^\top}, \underbrace{u_1, u_2, \ldots, u_m}_{\mathbf{u}^\top}) \, dt \qquad (3.31)$$

subject to n differential equation constraints

$$\dot{y}_i = g_i(\mathbf{y}, \dot{\mathbf{y}}, \mathbf{u}) \quad \text{or} \quad G_i \equiv -\dot{y}_i + g_i = 0; \quad i = 1, 2, \ldots, n \qquad (3.32)$$

with the n initial conditions

$$y_i(0) = y_{i,0}, \quad i = 1, 2, \ldots, n \qquad (3.33)$$

Using the Lagrange Multiplier Rule, the augmented functional is

$$J = \int_0^{t_f} \left(F + \sum_{i=1}^{m} \lambda_i G_i \right) dt \equiv \int_0^{t_f} f(\mathbf{y}, \dot{\mathbf{y}}, \mathbf{u}, \boldsymbol{\lambda}) \, dt$$

which was introduced in Section 2.5.2 (p. 44) using the same vector notation. At the optimum of J, δJ is zero and is given by

$$\int_0^{t_f} \left\{ \sum_{i=1}^{n} f_{y_i} \delta y_i + \sum_{i=1}^{n} f_{\dot{y}_i} \delta \dot{y}_i + \sum_{i=1}^{m} f_{u_i} \delta u_i + \sum_{i=1}^{n} f_{\lambda_i} \delta \lambda_i \right\} dt = 0$$

where

$$f_{y_i} = F_{y_i} + \sum_{j=1}^{n} \lambda_j G_{j y_i}, \quad f_{\dot{y}_i} = \sum_{j=1}^{n} \lambda_j G_{j \dot{y}_i} = -\sum_{j=1}^{n} \lambda_j, \quad \text{and}$$

$$f_{u_i} = F_{u_i} + \sum_{j=1}^{m} \lambda_j G_{j u_i}$$

The Necessary Conditions

Extending the optimal control analysis (Section 3.2.2, p. 59) to the generalized problem with multiple states and controls, and defining the Hamiltonian as

$$H = F + \sum_{i=1}^{n} \lambda_i g_i \tag{3.34}$$

we get the necessary conditions for the optimum of J, which are provided in Table 3.2.

Table 3.2 Necessary conditions for the optimum of I in the general problem defined by Equations (3.31)–(3.33)

1. The state equations	$\dot{y}_i = H_{\lambda_i}$	$0 \le t \le t_f$	(3.35)
with the initial conditions	$y_i(0) = y_{i,0},$		(3.36)
for $i = 1, 2, \ldots, n$			
2. The costate equations	$\dot{\lambda}_i = -H_{y_i},$	$0 \le t \le t_f$	(3.37)
with the final conditions	$\lambda_i(t_f) = 0$		(3.38)
for $i = 1, 2, \ldots, n$			
3. The stationarity condition	$H_{u_i} = 0,$	$0 \le t \le t_f$	(3.39)
for $i = 1, 2, \ldots, m$			

Example 3.4

Find the necessary conditions for the maximum of

$$I = \int_0^{t_f} \underbrace{k c x^a y^b}_{F} \, dt$$

subject to the satisfaction of the state equations

$$\dot{x} = \underbrace{-kax^ay^b}_{g_1} \tag{3.40}$$

$$\dot{y} = \underbrace{-kbx^ay^b}_{g_2} \tag{3.41}$$

where T is the control and

$$k = k_0 \exp\left(-\frac{E}{RT}\right) \tag{3.42}$$

The initial conditions are

$$x(0) = x_0 \quad \text{and} \quad y(0) = y_0$$

This is the batch reactor control problem (see p. 2) without the inequality constraints. Referring to Table 3.2, x and y are, respectively, the state variables y_1 and y_2. Furthermore, T is the control u. Using Equation (3.34) and the functions F, g_1, and g_2 as indicated above, the Hamiltonian for this problem is given by

$$H = \underbrace{kcx^ay^b}_{F} + \lambda_1\underbrace{(-kax^ay^b)}_{g_1} + \lambda_2\underbrace{(-kbx^ay^b)}_{g_2}$$

From Table 3.2, the necessary conditions for the maximum are

1. the state equations, $\dot{x} = H_{\lambda_1}$ and $\dot{y} = H_{\lambda_2}$, given respectively by Equations (3.40) and (3.41), and their initial conditions;

2. the costate equations given by $\dot{\lambda}_1 = -H_x$ and $\dot{\lambda}_2 = -H_y$, i.e.,

$$\dot{\lambda}_1 = -kax^{a-1}y^b(c - a\lambda_1 - b\lambda_2)$$
$$\dot{\lambda}_2 = -kx^aby^{b-1}(c - a\lambda_1 - b\lambda_2)$$

and their final conditions $\lambda_1(t_f) = \lambda_2(t_f) = 0$; and

3. the stationarity condition $H_T = 0$, i.e.,

$$\frac{kE}{RT^2}x^ay^b(c - a\lambda_1 - b\lambda_2) = 0$$

throughout the time interval $[0, t_f]$. ◻

Example 3.5
The change in reactant and catalyst concentrations x and y, and temperature T in a Constant (volume) Stirred Tank Reactor or CSTR without time delays

is given by (Ray and Soliman, 1970)

$$\dot{x} = \frac{Q}{V}(x_f - x) - kxy \tag{3.43}$$

$$\dot{y} = \frac{m - Qy}{V} \tag{3.44}$$

$$\dot{T} = \frac{Q}{V}(T_f - T) - \frac{\Delta H}{\rho C_p}kxy - \frac{hA + K(T - T_s)}{V\rho C_p}(T - T_c) \tag{3.45}$$

where V, ΔH, ρ, and C_p are, respectively, the reactor volume, heat of reaction, liquid density, and specific heat capacity. The subscript f denotes the feed property, k is given by Equation (3.42), and Q, m, and K are the three controls. They are, respectively, the volumetric flow rate, mass flow rate of the catalyst, and the gain of temperature controller as functions of time t. The initial conditions are

$$x(0) = x_0, \quad y(0) = y_0, \quad \text{and} \quad T(0) = T_0$$

It is desired to minimize the deviation of the state of the CSTR with minimum control action, i.e., to minimize

$$I = \int_0^{t_f} \left[(x - x_s)^2 + (y - y_s)^2 + (T - T_s)^2 + (Q - Q_s)^2 + (m - m_s)^2 \right] dt$$

where the subscript s denotes the steady state value. Find the necessary conditions for the minimum of I.

Using Equation (3.34), the Hamiltonian is given by

$$H = \left[(x - x_s)^2 + (y - y_s)^2 + (T - T_s)^2 + (Q - Q_s)^2 + (m - m_s)^2 \right]$$

$$+ \lambda_1 \left[\frac{Q}{V}(x_f - x) - kxy \right] + \lambda_2 \left[\frac{m - Qy}{V} \right]$$

$$+ \lambda_3 \left[\frac{Q}{V}(T_f - T) - \frac{\Delta H}{\rho C_p}kxy - \frac{hA + K(T - T_s)}{V\rho C_p}(T - T_c) \right]$$

In this example, x, y, T, Q, m, and K correspond, respectively, to y_1, y_2, y_3, u_1, u_2, and u_3 of Table 3.2. From the table, the necessary conditions for the minimum are

1. the state equations, Equations (3.43)–(3.45), respectively given by

$$\dot{x} = H_{\lambda_1}, \quad \dot{y} = H_{\lambda_2}, \quad \text{and} \quad \dot{T} = H_{\lambda_3}$$

with their initial conditions $x(0) = x_0$, $y(0) = y_0$, and $T(0) = T_0$.

2. the costate equations

$$\dot{\lambda}_1 = -H_x = -2(x - x_s) + \lambda_1 \left[\frac{Q}{V} + ky \right] + \lambda_3 \left[\frac{\Delta H}{\rho C_p} ky \right]$$

$$\dot{\lambda}_2 = -H_y = -2(y - y_s) + \lambda_1 kx + \lambda_2 \frac{Q}{V} + \lambda_3 \left[\frac{\Delta H}{\rho C_p} kx \right]$$

$$\dot{\lambda}_3 = -H_T = -2(T - T_s) + \lambda_1 \frac{kE}{RT^2} xy$$

$$+ \lambda_3 \left[\frac{Q}{V} + \frac{\Delta H}{\rho C_p} \frac{kE}{RT^2} xy + \frac{hA + K(2T - T_s - T_c)}{V \rho C_p} \right]$$

with the final conditions $\lambda_i(t_f) = 0$, $i = 1, 2, 3$.

3. the stationarity condition for each of Q, m, and K, i.e.,

$$H_Q = 2(Q - Q_s) + \lambda_1 \frac{x_f - x}{V} - \lambda_2 \frac{y}{V} + \lambda_3 \frac{T_f - T}{V} = 0$$

$$H_m = 2(m - m_s) + \frac{\lambda_2}{V} = 0$$

$$H_K = -\lambda_3 \frac{(T - T_s)(T - T_c)}{V \rho C_p} = 0$$

throughout the time interval $[0, t_f]$.   ▯

3.3 Solving an Optimal Control Problem

Although a little bit early, it is worthwhile to know how we can actually solve an optimal control problem. As indicated earlier, the answer lies in the necessary conditions we have established above. They must be satisfied by the optimal control functions.

It must be noted that the necessary conditions are frequently nonlinear and cannot be solved analytically. Therefore, optimal control problems are generally solved using numerical algorithms, the focus of Chapter 7.

At this moment, we outline a simple numerical algorithm, which begins by guessing the control functions, and involves the following steps:

1. forward integration of state equations using the initial conditions

2. backward integration of costate equations using the final conditions

3. improvement of each control function utilizing the gradient information from H_{u_i} — the variational derivative of the objective functional with respect to the control function u_i.

The improvement in control functions causes the objective functional value to get closer to the optimum. Therefore, iterative application of the above steps leads to the optimum, i. e., the optimal functional value, and the corresponding optimal control functions. While the state and costate equations are satisfied in each iteration, the variational derivatives are reduced successively. They vanish when the optimum is attained.

An application of the above algorithm to the batch reactor problem of Example 3.4 results in the optimal control temperature (T), states (x and y), and costates (λ_1 and λ_2) versus time, as shown in Figure 3.1 for the parameters listed in Table 3.3. The corresponding maximum value of the objective functional I is 6.67 kmol/m^3.

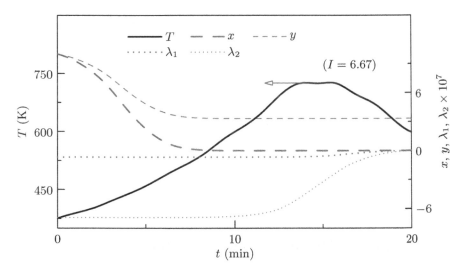

Figure 3.1 The optimal control, states, and costates versus time in Example 3.4

Table 3.3 Parameters for the problem in Example 3.4

a	1.5	t_f	20 min
b	1	E	7×10^7 J/kmol
c	1	k_0	6×10^6 (kmol/m^3)$^{-1.5}$/min

3.3.1 Presence of Several Local Optima

When solving an optimal control problem, it has to be kept in mind that several local optima may exist. Consider for example a problem with a single control function. The objective functional value may be locally optimal, i. e., optimal only in a vicinity of the obtained optimal control function. In another location within the space of all admissible* control functions, the objective functional may again be locally optimal corresponding to some other optimal control function. This new optimal objective functional value may be better or worse than, or, even the same as the previous one.

For example, in the problem of Example 3.4 solved above, the optimum obtained is local. Three different optima are obtained when the same problem is solved by initializing the algorithm with different guesses for the control function. Figure 3.2 shows the corresponding optimal control functions T_1, T_2, and T_3. They are quite different from not only each other but also the optimal T in Figure 3.1. While T_1 and T_2 yield the same maximum as obtained with T, T_3 provides a better maximum of $I = 6.70$ kmol/m^3.

The above outcome is not unexpected since it is based on the necessary conditions for the optimum. As shown in Figure 3.3, the necessary conditions are the consequence of an optimum, whether local or global. The satisfaction of these conditions does not guarantee the global optimum but merely provides a candidate for it. It is the satisfaction of the sufficient conditions that guarantees the global optimum.

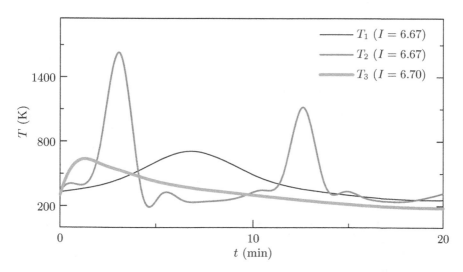

Figure 3.2 Three other optimal control functions for the batch reactor problem

---

* Satisfying the state equations and any other specified constraints.

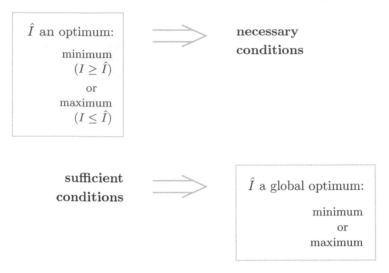

Figure 3.3 Necessary and sufficient conditions for the optimum of the objective functional I

Naturally, we would like to determine the optimum that is global, i. e., valid over the entire space of admissible control functions. This quest urges us to inquire into the conditions that are sufficient for the global optimum, i. e., the optimum over the entire space of admissible control functions.

3.4   Sufficient Conditions

For the minimum solution to be global, a sufficient condition must ensure that the objective functional value is the lowest possible. In the context of an optimal control problem, a sufficient condition for the minimum should lead to the inequality

$$I(\mathbf{y}, \mathbf{u}) \geq I(\hat{\mathbf{y}}, \hat{\mathbf{u}})$$

where I is the objective functional, $\mathbf{y}$ is the state vector corresponding to *any* admissible control vector $\mathbf{u}$, and $\hat{\mathbf{y}}$ is the optimal state vector corresponding to the optimal control vector $\hat{\mathbf{u}}$. The states, controls, and costates are functions of time t, which is the independent variable. The above inequality is opposite if the maximum of I is to be determined.

To yield the above inequality, a sufficient condition must include at least one inequality, which in fact is another optimal control problem. Thus, for the global minimum in the simplest optimal control problem, a set of sufficient conditions by Mangasarian (1966) requires that

1. the controls satisfy the necessary conditions for the minimum,

2. the costates at the minimum are non-positive if **g** [i. e., the vector of g_is in Equation (3.32) on p. 67] is nonlinear with respect to the states or controls, and

3. the functions F [Equation (3.31), p. 67] and **g** are convex jointly with respect to the states and corresponding controls

The proof of this result is provided in Appendix 3.C (p. 81).

The third requirement of Mangasarian gives rise to inequalities of the form $L(\mathbf{y}, \mathbf{u}) \geq 0$ [see Inequality (3.52), p. 82]. Proving it becomes a new optimal control problem in which zero needs to be established as the minimum of L for all admissible **u**. This task cannot be accomplished analytically except in rare, simple cases where the Hessian of L could be easily checked for positive definiteness. Because of this situation, the application of any sufficient condition in optimal control is very limited.

3.4.1 Weak Sufficient Condition

We will now examine the role of the second variation in identifying the optimum. Let us say we desire to find the minimum of I, and we have determined the minimum from the necessary conditions. Then from Equation (2.33) on p. 52, and noting that $\delta I = 0$ at the minimum, we get

$$I(\mathbf{y}, \mathbf{u}) - I(\hat{\mathbf{y}}, \hat{\mathbf{u}}) = \frac{1}{2}\delta^2 I(\hat{\mathbf{y}}; \delta \mathbf{y}) + \epsilon_2(\delta \mathbf{y}) \tag{3.46}$$

where $\delta^2 I$ is the second variation of I, and ϵ_2 is the error term, both depending on $\delta \mathbf{u} \equiv (\mathbf{u} - \hat{\mathbf{u}})$ and the corresponding $\delta \mathbf{y} \equiv (\mathbf{y} - \hat{\mathbf{y}})$. Observe that the non-negativity of $\delta^2 I$ does not assure the minimum. For example, there may exist a **u** for which $\delta^2 I$ is less than the absolute value of ϵ_2, which happens to be negative. In such a case, let $\epsilon_2 = -\beta$ where $\beta > 0$. Then the right-hand side of Equation (3.46) becomes

$$\frac{\delta^2 I}{2} + \epsilon_2 \; < \; \frac{|\epsilon_2|}{2} + \epsilon_2 = \frac{\beta}{2} - \beta \; < \; 0$$

The above result implies that $I(\mathbf{y}, \mathbf{u}) - I(\hat{\mathbf{y}}, \hat{\mathbf{u}}) < 0$ so that $I(\hat{\mathbf{y}}, \hat{\mathbf{u}})$ is not a minimum! We need to impose some condition on $\delta^2 I$ to ensure its minimum.

The Condition

If we place a condition that the error ratio $\epsilon_2/\|\delta\mathbf{u}\|^2$ vanishes with $\|\delta\mathbf{u}\|$, and $\delta^2 I \geq \alpha\|\delta\mathbf{u}\|^2$ where α is some positive number, then

$$I(\mathbf{y},\mathbf{u}) - I(\hat{\mathbf{y}},\hat{\mathbf{u}}) \geq \frac{\alpha}{2}\|\delta\mathbf{u}\|^2 + \epsilon_2$$

$$\geq \|\delta\mathbf{u}\|^2 \left(\frac{\alpha}{2} + \frac{\epsilon_2}{\|\delta\mathbf{u}\|^2} \right)$$

If we can find a sufficiently small neighborhood of $\hat{\mathbf{u}}$ such that all $\delta\mathbf{u}$ therein have norms sufficiently small to render any negative error ratio less than $\alpha/2$, then

$$I(\mathbf{y},\mathbf{u}) - I(\hat{\mathbf{y}},\hat{\mathbf{u}}) \geq 0$$

in that locality, and $I(\hat{\mathbf{y}},\hat{\mathbf{u}})$ is the local minimum.

The above sufficient condition is weak, since it is applicable only in a sufficiently small vicinity of the optimal control. The satisfaction of this condition does not guarantee a global minimum. Moreover, it is not easy to come up with a suitable α in an optimal control problem.

From a practical viewpoint, the second variation can be utilized

1. as another necessary condition for the optimum

2. to speed up the convergence of an optimal control algorithm near the optimum when the variational derivatives of I with respect to $\mathbf{u}$ approach zero.

Hence, to increase the confidence on the final optimal solution, the optimal control problem needs to be solved with different initial guesses to the numerical algorithm. If several optima are obtained, then the most optimal among them is adopted. There is no golden rule to ensure that the optimum is global.

3.5 Piecewise Continuous Controls

The optimal control analysis developed above is also applicable when a control function is piecewise continuous with a finite number of jump discontinuities, as shown in Figure 3.4. Observe that a jump discontinuity suddenly changes $\dot{y}$ or the slope of the state variable. The result is a corner in the graph of $y(t)$ at the time of discontinuity where $\dot{y}$ is not defined. Geometrically, there is no unique slope of the curve at a corner. One can draw an infinite number of tangents there.

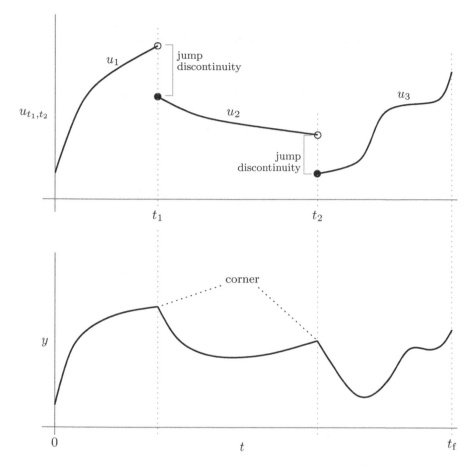

Figure 3.4 A piecewise continuous control u_{t_1,t_2} and the accompanying state $y(t)$

Let us use u_{t_1,t_2} to denote a piecewise continuous control with jump discontinuities at times t_1 and t_2. Thus, u_{t_1,t_2} is made up of three continuous controls — u_1, u_2, and u_3, as shown in the figure. Then the problem of finding the u_{t_1,t_2} that minimizes

$$I = \int_0^{t_f} F(y, u_{t_1,t_2}) \, dt \quad \text{or} \quad \underbrace{\int_0^{t_1} F(y, u_1) \, dt}_{I_1} + \underbrace{\int_{t_1}^{t_2} F(y, u_2) \, dt}_{I_2} + \underbrace{\int_{t_2}^{t_f} F(y, u_3) \, dt}_{I_3}$$

subject to

$$\dot{y} = g(y, u_{t_1,t_2}), \quad y(0) = y_0, \quad 0 \le t \le t_f$$

is the same as finding the minimum of I_1, I_2, and I_3 over the contiguous time intervals $[0, t_1)$, $[t_1, t_2)$, and $[t_1, t_f]$ during which the control has no jump discontinuities. Therefore, to find the control u_{t_1,t_2} that minimizes I, we apply the developed optimal control analysis to the following three subproblems:

1. Find the u_1 that minimizes I_1 subject to

$$\dot{y} = g(y, u), \quad y(0) = y_0, \quad 0 \le t \le t_1$$

2. Find the u_2 that minimizes I_2 subject to

$$\dot{y} = g(y, u), \quad t_1 \le t \le t_2$$

where $y(t_1)$ is obtained from the previous subproblem

3. Find the u_3 that minimizes I_3 subject to

$$\dot{y} = g(y, u), \quad t_2 \le t \le t_f$$

where $y(t_2)$ is obtained from the previous subproblem

At the minimum of I, additional conditions called, Weierstrass–Erdmann corner conditions, must be satisfied on the corners. These conditions require the continuity of the Hamiltonian and costate at the corners. The details will come later in Section 6.6.

3.A   Differentiability of λ

We need to show that λ is differentiable with respect to t in the following equation on p. 62:

$$\int_0^{t_f} (F_y + \lambda G_y)\delta y \, dt - \int_0^{t_f} \lambda \delta \dot{y} \, dt = 0 \qquad (3.16)$$

so that

$$\int_0^{t_f} (F_y + \lambda G_y)\delta y \, dt - [\lambda \delta y]_{t=t_f} + [\lambda \delta y]_{t=0} + \int_0^{t_f} \dot{\lambda} \delta y \, dt = 0 \qquad (3.17)$$

Applying integration by parts to the first integral in Equation (3.16), we obtain

$$\left[\delta y \int_0^t (F_y + \lambda G_y) \, dt \right]_0^{t_f} - \int_0^{t_f} \left[\delta \dot{y} \int_0^t (F_y + \lambda G_y) \, dt \right] dt - \int_0^{t_f} \lambda \delta \dot{y} \, dt = 0 \quad (3.47)$$

Introducing

$$\phi(t) \equiv \int_0^t (F_y + \lambda G_y)\, dt$$

and considering that δy is zero at $t = 0$ due to the initial condition [Equation (3.6), p. 58], we can write Equation (3.47) as

$$\delta y(t_f)\phi(t_f) - \int_0^{t_f} [\phi(t) + \lambda]\delta \dot{y}\, dt = 0 \qquad (3.48)$$

Note that the above equation is valid for the entire class of continuous δy functions that are zero at $t = 0$. This class includes the subclass of δy functions, which are zero at $t = t_f$ as well. Now if δy belongs to the subclass, then the coefficient of $\delta \dot{y}$ in Equation (3.48) is a constant, as we will show shortly. This is a very important result, which implies that the coefficient $[\phi(t) + \lambda]$ has to be the same constant for the entire class of δy functions. If that is not so, then Equation (3.48) will obviously not apply to the subclass and therefore will not be valid for the entire class of δy functions.

Before examining the differentiability of λ, let us show the constancy of the coefficient of $\delta \dot{y}$, i.e., $[\phi(t) + \lambda]$. This result is known as the **Lemma of Dubois–Reymond**.

Constancy of the coefficient of $\delta \dot{y}$

With δy additionally being zero at $t = t_f$, Equation (3.48) becomes

$$\int_0^{t_f} \underbrace{[\phi(t) + \lambda]}_{c(t)} \delta \dot{y}\, dt \equiv \int_0^{t_f} c(t)\delta \dot{y}\, dt = 0 \qquad (3.49)$$

where $c(t)$ stands for the coefficient of $\delta \dot{y}$. Let c_0 be the average of $c(t)$ in the interval $[0, t_f]$. Then

$$\int_0^{t_f} c(t)\, dt = c_0(t_f - 0) = \int_0^{t_f} c_0\, dt$$

which yields upon rearrangement

$$\int_0^{t_f} [c(t) - c_0]\, dt = 0$$

Thus, the choice of

$$\delta y = \int_0^t [c(t) - c_0]\, dt$$

conforms to the current conditions of δy being zero at $t = 0$ and t_f. For this choice, $\delta \dot{y}$ is given by

$$\delta \dot{y} = c(t) - c_0 \tag{3.50}$$

and is admissible. We will use it in the following result

$$\int_0^{t_f} [c(t) - c_0] \delta \dot{y} \, dt = \int_0^{t_f} c(t) \delta \dot{y} \, dt - c_0 \left[\delta y \right]_0^{t_f} = 0 \tag{3.51}$$

which is due to Equation (3.49) and the conditions $\delta y(0) = \delta y(t_f) = 0$. Substituting the $\delta \dot{y}$ from Equation (3.50) in Equation (3.51), we obtain

$$\int_0^{t_f} [c(t) - c_0]^2 \, dt = 0$$

Observe that the above integrand, being a squared term, cannot be negative. It must be zero throughout the interval in order to satisfy the above equation. Thus,

$$c(t) = c_0, \quad 0 \le t \le t_f$$

showing that $c(t)$ or $[\phi(t) + \lambda]$, the coefficient of $\delta \dot{y}$ in Equation (3.49), is a constant. As explained earlier in the paragraph following Equation (3.48), this result extends to the entire class of δy functions. Thus, $[\phi(t) + \lambda]$ is a constant in Equation (3.48) for all δy functions. Let us now examine the differentiability of λ by expressing the above result, i. e.,

$$\phi(t) + \lambda = \int_0^t (F_y + \lambda G_y) \, dt + \lambda = c_0$$

or

$$\lambda = - \int_0^t (F_y + \lambda G_y) \, dt + c_0$$

The above equation shows that λ is differentiable with respect to t at the optimum for which Equation (3.16) is the necessary condition. At the optimum, the derivative of λ with respect to t is

$$\dot{\lambda} = -(F_y + \lambda G_y)$$

3.B   Vanishing of $(F_y + \lambda G_y + \dot{\lambda})$ at $t = 0$

We need to show that the coefficient $(F_y + \lambda G_y + \dot{\lambda})$ of δy in the following equation on p. 63:

$$\int_0^{t_f} \left(F_y + \lambda G_y + \dot{\lambda}\right) \delta y \, dt = 0 \qquad (3.20)$$

has to be zero at $t = 0$ even though δy there is zero.

Suppose that the coefficient $(F_y + \lambda G_y + \dot{\lambda})$ is greater than zero at $t = 0$. This supposition means that $(F_y + \lambda G_y + \dot{\lambda})$, which is a continuous function,[*] has to be greater than zero in a *finite* subinterval $[0, t_1]$ *however* small. Now δy could be any continuous function provided it is zero at $t = 0$. Satisfying this provision, we choose a δy to be $t(t_1 - t)$ in the subinterval but zero outside it. Figure 3.5 shows these two functions as well as their product, the integral of which is

$$\int_0^{t_f} \left(F_y + \lambda G_y + \dot{\lambda}\right) \delta y \, dt \; = \; \int_0^{t_1} \left(F_y + \lambda G_y + \dot{\lambda}\right) t(t_1 - t) \, dt \; > \; 0$$

in contradiction to Equation (3.20). The above integral turns out to be positive because $(F_y + \lambda G_y + \dot{\lambda})$ is positive in the subinterval $[0, t_1]$, and $t(t_1 - t)$ is zero at $t = 0$ but positive otherwise.

Supposing the coefficient $(F_y + \lambda G_y + \dot{\lambda})$ at $t = 0$ to be less than zero contradicts Equation (3.20) again. The integral turns out to be negative this time.

Thus, neither greater nor less than zero as supposed, the coefficient of δy, i. e., $(F_y + \lambda G_y + \dot{\lambda})$, has to be zero at $t = 0$ where δy is already zero. This result is due to Du Bois-Reymond (1879) and holds for any t where δy is zero.

3.C   Mangasarian Sufficiency Condition

For the minimum (maximum) of the optimal control problem given by Equations (3.31)–(3.33) on p. 67, it is sufficient to have

1. the control $\hat{\mathbf{u}}$ satisfy the necessary conditions in Table 3.2,

---

[*] Because of the assumption made at the outset that the partial derivatives of F and g with respect to y are continuous.

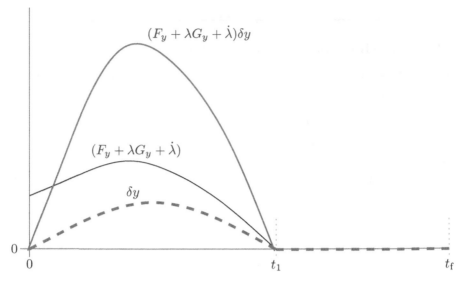

Figure 3.5 The functions $\left(F_y + \lambda G_y + \dot{\lambda}\right)$, δy, and their product, i. e., the integrand of Equation (3.20)

2. all non-positive (non-negative) elements of the costate vector $\boldsymbol{\lambda}$ in the interval $[0, t_f]$ at the minimum (maximum) if

$$\mathbf{g} = \begin{bmatrix} g_1 & g_2 & \cdots & g_n \end{bmatrix}^\top$$

is nonlinear with respect to $\mathbf{u}$ or $\mathbf{y}$, and

3. a convex (concave) F as well as $\mathbf{g}$ jointly with respect to $\mathbf{u}$ and the corresponding $\mathbf{y}$ in the interval $[0, t_f]$

Note that the entities H, F, $\mathbf{g}$, and their partial derivatives are functions of $\mathbf{y}$ and $\mathbf{u}$. Such an entity, say, H, is denoted by $\hat{H}$ when evaluated at the optimum, i. e., for the vector of optimal controls $\hat{\mathbf{u}}$ and the corresponding vector of optimal states $\hat{\mathbf{y}}$.

Let us show the result for the minimum of I or $\hat{I}$.[*] Because F and $\mathbf{g}$ are specified to be convex and continuously differentiable,

$$F - \hat{F} \geq \hat{F}_\mathbf{y}^\top \left(\mathbf{y} - \hat{\mathbf{y}}\right) + \hat{F}_\mathbf{u}^\top \left(\mathbf{u} - \hat{\mathbf{u}}\right) \tag{3.52}$$

$$\mathbf{g} - \hat{\mathbf{g}} \geq \hat{\mathbf{g}}_\mathbf{y}^\top \left(\mathbf{y} - \hat{\mathbf{y}}\right) + \hat{\mathbf{g}}_\mathbf{u}^\top \left(\mathbf{u} - \hat{\mathbf{u}}\right) \tag{3.53}$$

throughout the time interval $[0, t_f]$.

---

[*] To show it for the maximum, reverse the inequalities that follow.

Since $\hat{\mathbf{u}}$ satisfies the necessary optimality conditions (see Table 3.2, p. 68),

$$\boldsymbol{\lambda} = -\hat{H}_{\mathbf{y}} \quad \Rightarrow \quad \hat{F}_{\mathbf{y}} = -\dot{\boldsymbol{\lambda}} - \hat{\mathbf{g}}_{\mathbf{y}}^{\top} \boldsymbol{\lambda} \quad \Rightarrow \quad \hat{F}_{\mathbf{y}}^{\top} = -\dot{\boldsymbol{\lambda}}^{\top} - \boldsymbol{\lambda}^{\top} \hat{\mathbf{g}}_{\mathbf{y}} \quad (3.54)$$

$$\hat{H}_{\mathbf{u}} = 0 \quad \Rightarrow \quad \hat{F}_{\mathbf{u}} = -\hat{\mathbf{g}}_{\mathbf{u}}^{\top} \boldsymbol{\lambda} \quad \Rightarrow \quad \hat{F}_{\mathbf{u}}^{\top} = -\boldsymbol{\lambda}^{\top} \hat{\mathbf{g}}_{\mathbf{u}} \quad (3.55)$$

Using Equations (3.54) and (3.55) in Inequality (3.52), and integrating both sides provides

$$I - \hat{I} \geq -\int_0^{t_f} \dot{\boldsymbol{\lambda}}^{\top} (\mathbf{y} - \hat{\mathbf{y}}) \, dt - \int_0^{t_f} [\boldsymbol{\lambda}^{\top} \hat{\mathbf{g}}_{\mathbf{y}} (\mathbf{y} - \hat{\mathbf{y}}) + \boldsymbol{\lambda}^{\top} \hat{\mathbf{g}}_{\mathbf{u}} (\mathbf{u} - \hat{\mathbf{u}})] \, dt \quad (3.56)$$

Integrating by parts the first integral in the above inequality, we get

$$\int_0^{t_f} \dot{\boldsymbol{\lambda}}^{\top} (\mathbf{y} - \hat{\mathbf{y}}) \, dt = -\int_0^{t_f} \boldsymbol{\lambda}^{\top} (\dot{\mathbf{y}} - \dot{\hat{\mathbf{y}}}) \, dt = -\int_0^{t_f} \boldsymbol{\lambda}^{\top} (\mathbf{g} - \hat{\mathbf{g}}) \, dt \quad (3.57)$$

where we have applied the final costate condition $\boldsymbol{\lambda}(t_f) = \mathbf{0}$, the non-variation of the state at $t = 0$, i.e., $\mathbf{y}(0) - \hat{\mathbf{y}}(0) = \delta \mathbf{y}(0) = \mathbf{0}$, and state equations in the last step. With the help of Equation (3.57), Inequality (3.56) becomes

$$I - \hat{I} \geq \int_0^{t_f} \boldsymbol{\lambda}^{\top} \Big[\mathbf{g} - \hat{\mathbf{g}} - \hat{\mathbf{g}}_{\mathbf{y}} (\mathbf{y} - \hat{\mathbf{y}}) - \hat{\mathbf{g}}_{\mathbf{u}} (\mathbf{u} - \hat{\mathbf{u}}) \Big] \, dt$$

in which the square-bracketed term is greater than zero due to Inequality (3.53). Because the costate at the minimum is specified to be non-positive, i.e., $\boldsymbol{\lambda} \geq \mathbf{0}$, we obtain $I - \hat{I} \geq 0$.

If $\hat{\mathbf{g}}$ is linear with respect to $\mathbf{u}$ and $\mathbf{y}$, then Inequality (3.53) becomes an equation so that the specification $\boldsymbol{\lambda} \geq \mathbf{0}$ is not required to obtain $I - \hat{I} \geq 0$.

Bibliography

P. Du Bois-Reymond. Erläuterungen zu der anfangsgründen der variationsrechnung. *Math. Ann.*, 15:283–314, 1879.

T. Jensen. *Dynamic Control of Large Dimension Nonlinear Chemical Processes.* PhD thesis, Princeton University, 1964.

M.I. Kamien and N.L. Schwartz. *Dynamic Optimization*, Chapter 6, pages 124–132. Elsevier, Amsterdam, 1991.

O.L. Mangasarian. Sufficient conditions for the optimal control of nonlinear systems. *SIAM J. Control*, 4:139–152, 1966.

W.H. Ray and M.A. Soliman. The optimal control of processes containing pure time delays — I. Necessary conditions for an optimum. *Chem. Eng. Sci.*, 25:1911–1925, 1970.

H. Sagan. *Introduction to the Calculus of Variations*, Chapter 1, pages 32–43. Dover Publications Inc., New York, 1969.

D.R. Smith. *Variational Methods in Optimization*, Chapter 1, pages 31–41. Dover Publications Inc., New York, 1998.

Exercises

3.1 Find the necessary conditions for the minimum of

$$I = \int_0^{t_f} (u^2 - y_2)\, \mathrm{d}t$$

subject to

$$\dot{y}_1 = -y_1 u + y_1 u^2, \qquad y_1(0) = y_{1,0}$$
$$\dot{y}_2 = y_2 u, \qquad\qquad y_2(0) = 0$$

3.2 For the plug flow reactor problem on p. 6, find the necessary conditions for the minimum of

$$I = \int_0^Z \frac{\mathrm{d}y}{\mathrm{d}z}\, \mathrm{d}z \tag{1.12}$$

subject to

$$\frac{\mathrm{d}y}{\mathrm{d}z} = \tau S \left[\frac{-k_1 y P}{2y_0 - y} + \frac{4k_2(y_0 - y)^2 P^2}{(2y_0 - y)^2} \right] \tag{3.58}$$

with $y(0) = y_0$.

3.3 Find the necessary conditions to maximize the profits

$$I = y_8(t_f)$$

from an isothermal CSTR carrying out four simultaneous reactions described by (Jensen, 1964)

$$\dot{y}_1 = u_4 - (u_1 + u_2 + u_4)y_1 - a_1 y_1 y_2 - a_2 y_1 y_6 u_3$$
$$\dot{y}_2 = u_1 - (u_1 + u_2 + u_4)y_2 - a_1 y_1 y_2 - 2a_3 y_2 y_3$$
$$\dot{y}_3 = u_2 - (u_1 + u_2 + u_4)y_3 - a_3 y_2 y_3$$
$$\dot{y}_4 = -(u_1 + u_2 + u_4)y_4 + 2a_1 y_1 y_2 - a_4 y_4 y_5$$
$$\dot{y}_5 = -(u_1 + u_2 + u_4)y_5 + a_5 y_2 y_3 - a_4 y_4 y_5$$
$$\dot{y}_6 = -(u_1 + u_2 + u_4)y_6 + 2a_4 y_4 y_5 - a_2 y_1 y_6 u_3$$
$$\dot{y}_7 = -(u_1 + u_2 + u_4)y_7 + 2a_2 y_1 y_6 u_3$$
$$\dot{y}_8 = a_5[(u_1 + u_2 + u_4)y_1 - u_4] - a_6 u_1 - a_7 u_2$$
$$+ (u_1 + u_2 + u_4)(a_2 y_4 + a_8 y_5 + a_9 y_6 + a_{10} y_7) - a_{11} u_3^2 - a_{12}$$

with initial conditions $y_i(0) = y_{i,0}$ for $i = 1, 2, \ldots, 8$ over the time interval $[0, t_f]$. The coefficients a_1 to a_{12} are constants. The controls u_1 to u_4 are, respectively, three feed flow rates and electrical energy input to the CSTR as functions of time.

3.4 Show that the non-autonomous problem of finding the minimum of

$$I = \int_0^{t_f} F[y(t), u(t), t] \, dt$$

subject to

$$G[y(t), \dot{y}, u(t), t] \equiv -\dot{y} + g[y(t), u(t), t] = 0$$

is similar to the autonomous problem in which t does not appear explicitly as an argument of F and g.

Hint: Propose t as a new state variable, say, y_0 (see Section 9.7.1, p. 270).

Chapter 4

Lagrange Multipliers

In this chapter, we introduce the concept of Lagrange multipliers. We show how the Lagrange Multiplier Rule and the John Multiplier Theorem help us handle the equality and inequality constraints in optimal control problems.

4.1   Motivation

Consider the simplest optimal control problem, in which we wish to find the control function u that optimizes

$$I = \int_0^{t_f} F(y, u)\, dt \tag{3.4}$$

subject to the differential equation constraint

$$\dot{y} = g(y, u) \quad \text{or} \quad G(y, \dot{y}, u) \equiv -\dot{y} + g(y, u) = 0 \tag{3.5}$$

with the initial condition

$$y(0) = y_0 \tag{3.6}$$

At the optimum, it is necessary that the variation given by

$$\delta I = \int_0^{t_f} (F_y \delta y + F_u \delta u)\, dt$$

is zero, while satisfying the differential equation constraint. Because the constraint ties y and u together, we cannot have δy or δu arbitrary and thus independent of each other in the above equation. Recall from the last chapter that when the variations are arbitrary, their coefficients are individually zero, thereby leading to the necessary conditions for the optimum. This simplification is, however, not possible with the above equation.

Dealing with this problem is easy if Equation (3.5) could be integrated to provide an explicit solution for y in terms of u. Then one could substitute

$y = y(u)$ into Equation (3.4) and obtain I in terms of u alone. However, this approach fails in most problems where analytical solutions of the involved constraints are simply not possible.

4.2   Role of Lagrange Multipliers

The above difficulty is surmounted by introducing an undetermined function, $\lambda(t)$, called Lagrange multiplier, in the augmented objective functional defined as

$$J \equiv \int_0^{t_f} \left[F(y, u) + \lambda G(y, \dot{y}, u) \right] \mathrm{d}t \qquad (3.7)$$

At the optimum, the variation of J is

$$\delta J = \int_0^{t_f} [F_y \delta y + F_u \delta u + \lambda(G_y \delta y + G_{\dot{y}} \delta \dot{y} + G_u \delta u) + G \delta \lambda]\, \mathrm{d}t = 0 \qquad (4.1)$$

where the role of λ is to untie y from u by assuming certain values in the interval $[0, t_f]$. Given such a λ, we are then able to vary δy and δu arbitrarily and independently of each other. This ability leads to the simplified necessary conditions for the optimum of J and, equivalently, for the constrained optimum of I. The conditions include an additional equation for λ, the satisfaction of which enables the arbitrary variations in the first place.

In Section 3.2.1 (p. 59), we had asserted the Lagrange Multiplier Rule that the optimum of the augmented J is equivalent to the constrained optimum of I. This rule is based on the Lagrange Multiplier Theorem, which provides the necessary conditions for the constrained optimum. We will first prove this theorem and then apply it to optimal control problems subject to different types of constraints.

4.3   Lagrange Multiplier Theorem

The theorem states that if a functional $I(y)$ subject to the constraint

$$K(y) = k_0$$

is optimal at $y = \hat{y}$ near which δI and δK are *weakly continuous*, and there exists a variation δz at $\hat{y}$ along which

$$\delta K(\hat{y}; \delta z) \neq 0$$

then the following conditions are necessary:

1. The constraint is satisfied, i. e.,

$$K(\hat{y}) = k_0 \tag{4.2}$$

2. There exists a constant multiplier λ such that

$$\delta I(\hat{y}; \delta y) + \lambda \delta K(\hat{y}; \delta y) = 0 \tag{4.3}$$

for all variations δy.

Remarks

1. The provision that $\delta K(\hat{y}; \delta z) \neq 0$ along at least one variation δz is called the **normality condition** or the **constraint qualification**.

2. Equations (4.2) and (4.3) are the necessary conditions for the constrained optimum of I at $\hat{y}$.

3. Both I and K in a general case are functionals of a function y depending on an independent variable.

4. **Weak continuity** of δI means the continuity of δI with respect to y for each fixed δy. This is a relaxed requirement. Not fixing δy would have imposed the demand for the Fréchet differential instead of the variation of I.

5. If δI is weakly continuous near $\hat{y}$, it means at any y near $\hat{y}$, δI is continuous for each fixed δy. In other words, if y_1 and y_2 are two functions in the vicinity of $\hat{y}$, then $\delta I(y_1; \delta y)$ approaches $\delta I(y_2; \delta y)$ as y_1 approaches y_2 for each fixed δy.

Example 4.1

Consider the functional

$$I = \int_0^{t_f} F(y) \, dt$$

whose variation is given by

$$\delta I = \int_0^{t_f} F_y(y) \, \delta y \, dt$$

Then the weak continuity of δI means the continuity of the partial derivative F_y. $\square$

Note that most of the optimal control problems have continuous partial derivatives of the involved integrands and, therefore, satisfy the requirement of weak continuity.

Outline of the Proof

As shown in Figure 4.1, together with the givens and an assumption of a non-zero Jacobian of I and K at the optimum, we will invoke the Inverse Function Theorem (Appendix 4.A, p. 115) to contradict the optimum of I. This step will entail the Jacobian being zero and lead to the desired conclusion.

at $\hat{y}$:

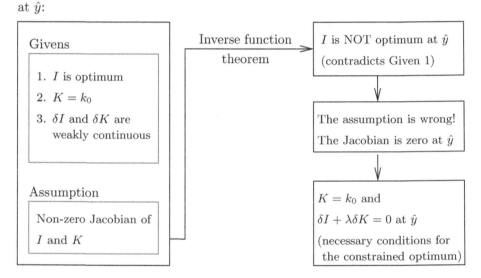

Figure 4.1 Outline of the proof for the Lagrange Multiplier Theorem

The Proof

Let $\hat{y}$ be the optimal function. Since any admissible function must satisfy the the constraint $K(y) = k_0$, we have

$$\boxed{K(\hat{y}) = k_0} \tag{4.2}$$

as **the first necessary condition** for the constrained optimum of I.

Consider this $\hat{y}$ and its arbitrary variations, δy and δz. Keeping these three functions fixed, the functional values $I(\hat{y}+\mu\delta y+\nu\delta z)$ and $K(\hat{y}+\mu\delta y+\nu\delta z)$ are functions of μ and ν, which are some real numbers. Let these two functions be, respectively,

$$i(\mu, \nu) \equiv I(\hat{y} + \mu \delta y + \nu \delta z) \tag{4.4}$$

$$\text{and} \quad k(\mu, \nu) \equiv K(\hat{y} + \mu \delta y + \nu \delta z) \tag{4.5}$$

for all μ and ν within a region R that is sufficiently small to allow for the weak continuity of δI as well as δK at $y = (\hat{y} + \mu y + \nu z)$. Figure 4.2 shows the (μ, ν) coordinates in R mapped to the (i, k) coordinates by $i(\mu, \nu)$ and $k(\mu, \nu)$.

Jacobian Determinant

We now introduce the Jacobian determinant of i and k

$$D(\mu, \nu) \equiv \begin{vmatrix} i_\mu & i_\nu \\ k_\mu & k_\nu \end{vmatrix}$$

where i_μ, for example, is the partial derivative of i with respect to μ and is given by

$$i_\mu = \lim_{\alpha \to 0} \frac{i(\mu + \alpha, \nu) - i(\mu, \nu)}{\alpha}$$

$$= \lim_{\alpha \to 0} \frac{I[(\hat{y} + \mu \delta y + \nu \delta z) + \alpha \delta y] - I[(\hat{y} + \mu \delta y + \nu \delta z)]}{\alpha}$$

$$- \frac{\mathrm{d}}{\mathrm{d}\alpha} I(\hat{y} + \mu \delta y + \nu \delta z)_{\alpha=0}$$

$$= \delta I(\hat{y} + \mu \delta y + \nu \delta z; \delta y) \tag{4.6}$$

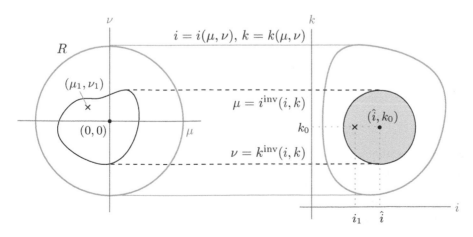

Figure 4.2 Mappings between (μ, ν) and (i, k). The shaded region has the inverse mappings.

The last equation follows from the definition of the variation [see Equation (2.12), p. 36]. Similar results are obtained for the remaining partial derivatives. Using them in the definition of D, we obtain

$$D(\mu, \nu) = \begin{vmatrix} \delta I(\hat{y} + \mu \delta y + \nu \delta z; \delta y) & \delta I(\hat{y} + \mu \delta y + \nu \delta z; \delta z) \\ \delta K(\hat{y} + \mu \delta y + \nu \delta z; \delta y) & \delta K(\hat{y} + \mu \delta y + \nu \delta z; \delta z) \end{vmatrix}$$

Note that the given weak continuity of δI and δK gets translated into the continuity of their equivalents, i.e., the partial derivatives of i and k. In other words, the functions i and k are differentiable in the sufficiently small region containing μ and ν.

When $\mu = \nu = 0$, the Jacobian becomes

$$D(0, 0) = \begin{vmatrix} \delta I(\hat{y}; \delta y) & \delta I(\hat{y}; \delta z) \\ \delta K(\hat{y}; \delta y) & \delta K(\hat{y}; \delta z) \end{vmatrix}$$

which is at the optimum. Let us assume that $D(0, 0)$ is not zero. This assumption rules out δy or δz being zero throughout the t-interval $[0, t_f]$. Otherwise, say, if δy is zero throughout the interval, then both terms of the first column of the Jacobian would be zero and result in $D = 0$.

Application of Inverse Function Theorem

Observe from the definitions in Equations (4.4) and (4.5) that $\mu = 0$ and $\nu = 0$ correspond to $i(0, 0) = I(\hat{y}) \equiv \hat{i}$, and $k(0, 0) = K(\hat{y}) = k_0$ using Equation (4.2). Figure 4.2 shows these coordinates. The functions $i(\mu, \nu)$ and $k(\mu, \nu)$ are differentiable in the region R around $\mu = \nu = 0$.

Since $D(0, 0)$ is not zero, the Inverse Function Theorem (Appendix 4.A, p. 115) guarantees the existence of inverse functions

$$\mu = i^{\text{inv}}(i, k) \tag{4.7}$$

$$\nu = k^{\text{inv}}(i, k) \tag{4.8}$$

which are continuous in some region shown shaded around $(\hat{i}, k_0)$ in the figure. In this region, we can always pick a coordinate (i_1, k_0) where i_1 is less than $I(\hat{y})$ and the constraint $K = k_0$ is satisfied. Then Equations (4.7) and (4.8) provide

$$\mu_1 = i^{\text{inv}}(i_1, k_0) \quad \text{and} \quad \nu_1 = k^{\text{inv}}(i_1, k_0)$$

which contradict the specification that $\hat{i}$ is the minimum. The presence of μ_1 and ν_1 implies that while satisfying the constraint $K(y) = k_0$, a control function $y = (\hat{y} + \mu_1 \delta y + \nu_1 \delta z)$ provides $i_1 = I(y)$, which is *less* than $\hat{i} = I(\hat{y})$. Since this is impossible, our assumption cannot be true, so that the $D(0, 0)$ has to be zero. Therefore,

$$D(0, 0) = \begin{vmatrix} \delta I(\hat{y}; \delta y) & \delta I(\hat{y}; \delta z) \\ \delta K(\hat{y}; \delta y) & \delta K(\hat{y}; \delta z) \end{vmatrix} = 0 \tag{4.9}$$

which leads to the second necessary condition for the minimum of $I(y)$ at $\hat{y}$ subject to the constraint $K(y) = k_0$. The same condition is obtained for the constrained maximum of I by considering i_1 greater than $I(\hat{y})$.

Expanding Equation (4.9), we get

$$\delta I(\hat{y}; \delta y)\delta K(\hat{y}; \delta z) - \delta I(\hat{y}; \delta z)\delta K(\hat{y}; \delta y) = 0 \qquad (4.10)$$

for all functions δy and δz. With the introduction of

$$\lambda \equiv -\frac{\delta I(\hat{y}; \delta z)}{\delta K(\hat{y}; \delta z)}$$

where $\delta K(\hat{y}; \delta z)$ is not zero as per the given constraint qualification, Equation (4.10) becomes

$$\boxed{\delta I(\hat{y}; \delta y) + \lambda \delta K(\hat{y}; \delta y) = 0} \qquad (4.3)$$

for all δy and a Lagrange multiplier λ. The above equation is the **second necessary condition** for the constrained optimum of $I(y)$ at $\hat{y}$ subject to the constraint $K(y) = k_0$. This completes the proof of the Lagrange Multiplier Theorem.

Lagrange Multiplier Rule

This rule is based on the Lagrange Multiplier Theorem. Consider the augmented functional

$$J(y) \equiv I(y) + \lambda[K(y) - k_0]$$

where both y and λ are functions of an independent variable t. According to the theorem, the necessary condition for the optimum of J at $\hat{y}$ is that

$$\delta J(\hat{y}; \delta y) = 0, \quad \text{or}$$

$$\underbrace{\delta I(\hat{y}; \delta y) + \lambda \delta K(\hat{y}; \delta y)}_{J_1} + \underbrace{\delta\lambda[K(\hat{y}) - k_0]}_{J_2} = 0$$

Since $\delta\lambda$ is arbitrary and appears only in the last term J_2 of the above equation, both J_1 and J_2 should be individually zero. Thus,

$$J_2 = \delta\lambda[K(\hat{y}) - k_0] = 0$$

implies that

$$\boxed{K(\hat{y}) - k_0} \qquad (4.2)$$

because $\delta\lambda$ is an arbitrary function of t. Next, $J_1 = 0$ yields

$$\boxed{\delta I(\hat{y}; \delta y) + \lambda \delta K(\hat{y}; \delta y) = 0} \qquad (4.3)$$

According to the Lagrange Multiplier Theorem, Equations (4.2) and (4.3) are the necessary conditions for the optimum of I subject to the following preconditions:

1. The variations of both I and K are weakly continuous near $\hat{y}$.

2. The constraint qualification that the variation of K at $y = \hat{y}$ is non-zero for at least one variation of y.

Observe that Equations (4.2) and (4.3), which arise from $\delta J(\hat{y}; \delta y) = 0$, are also the necessary conditions for the optimum of J. This fact gives rise to the following **Lagrange Multiplier Rule**:

The optimum of a functional $I(y)$ subject to a constraint $K(y) = k_0$ is equivalent to the optimum of the augmented functional

$$J(y) = I(y) + \lambda[K(y) - k_0]$$

where λ is an undetermined Lagrange multiplier, both δI and δK are weakly continuous near the optimal y, and the constraint qualification $\delta K(y; \delta z) \neq 0$ holds along at least one variation δz at the optimal y.

Before considering the applications of the Lagrange Multiplier Theorem, it is worthwhile to generalize the theorem to handle several equality constraints and functions.

4.3.1   Generalization to Several Equality Constraints

Let us derive in terms of Lagrange multipliers the necessary conditions for the optimum of $I(y)$ at $y = \hat{y}$ subject to two equality constraints

$$K_1(y) = k_1 \quad \text{and} \quad K_2(y) = k_2$$

and weak continuity of δI, δK_1, and δK_2 near $\hat{y}$ with the following constraint qualification:

There exists at least a pair of variations δz_1 and δz_2 corresponding to which the determinant

$$\begin{vmatrix} \delta K_1(\hat{y}; \delta z_1) & \delta K_1(\hat{y}; \delta z_2) \\ \delta K_2(\hat{y}; \delta z_1) & \delta K_2(\hat{y}; \delta z_2) \end{vmatrix} \neq 0$$

If $\hat{y}$ is optimal, then it must satisfy the above constraints. Therefore, the first set of two necessary conditions is

$$K_1(\hat{y}) = k_1 \quad \text{and} \quad K_2(\hat{y}) = k_2$$

Now consider a function

$$y \equiv \hat{y} + \mu \delta y + \nu_1 \delta z_1 + \nu_2 \delta z_2$$

in the vicinity of $\hat{y}$ with three arbitrary but fixed variations, δy, δz_1, and δz_2. The variables μ, ν_1, and ν_2 are some real numbers in a region R. Then the functional values $I(y)$, $K_1(y)$, and $K_2(y)$ are the functions $i(\mu, \nu_1, \nu_2)$, $k_1(\mu, \nu_1, \nu_2)$, and $k_2(\mu, \nu_1, \nu_2)$, respectively. If R is small enough to allow for weak continuity of δI, δK_1, and δK_2 near $\hat{y}$, then from Equation (4.6) it follows that

$$
D(\mu, \nu_1, \nu_2) \equiv
\begin{vmatrix}
\dfrac{\partial i}{\partial \mu} & \dfrac{\partial i}{\partial \nu_1} & \dfrac{\partial i}{\partial \nu_2} \\[2ex]
\dfrac{\partial k_1}{\partial \mu} & \dfrac{\partial k_1}{\partial \nu_1} & \dfrac{\partial k_1}{\partial \nu_2} \\[2ex]
\dfrac{\partial k_2}{\partial \mu} & \dfrac{\partial k_2}{\partial \nu_1} & \dfrac{\partial k_2}{\partial \nu_2}
\end{vmatrix}
$$

$$
=
\begin{vmatrix}
\delta I(y; \delta y) & \delta I(y; \delta z_1) & \delta I(y; \delta z_2) \\[1ex]
\delta K_1(y; \delta y) & \delta K_1(y; \delta z_1) & \delta K_1(y; \delta z_2) \\[1ex]
\delta K_2(y; \delta y) & \delta K_2(y; \delta z_1) & \delta K_2(y; \delta z_2)
\end{vmatrix}
$$

Applying the Inverse Function Theorem as before, we establish

$$
D(0,0,0) =
\begin{vmatrix}
\delta I(\hat{y}; \delta y) & \delta I(\hat{y}; \delta z_1) & \delta I(\hat{y}; \delta z_2) \\[1ex]
\delta K_1(\hat{y}; \delta y) & \delta K_1(\hat{y}; \delta z_1) & \delta K_1(\hat{y}; \delta z_2) \\[1ex]
\delta K_2(\hat{y}; \delta y) & \delta K_2(\hat{y}; \delta z_1) & \delta K_2(\hat{y}; \delta z_2)
\end{vmatrix}
= 0
$$

which is the remaining necessary condition for the optimum of $I(\hat{y})$ subject to $K_1(\hat{y}) = k_1$ and $K_2(\hat{y}) = k_2$. This condition expands to

$$
\delta I(\hat{y}; \delta y) \underbrace{\begin{vmatrix} \delta K_1(\hat{y}; \delta z_1) & \delta K_1(\hat{y}; \delta z_2) \\ \delta K_2(\hat{y}; \delta z_1) & \delta K_2(\hat{y}; \delta z_2) \end{vmatrix}}_{\eta_0} - \delta K_1(\hat{y}; \delta y) \underbrace{\begin{vmatrix} \delta I(\hat{y}; \delta z_1) & \delta I(\hat{y}; \delta z_2) \\ \delta K_2(\hat{y}; \delta z_1) & \delta K_2(\hat{y}; \delta z_2) \end{vmatrix}}_{\eta_1}
$$

$$
+ \delta K_2(\hat{y}; \delta y) \underbrace{\begin{vmatrix} \delta I(\hat{y}; \delta z_1) & \delta I(\hat{y}; \delta z_2) \\ \delta K_1(\hat{y}; \delta z_1) & \delta K_1(\hat{y}; \delta z_2) \end{vmatrix}}_{\eta_2} = 0
$$

where η_0, η_1, and η_2 are the determinants as indicated above.

Because of the provision of constraint qualification, $\eta_0 \neq 0$. Thus, we can introduce the Lagrange multipliers

$$
\lambda_1 \equiv -\eta_1/\eta_0 \quad \text{and} \quad \lambda_2 \equiv \eta_2/\eta_0
$$

to obtain

$$
\boxed{\delta I(\hat{y}; \delta y) + \lambda_1 \delta K_1(\hat{y}; \delta y) + \lambda_2 \delta K_2(\hat{y}; \delta y) = 0}
$$

as the final necessary condition for the constrained optimum of I.

Lagrange Multiplier Theorem for Several Equality Constraints

Generalizing the above results for m constraints, $K_i(y) = k_i$, $i = 1, 2, \ldots, m$, the necessary conditions for the optimum of $I(y)$ at $\hat{y}$

$$K_i(\hat{y}) = k_i, \quad i = 1, 2, \ldots, m$$

$$\delta I(\hat{y}; \delta y) + \sum_{i=1}^{m} \lambda_i \delta K_i(\hat{y}; \delta y) = 0$$

subject to the following preconditions:

1. The variations of I and K_i, $i = 1, 2, \ldots, m$ are weakly continuous near the optimum at $\hat{y}$.

2. The constraint qualification that there exists a set of m variations

$$(\delta z_1, \ \delta z_2, \ \ldots, \ \delta z_m)$$

of **y** such that the determinant

$$\eta_0 = \begin{vmatrix} \delta K_1(\hat{y}; \delta z_1) & \delta K_1(\hat{y}; \delta z_2) & \ldots & \delta K_1(\hat{y}; \delta z_m) \\ \delta K_2(\hat{y}; \delta z_1) & \delta K_2(\hat{y}; \delta z_2) & \ldots & \delta K_2(\hat{y}; \delta z_m) \\ \vdots & \vdots & \vdots & \vdots \\ \delta K_m(\hat{y}; \delta z_1) & \delta K_m(\hat{y}; \delta z_2) & \ldots & \delta K_m(\hat{y}; \delta z_m) \end{vmatrix} \neq 0$$

Lagrange Multiplier Rule for Several Equality Constraints

Similar to the case of a single constraint in the last section, the rule is as follows:

The optimum of a functional $I(y)$ subject to constraints

$$K_i(y) = k_i, \quad i = 1, 2, \ldots, m$$

is equivalent to the optimum of the augmented functional

$$J(y) = I(y) + \sum_{i=1}^{m} \lambda_i [K_i(y) - k_i]$$

where λ_is are undetermined Lagrange multipliers, δI and all δK_is are weakly continuous near the optimum, and the aforementioned constraint qualification $\eta_0 \neq 0$ holds for at least one set of m variations of y at the optimum.

4.3.2   Generalization to Several Functions

The result of the above section is easily extensible to the general case in which the objective functional and equality constraints depend on several functions.

Let I be a functional dependent on the vector of n functions

$$\mathbf{y} = \begin{bmatrix} y_1 & y_2 & \cdots & y_n \end{bmatrix}^\top$$

and subject to m constraints

$$K_i(\mathbf{y}) = k_i, \quad i = 1, 2, \ldots, m$$

If satisfying the above constraints I is optimum at $\mathbf{y} = \hat{\mathbf{y}}$ such that

1. the functionals δI and all δK_is are weakly continuous near $\hat{\mathbf{y}}$, and

2. there exists a set of m variations

$$(\delta \mathbf{z}_1, \ \delta \mathbf{z}_2, \ \ldots, \ \delta \mathbf{z}_m)$$

where $\delta \mathbf{z}_i$ is the i-th variation vector at $\hat{\mathbf{y}}$ given by

$$\delta \mathbf{z}_i = \begin{bmatrix} \delta z_{i1} & \delta z_{i2} & \cdots & \delta z_{in} \end{bmatrix}^\top$$

for which the following constraint qualification is satisfied:

$$\eta_0 = \begin{vmatrix} \delta K_1(\hat{\mathbf{y}}; \delta \mathbf{z}_1) & \delta K_1(\hat{\mathbf{y}}; \delta \mathbf{z}_2) & \cdots & \delta K_1(\hat{\mathbf{y}}; \delta \mathbf{z}_m) \\ \delta K_2(\hat{\mathbf{y}}; \delta \mathbf{z}_1) & \delta K_2(\hat{\mathbf{y}}; \delta \mathbf{z}_2) & \cdots & \delta K_2(\hat{\mathbf{y}}; \delta \mathbf{z}_m) \\ \vdots & \vdots & \vdots & \vdots \\ \delta K_m(\hat{\mathbf{y}}; \delta \mathbf{z}_1) & \delta K_m(\hat{\mathbf{y}}; \delta \mathbf{z}_2) & \cdots & \delta K_m(\hat{\mathbf{y}}; \delta \mathbf{z}_m) \end{vmatrix} \neq 0$$

then the necessary conditions for the optimum are

$$K_i(\hat{\mathbf{y}}) = k_i, \quad i = 1, 2, \ldots, m$$

$$\delta I(\hat{\mathbf{y}}; \delta \hat{\mathbf{y}}) + \sum_{i=1}^{m} \lambda_i \delta K_i(\hat{\mathbf{y}}; \delta \hat{\mathbf{y}}) = 0$$

where λ_is are the m undetermined Lagrange multipliers.

4.3.2.1 Simplification of Constraint Qualification

The constraint qualification $\eta_0 \neq 0$ can be simplified by expanding the terms of the determinant. In general

$$\delta K_i(\hat{\mathbf{y}}; \delta \mathbf{z}_j) = \begin{bmatrix} \frac{\partial K_i}{\partial y_1} \delta z_{j1} + \frac{\partial K_i}{\partial y_2} \delta z_{j2} + \cdots + \frac{\partial K_i}{\partial y_n} \delta z_{jn} \end{bmatrix}_{\hat{\mathbf{y}}}$$

$$= K_{i\hat{\mathbf{y}}}^\top \delta \mathbf{z}_j$$

where $K_{i\hat{y}}$ is the vector of partial derivatives of K_i with respect to $\mathbf{y}$ and is evaluated at $\mathbf{y} = \hat{\mathbf{y}}$, i. e.,

$$K_{i\hat{y}} = \begin{bmatrix} \dfrac{\partial K_i}{\partial y_1} & \dfrac{\partial K_i}{\partial y_2} & \cdots & \dfrac{\partial K_i}{\partial y_n} \end{bmatrix}_{\hat{y}}^{\top}$$

Thus, the constraint qualification can be written as

$$\eta_0 = \begin{vmatrix} K_{1\hat{y}}^{\top}\delta\mathbf{z}_1 & K_{1\hat{y}}^{\top}\delta\mathbf{z}_2 & \cdots & K_{1\hat{y}}^{\top}\delta\mathbf{z}_m \\ K_{2\hat{y}}^{\top}\delta\mathbf{z}_1 & K_{2\hat{y}}^{\top}\delta\mathbf{z}_2 & \cdots & K_{2\hat{y}}^{\top}\delta\mathbf{z}_m \\ \vdots & \vdots & \vdots & \vdots \\ K_{m\hat{y}}^{\top}\delta\mathbf{z}_1 & K_{m\hat{y}}^{\top}\delta\mathbf{z}_2 & \cdots & K_{m\hat{y}}^{\top}\delta\mathbf{z}_m \end{vmatrix} \neq 0$$

If a determinant is not zero, it means that its rows are linearly independent. In other words, none of the rows can be expressed as a linear combination of other rows. This result in case of the above determinant η_0 means that there is no $K_{i\hat{y}}$ that can be expressed as a linear combination of other $K_{j\hat{y}}$ $(j = 1, 2, \ldots, m, \ j \neq i)$.

As a consequence, the simplified constraint qualification is that the vectors

$$K_{i\hat{y}} = \begin{bmatrix} \dfrac{\partial K_i}{\partial y_1} & \dfrac{\partial K_i}{\partial y_2} & \cdots & \dfrac{\partial K_i}{\partial y_n} \end{bmatrix}_{\hat{y}}^{\top}, \quad i = 1, 2, \ldots, m$$

are linearly independent.

Generalized Lagrange Multiplier Rule

The rule for this general case of several constraints and functions is as follows: The optimum of a functional $I(\mathbf{y})$ where

$$\mathbf{y} = \begin{bmatrix} y_1 & y_2 & \cdots & y_n \end{bmatrix}^{\top}$$

subject to m constraints

$$K_i(\mathbf{y}) = k_i, \quad i = 1, 2, \ldots, m$$

is equivalent to the optimum of the augmented functional

$$J(\mathbf{y}) = I(\mathbf{y}) + \sum_{i=1}^{m} \lambda_i [K_i(\mathbf{y}) - k_i]$$

where λ_is are undetermined Lagrange multipliers, δI and all δK_is are weakly continuous near the optimal y, and the vectors of partial derivatives $K_{i\hat{y}}$ $(i = 1, 2, \ldots m)$ are linearly independent.

4.3.3 Application to Optimal Control Problems

We will apply the Lagrange Multiplier Rule to obtain the set of necessary conditions for the optimum in an optimal control problem constrained by a differential equation. In Section 3.2, we asserted the rule and obtained the following necessary conditions (see p. 60):

$$G = 0 \tag{3.11}$$

and

$$\int\limits_0^{t_f} (\delta F + \lambda \delta G)\, \mathrm{d}t = 0 \tag{3.12}$$

for the optimum of

$$I = \int\limits_0^{t_f} F(y, u)\, \mathrm{d}t \tag{3.4}$$

subject to the differential equation constraint

$$\dot{y} = g(y, u) \quad \text{or} \quad G \equiv -\dot{y} + g(y, u) = 0 \tag{3.5}$$

over the interval $[0, t_f]$ with the initial condition $y(0) = y_0$.

Series of Equality Constraints

We will first show that the differential equation poses a series of equality constraints along the t-direction. Then we will apply the Lagrange Multiplier Rule for the optimum of I subject to those constraints.

As shown in Figure 4.3, let y_i and u_i be the values of y and u at a point t_i in the interval $[0, t_f]$. Then at each successive t greater than 0, the differential

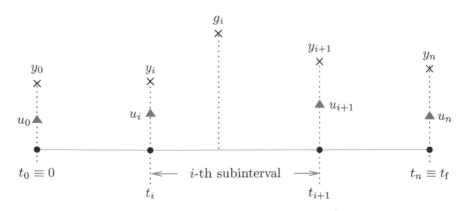

Figure 4.3 Values of y, u, and g in the i-th subinterval

equation constrains the corresponding values of y to

$$\underbrace{y_0 + \int_{t_0=0}^{t_1} g\,\mathrm{d}t}_{y_1}, \quad \underbrace{y_1 + \int_{t_1}^{t_2} g\,\mathrm{d}t}_{y_2}, \quad \ldots, \quad \underbrace{y_{n-2} + \int_{t_{n-2}}^{t_{n-1}} g\,\mathrm{d}t}_{y_{n-1}}, \quad \text{and} \quad \underbrace{y_{n-1} + \int_{t_{n-1}}^{t_n=t_f} g\,\mathrm{d}t}_{y_n}$$

where for each i-th subinterval $[t_i, t_{i+1}]$, the length $\Delta t_i \equiv (t_{i+1} - t_i)$ tends to zero. The i-th such constraint is

$$\int_{t_i}^{t_{i+1}} g(y, u)\,\mathrm{d}t = y_{i+1} - y_i$$

which can be written as

$$\lim_{\Delta t_i \to 0} g_i \Delta t_i = y_{i+1} - y_i$$

where from the Mean Value Theorem for integrals (Section 9.14.2, p. 276), g_i is the value of g at some t in the i-th subinterval. Rearranging the above equation,

$$\underbrace{\lim_{\Delta t_i \to 0} \left[-\frac{y_{i+1} - y_i}{\Delta t_i} + g_i \right] \Delta t_i}_{K_i(\mathbf{z}_i)} = \underbrace{0}_{k_i}$$

where we have denoted the left-hand side as the functional K_i. In the limit Δt_i tending to zero, K_i depends on the vector

$$\mathbf{z}_i = \begin{bmatrix} \dot{y}_i & y_i & u_i \end{bmatrix}^{\mathsf{T}}$$

Hence, for all $t > 0$ in the interval $[0, t_f]$, the differential equation constraint is equivalent to the following series of equality constraints:

$$K_i(\mathbf{z}_i) = 0, \quad i = 1, 2, \ldots, n \tag{4.11}$$

where n tends to infinity. Note that there is only *one* constraint $K_i(\mathbf{z}_i) = 0$ at a given $t = t_i$ in the interval $(0, t_f]$.

4.3.3.1 Serial Application of the Lagrange Multiplier Rule

The optimum of I subject to the first constraint $K_1 = 0$ is equivalent to the optimum of

$$J_1 = I + \lambda_1 K_1$$

In turn, the optimum of J_1 subject to the second constraint $K_2 = 0$ is equivalent to the optimum of

$$J_2 = J_1 + \lambda_2 K_2 = I + \lambda_1 K_1 + \lambda_2 K_2$$

Thus, the optimum of J_2 is equivalent to the optimum of I subject to the first two constraints. In this way, we sequentially arrive at

$$J \equiv J_n = I + \sum_{i=1}^{n} \lambda_i K_i, \quad \lim n \to \infty$$

whose optimum is equivalent to the optimum of I subject to all constraints

$$K_i(\mathbf{z}_i) = 0, \quad i = 1, 2, \ldots, n, \quad \lim n \to \infty \tag{4.11}$$

Consequently, the necessary conditions for the optimum at $\hat{\mathbf{z}}$ are

1. the constraints given by Equation (4.11) or $G = 0$ along with $y(0) = y_0$, and

2. the equation

$$\delta I + \sum_{i=0}^{n \to \infty} \lambda_i \delta K_i = 0 \tag{4.12}$$

where λ_i is the Lagrange multiplier corresponding to the i-th constraint, $K_i(y) = 0$, at $t = t_i$.

We now need to simplify the last condition. The variations δI and δK_i are respectively given by

$$\delta I = \int_0^{t_f} \delta F \, dt \quad \text{and}$$

$$\delta K_i = \lim_{\Delta t_i \to 0} \left[-\frac{\delta y_{i+1} - \delta y_i}{\Delta t_i} + \underbrace{g_y\big|_{t_i} \delta y_i + g_u\big|_{t_i} \delta u_i}_{\delta g} \right] \Delta t_i$$

Inserting them in Equation (4.12), we get

$$\int_0^{t_f} \delta F \, dt + \sum_{i=0}^{n \to \infty} \lim_{\Delta t_i \to 0} \lambda_i \left[-\frac{\delta y_{i+1} - \delta y_i}{\Delta t_i} + \right.$$

$$\left. g_y\big|_{t_i} \delta y_i + g_u\big|_{t_i} \delta u_i \right] \Delta t_i = 0 \tag{4.13}$$

With Δt_i tending to zero as n increases to infinity in the above equation,

$$\frac{\delta y_{i+1} - \delta y_i}{\Delta t_i} = \frac{d}{dt} \delta y \bigg|_{t_i} = \delta \dot{y} \big|_{t_i}, \quad i = 1, 2, \ldots, n$$

and the summation term becomes an integral. Incorporating these two results, Equation (4.13) becomes

$$\int_0^{t_f} \left[\delta F + \underbrace{\lambda(-\delta \dot{y} + g_y \delta y + g_u \delta u)}_{\delta G} \right] dt = 0$$

where λ is an undetermined function of the independent variable t. The bracketed term multiplied by λ is δG obtained from the definition of G. Hence, the above equation can be expressed as

$$\int_0^{t_f} (\delta F + \lambda \delta G)\, dt = 0 \qquad (3.12)$$

To summarize, the above equation and

$$G = 0 \qquad (3.11)$$

along with $y(0) = y_0$ are necessary for the constrained optimum of I.

Preconditions for the Optimum

It can be easily verified that the above result is subject to the following preconditions at each t in $(0, t_f]$:

1. The variations of I and G are weakly continuous near the optimum.

   Given that $F_{\dot{y}} = 0$ and $G_{\dot{y}} = -1$ are constant and therefore continuous, the weak continuity condition means that the partial derivatives of F and g are continuous with respect to y and u in the vicinity of the optimal pair $(\hat{y}, \hat{u})$.

2. The following constraint qualification exists. There exists at least one set of variations $\delta z = (\delta \dot{y}, \delta y, \delta u)$ for which the variation of G at each t in $(0, t_f]$ is not zero at the optimum.

Example 4.2
Consider the problem in Example 3.1 (p. 65). According to the Lagrange Multiplier Rule, the minimum of

$$I = \int_0^{t_f} (u^2 - yu)\, dt$$

subject to

$$\dot{y} = -\sqrt{y} + u \qquad (3.24)$$

with the initial condition $y(0) = 0$ is equivalent to the minimum of the augmented functional

$$J = \int_0^{t_f} \Big[\underbrace{(u^2 - yu)}_{F} + \lambda\, \underbrace{(-\dot{y} - \sqrt{y} + u)}_{g} \Big] dt \qquad (4.14)$$

subject to the initial condition. The Lagrange multiplier λ is an undetermined function of the independent variable t. This result is subject to the following preconditions:

1. The partial derivatives of F and g with respect to y and u are continuous in the vicinity of the minimum.

2. The following constraint qualification is satisfied for $G \equiv -\dot{y} + g$. There exists at least one set of variations $(\delta\dot{y}, \delta y, \delta u)$ at each t in $(0, t_f]$ for which the variation of G is not zero at the minimum.

☐

Example 4.3
Similarly in Example 3.2 (p. 65), the maximum of

$$I = \int_0^{t_f} ckx^a \, dt, \quad k = k_o \exp\left(-\frac{E}{RT}\right)$$

subject to

$$\dot{x} = -akx^a, \quad x(0) = x_0$$

with T as the control function is equivalent to the maximum of the augmented functional

$$J = \int_0^{t_f} \left[\underbrace{ckx^a}_{F} + \lambda(\underbrace{-\dot{x} - akx^a}_{g}) \right] dt, \quad x(0) = x_0$$

where λ is an undetermined function of the independent variable t. The vector $\hat{\mathbf{z}} = \begin{bmatrix} \hat{x} & \hat{T} \end{bmatrix}^\top$ provides the maximum under the following preconditions:

1. The partial derivatives of F and g with respect to x and T are continuous in the vicinity of $\hat{\mathbf{z}}$.

2. The constraint qualification is satisfied for $G \equiv -\dot{y} + g$. Thus, there exists at least one set of variations $(\delta\dot{y}, \delta y, \delta u)$ at each t in $(0, t_f]$ for which the variation of G is not zero at the maximum.

☐

4.3.3.2 Generalization to Several States and Controls

Consider the optimization of the functional

$$I = \int_0^{t_f} F(\mathbf{y}, \mathbf{u}) \, dt$$

where $\mathbf{y}$ and $\mathbf{u}$ are, respectively, the state and control vectors

$$\mathbf{y} = \begin{bmatrix} y_1(t) & y_2(t) & \cdots & y_n(t) \end{bmatrix}^\top \quad \text{and}$$

$$\mathbf{u} = \begin{bmatrix} u_1(t) & u_2(t) & \cdots & u_m(t) \end{bmatrix}^\top$$

subject to the autonomous ordinary differential equations

$$\dot{\mathbf{y}} = \mathbf{g}(\mathbf{y}, \mathbf{u})$$

with the initial conditions

$$\mathbf{y}(0) = \mathbf{y}_0$$

where $\mathbf{g}$ is the function vector

$$\begin{bmatrix} g_1(\mathbf{y}, \mathbf{u}) & g_2(\mathbf{y}, \mathbf{u}) & \cdots & g_n(\mathbf{y}, \mathbf{u}) \end{bmatrix}^\top$$

The above problem is equivalent to optimizing the augmented functional

$$J = \int_0^{t_f} \left[F + \boldsymbol{\lambda}^\top \underbrace{(-\dot{\mathbf{y}} + \mathbf{g})}_{\mathbf{G}} \right] dt$$

subject to the initial conditions where $\boldsymbol{\lambda}$ is the vector of time dependent Lagrange multipliers

$$\begin{bmatrix} \lambda_1(t) & \lambda_2(t) & \cdots & \lambda_n(t) \end{bmatrix}^\top$$

The preconditions for the optimum are as follows:

1. The partial derivatives of F and $\mathbf{g}$ with respect to $\mathbf{y}$ and $\mathbf{u}$ are continuous in the vicinity of the optimal pair $(\hat{\mathbf{y}}, \hat{\mathbf{u}})$.

2. The following constraint qualification is satisfied at each t in $(0, t_f]$:

$$\eta_0 = \begin{vmatrix} \delta G_1(\hat{\mathbf{z}}; \delta\mathbf{z}_1) & \delta G_1(\hat{\mathbf{z}}; \delta\mathbf{z}_2) & \cdots & \delta G_1(\hat{\mathbf{z}}; \delta\mathbf{z}_m) \\ \delta G_2(\hat{\mathbf{z}}; \delta\mathbf{z}_1) & \delta G_2(\hat{\mathbf{z}}; \delta\mathbf{z}_2) & \cdots & \delta G_2(\hat{\mathbf{z}}; \delta\mathbf{z}_m) \\ \vdots & \vdots & \vdots & \vdots \\ \delta G_m(\hat{\mathbf{z}}; \delta\mathbf{z}_1) & \delta G_m(\hat{\mathbf{z}}; \delta\mathbf{z}_2) & \cdots & \delta G_m(\hat{\mathbf{z}}; \delta\mathbf{z}_m) \end{vmatrix} \neq 0$$

where $\mathbf{z}$ is the coordinate set $(\dot{\mathbf{y}}, \mathbf{y}, \mathbf{u})$, $\hat{\mathbf{z}}$ is the optimal $\mathbf{z}$, and $\delta\mathbf{z}_i$ the i-th variation $(\delta\dot{\mathbf{y}}_i, \delta\mathbf{y}_i, \delta\mathbf{u}_i)$ in $\mathbf{z}$ at $\hat{\mathbf{z}}$.

Example 4.4

Consider the isothermal operation of the CSTR in Example 3.5 (p. 69). The reactant and catalyst concentrations, x_1 and x_2, are governed by

$$\dot{x}_1 = \frac{u_1}{V}(x_f - x_1) - kx_1x_2, \quad x_1(0) = x_{1,0}$$

$$\dot{x}_2 = \frac{u_2 - u_1x_2}{V}, \qquad\qquad x_2(0) = x_{2,0}$$

where the controls u_1 and u_2 are the volumetric flow rate of the CSTR and the catalyst mass flow rate, respectively. The aim is to minimize the deviation of the state of the CSTR with minimum control action, i. e., to minimize

$$I = \int_0^{t_f} \sum_{i=1}^2 \left[(x_i - x_i^s)^2 + (u_i - u_i^s)^2 \right] dt$$

where the superscript s denotes the steady state value.

The above problem is equivalent to minimizing the augmented objective functional

$$J = \int_0^{t_f} \left\{ \underbrace{\sum_{i=1}^2 \left[(x_i - x_i^s)^2 + (u_i - u_i^s)^2 \right]}_{F} + \lambda_1 \left[\underbrace{-\dot{x}_1 + \frac{u_1}{V}(x_f - x_1) - kx_1x_2}_{g_1} \right] \right.$$

$$\left. + \lambda_2 \left[\underbrace{-\dot{x}_2 + \frac{u_2 - u_1x_2}{V}}_{g_2} \right] \right\} dt \tag{4.15}$$

where $x_1(0) = x_{1,0}$, $x_2 = x_{2,0}$, and λ_1 and λ_2 are time dependent Lagrange multipliers. The preconditions are as follows:

1. The partial derivatives of F, g_1, and g_2 are continuous with respect to x_1, x_2, u_1, and u_2 in the vicinity of the minimum.

2. With G_i defined as $(-\dot{x}_i + g_i)$ for $i = 1$ and 2, the following constraint qualification holds at each t in the interval $(0, t_f]$:

$$\eta_0 = \begin{vmatrix} \delta G_1(\hat{z}; \delta z_1) & \delta G_1(\hat{z}; \delta z_2) \\ \delta G_2(\hat{z}; \delta z_1) & \delta G_2(\hat{z}; \delta z_2) \end{vmatrix} \neq 0$$

where z is the coordinate set

$$(\dot{x}_1, \dot{x}_2, x_1, x_2, u_1, u_2)$$

$\hat{z}$ is the optimal z and δz_i the i-th variation set

$$(\delta \dot{x}_{1i}, \delta \dot{x}_{2i}, \delta x_1, \delta x_2, \delta u_{1i}, \delta u_{2i})$$

▯

4.3.3.3 Presence of Algebraic Constraints

Consider the "base" problem to optimize

$$I = \int_0^{t_f} F(\mathbf{y}, \mathbf{u}) \, dt$$

subject to

$$\dot{\mathbf{y}} = \mathbf{g}(\mathbf{y}, \mathbf{u}), \quad \mathbf{y}(0) = \mathbf{y}_0$$

To the base problem we add the following algebraic equality constraints:

$$h_i(\mathbf{y}, \mathbf{u}) = 0, \quad i = 1, 2, \ldots, l \quad \text{or} \quad \mathbf{h}(\mathbf{y}, \mathbf{u}) = \mathbf{0}; \quad 0 \leq t \leq t_\mathrm{f}$$

where l is less than m, the dimension of $\mathbf{u}$. From Section 4.3.3.2 (p. 103), the base problem is equivalent to the optimization of

$$J = \int_0^{t_\mathrm{f}} \left[F + \boldsymbol{\lambda}^\top (-\dot{\mathbf{y}} + \mathbf{g}) \right] \mathrm{d}t$$

subject to the usual preconditions of weak continuity and constraint qualification. Thus, the current problem is to find the optimum of J subject to $\mathbf{h} = \mathbf{0}$. Observe that the optimum of J already implies the satisfaction of state equations. Then, according to the Lagrange Multiplier Rule, the current problem is equivalent to finding the optimum of the further-augmented functional

$$M(\hat{\mathbf{y}}, \mathbf{u}) = J(\hat{\mathbf{y}}, \mathbf{u}) + \int_0^{t_\mathrm{f}} \sum_{i=1}^l \nu_i h_i(\hat{\mathbf{y}}, \mathbf{u}) \, \mathrm{d}t = J + \int_0^{t_\mathrm{f}} \boldsymbol{\nu}^\top \mathbf{h}(\hat{\mathbf{y}}, \mathbf{u}) \, \mathrm{d}t$$

where $\hat{\mathbf{y}}$ denotes the state vector that satisfies state equations for any admissible control vector $\mathbf{u}$ and $\boldsymbol{\nu}$ is the vector of time dependent Lagrange multipliers

$$\begin{bmatrix} \nu_1(t) & \nu_2(t) & \ldots & \nu_l(t) \end{bmatrix}^\top$$

The additional preconditions for the optimum are as follows:

1. The partial derivatives of $\mathbf{h}$ are continuous with respect to $\mathbf{y}$ and $\mathbf{u}$ in the vicinity of the optimum at $(\hat{\mathbf{y}}, \hat{\mathbf{u}})$.

2. The following constraint qualification is satisfied: there exists a set of m control variations $(\delta\mathbf{u}_1, \delta\mathbf{u}_2, \ldots, \delta\mathbf{u}_m)$ for which at each t in $(0, t_\mathrm{f}]$

$$\eta_0 = \begin{vmatrix} \delta h_1(\hat{\mathbf{z}}; \delta\mathbf{u}_1) & \delta h_1(\hat{\mathbf{z}}; \delta\mathbf{u}_2) & \ldots & \delta h_1(\hat{\mathbf{z}}; \delta\mathbf{u}_m) \\ \delta h_2(\hat{\mathbf{z}}; \delta\mathbf{u}_1) & \delta h_2(\hat{\mathbf{z}}; \delta\mathbf{u}_2) & \ldots & \delta h_2(\hat{\mathbf{z}}; \delta\mathbf{u}_m) \\ \vdots & \vdots & \vdots & \vdots \\ \delta h_m(\hat{\mathbf{z}}; \delta\mathbf{u}_1) & \delta h_m(\hat{\mathbf{z}}; \delta\mathbf{u}_2) & \ldots & \delta h_m(\hat{\mathbf{z}}; \delta\mathbf{u}_m) \end{vmatrix} \neq 0$$

where $\hat{\mathbf{z}}$ is the coordinate set $(\hat{\mathbf{y}}, \hat{\mathbf{u}})$ at the optimum.

Example 4.5

In the last example, let us say we want u_2 to be increasingly higher for a smaller amount of the catalyst-to-reactant ratio. For this purpose we enforce the following equality constraint at all times:

$$\frac{x_2}{x_1} = a\left(\frac{1}{b^{u_2-c}} - 1\right)$$

where a, b, and c are some suitable parameters. The objective is to minimize I subject to state equations and the above algebraic constraint. In turn, the equivalent objective is to minimize J given by Equation (4.15) on p. 105 subject to the algebraic constraint. Provided that the preconditions for the minimum of J are satisfied, the objective is to find the minimum of

$$M = J + \int_0^{t_f} \mu \underbrace{\left[\frac{x_2}{x_1} - a\left(\frac{1}{b^{u_2-c}} - 1\right)\right]}_{h} dt \qquad (4.16)$$

where μ is another time dependent Lagrange multiplier.

The additional preconditions are as follows:

1. The partial derivatives of h are continuous with respect to $\mathbf{x}$ and $\mathbf{u}$ in the vicinity of the minimum at $(\hat{\mathbf{x}}, \hat{\mathbf{u}})$.

2. The following constraint qualification is satisfied at each t in $(0, t_f]$. There exists a control variation $\delta\mathbf{u}$ for which $\delta h(\hat{\mathbf{x}}, \hat{\mathbf{u}}; \delta\mathbf{u}) \neq 0$.

□

4.4 Lagrange Multiplier and Objective Functional

In most problems, a Lagrange multiplier can be shown to be related to the rate of change of the optimal objective functional with respect to the constraint value. This is an important result, which will be utilized in developing the necessary conditions for optimal control problems having inequality constraints.

For simplicity, consider an objective functional J dependent on a control function $u(t)$ and subject to the constraint

$$K(u) = k_0$$

We assume that both J and K are Gâteaux differentiable. This is a modest assumption, which is valid in most optimal control problems we encounter. The reason for this assumption is the need for the linearity of the differentials in the following three-step derivation of the relation between a Lagrange multiplier μ and the objective functional J:

Step 1 Let J be optimal at $\hat{u}$, which depends on k_0, the value of the constraint. Then at any t in the t-interval, Taylor's first order expansion gives

$$\hat{u}(k_0 + \Delta k_0) = \hat{u}(k_0) + \hat{u}_{k_0}\Delta k_0 + \epsilon$$

where $\epsilon/\Delta k_0$ vanishes with Δk_0. We can rewrite the above equation as

$$\hat{u}(k_0 + \Delta k_0) = \hat{u}(k_0) + \Delta k_0 \underbrace{(\hat{u}_{k_0} + \epsilon/\Delta k_0)}_{\delta u}$$

Step 2 Using δu as indicated in the above equation,

$$J[\hat{u}(k_0 + \Delta k_0)] = J[\hat{u}(k_0) + \Delta k_0 \delta u]$$
$$= J[\hat{u}(k_0)] + \mathrm{d}J[\hat{u}(k_0); \Delta k_0 \delta u] + \epsilon_1(\Delta k_0 \delta u)$$

where $\mathrm{d}J$ is the Gâteaux differential of J [from Equation (2.8), p. 30] and $\epsilon_1/\Delta k_0$ vanishes with Δk_0. Since $\mathrm{d}J$ is linear with respect to the second argument, i. e., $\Delta k_0 \delta y$, we obtain, upon expanding δu,

$$J[\hat{u}(k_0 + \Delta k_0]) = J[\hat{u}(k_0)] + \mathrm{d}J[\hat{u}(k_0); \hat{u}_{k_0}\Delta k_0] + \mathrm{d}J[\hat{u}(k_0); \epsilon] + \epsilon_1(\Delta k_0 \delta u)$$
$$= J[\hat{u}(k_0)] + \Delta k_0 \mathrm{d}J[\hat{y}(k_0); \hat{u}_{k_0}] + \Delta k_0 \mathrm{d}J[\hat{u}(k_0); \epsilon/\Delta k_0] + \epsilon_1$$

The above equation can be rearranged as

$$\frac{J[\hat{u}(k_0 + \Delta k_0)] - J[\hat{u}(k_0)]}{\Delta k_0} = \mathrm{d}J[\hat{y}(k_0); \hat{u}_{k_0}] + \mathrm{d}J[\hat{u}(k_0); \epsilon/\Delta k_0] + \frac{\epsilon_1}{\Delta k_0} \quad (4.17)$$

In the limit, $\Delta k_0 \to 0$, we have

$$\frac{J[\hat{u}(k_0 + \Delta k_0)] - J[\hat{u}(k_0)]}{\Delta k_0} = J_{k_0}(\hat{u}) \quad \text{(i. e., the partial derivative)}$$
$$\mathrm{d}J[\hat{u}(k_0); \epsilon/\Delta k_0] = \mathrm{d}J[\hat{u}(k_0); 0] = 0$$
$$\epsilon_1/\Delta k_0 = 0$$

so that Equation (4.17) becomes

$$J_{k_0}(\hat{u}) = \mathrm{d}J[\hat{u}(k_0); \hat{u}_{k_0}]$$

We already know that if J is Gâteaux differentiable, then its variation δJ exists and is equal to $\mathrm{d}J$. Therefore,

$$\delta J[\hat{u}(k_0); \hat{u}_{k_0}] = J_{k_0}(\hat{u})$$

Step 3 The augmented objective functional is $M \equiv J + \mu K$ where μ is a Lagrange multiplier. From the Lagrange Multiplier Theorem, assuming that $\delta K[\hat{u}(k_0); \hat{u}_{k_0}] \neq 0$, we have

$$\delta M[\hat{u}(k_0); \hat{u}_{k_0}] = \delta J[\hat{u}(k_0); \hat{u}_{k_0}] + \mu \delta K[\hat{u}(k_0); \hat{u}_{k_0}] = 0$$

The last two equations yield

$$J_{k_0}(\hat{u}) = -\mu \delta K[\hat{u}(k_0); \hat{u}_{k_0}] \tag{4.18}$$

Repeating Step 2 for K in place of J, we obtain

$$\delta K[\hat{u}(k_0); \hat{u}_{k_0}] = K_{k_0}(\hat{u})$$

Expanding the right-hand side of the above equation,

$$\delta K[\hat{u}(k_0); \hat{u}_{k_0}] = \frac{(k_0 + \Delta k_0) - k_0}{\Delta k_0} = 1 \tag{4.19}$$

From Equations (4.18) and (4.19), we finally obtain

$$\boxed{\mu = -J_{k_0}(\hat{u}) = -\frac{\partial}{\partial k_0} J[\hat{u}(k_0)]} \tag{4.20}$$

4.4.1 General Relation

The above result can be readily generalized for the optimal control problem in which J is dependent on vectors $\mathbf{y}$ and $\mathbf{u}$ of state and control functions and is subject to m constraints, $K_i = k_i$, $i = 1, 2, \ldots, m$. In this case, the Lagrange multipliers are given by

$$\mu_i = -\frac{\partial}{\partial k_i} J[\hat{\mathbf{y}}(\mathbf{k}), \hat{\mathbf{u}}(\mathbf{k})]; \quad i = 1, 2, \ldots, m \tag{4.21}$$

where $\mathbf{k} = \begin{bmatrix} k_1 & k_2 & \cdots & k_m \end{bmatrix}^\top$.

4.5 John Multiplier Theorem for Inequality Constraints

In this section, we will derive the John Multiplier Theorem, which is a set of necessary conditions for the minimum of an objective functional constrained by inequalities.

We begin with an objective functional J dependent on a control function $u(t)$ and subject to the constraint $K(u) \leq k_0$. As before, we assume that both

J and K are Gâteaux differentiable since we will need the continuity of the differentials.

Let J be minimum at $\hat{u}$ among all u satisfying the inequality constraint. Then $K(\hat{u})$ could be either k_0 or less. We will consider these two cases as follows:

Case 1   Here $K(\hat{u}) = k_0$ and the inequality constraint is said to be **active**. The augmented objective functional is $M \equiv J + \mu K$ where μ is a Lagrange multiplier. The Lagrange Multiplier Theorem yields

$$\delta M = \delta J + \mu \delta K = 0$$

where, from Equation (4.20),

$$\mu = -\frac{\partial}{\partial k_0} J[\hat{u}(k_0)]$$

Now, any change in K from $K(\hat{u})$ has to be negative since $K(u) \leq k_0$. This change, which is Δk_0, cannot decrease the value of J because $J(\hat{u})$ is already the minimum subject to the constraint $K(u) \leq k_0$. Hence, the change ΔJ has to be positive or zero, corresponding to the negative Δk_0. In other words, the partial derivative in Equation (4.20) has to be either negative or zero. This finally means that

$$\mu \geq 0$$

Case 2   Here the strict inequality $K(\hat{u}) < k_0$ is in effect and the inequality constraint is said to be **inactive**. Let $K(\hat{u}) = k_c$ where k_c is some real number less than k_0. Then Equation (4.20) for $K(\hat{u}) = k_c$ is

$$\mu = -\frac{\partial}{\partial k_c} J[\hat{u}(k_c)]$$

The continuity of K implies that there is an interval around k_c in which $K < k_0$ for all u in a region around $\hat{u}$. In that interval, Δk_c can be positive or negative, but ΔJ has to be positive or zero since $J(\hat{u})$ is already the minimum. Thus, the partial derivative in the above equation can only be zero, thereby leading to

$$\mu = 0$$

In this case, the Lagrange Multiplier Theorem yields $\delta M = 0$, which is the necessary condition for the minimum.

The John Multiplier Theorem can now be expressed by combining the results for both the cases as follows:

The necessary condition for the minimum at $\hat{u}(t)$ is

$$\delta M(\hat{u}) = \delta J(\hat{u}) + \mu \delta K(\hat{u}) = 0$$
$$K(\hat{u}) \leq k_0$$
$$\mu \geq 0, \qquad \underbrace{\mu[K(\hat{u}) - k_0] = 0}$$
$$\text{complementary slackness condition}$$

The following preconditions must be satisfied:

1. The Gâteaux differentials of both J and K are weakly continuous near $\hat{u}(t)$.

2. The constraint qualification — There exists a δu for which $K(\hat{u}; \delta u) \neq 0$ whenever the constraint is active.

Observe that the complementary slackness condition, $\mu[K(\hat{u}) - k_0] = 0$, requires μ to be zero when the constraint is inactive. Otherwise, μ could be zero or greater.

Example 4.6

Let the problem in Example 4.2 (p. 102) be constrained by the inequality

$$u \leq cy, \quad \text{at} \quad t = t_f$$

where c is some constant. Thus, the minimum of the augmented functional J given by Equation (4.14) is now subject to the inequality constraint

$$\underbrace{u - cy}_{K} \leq \underbrace{0}_{k_0} \quad \text{at } t = t_f$$

We assume that the preconditions already hold for the minimum of J.

Now the minimum of J implies that the state equation is already satisfied for the given initial condition $y(0) = 0$. The augmented functional is given by

$$M(\hat{y}, u) = J(\hat{y}, u) + \mu K = J + \mu \Big[u - c\hat{y} \Big]_{t=t_f}$$

where $\hat{y}$ is the state satisfying the state equation for any admissible control u and μ is an undetermined multiplier. Hence, from the John Multiplier Theorem, the necessary conditions for the minimum of $M(\hat{y}, u)$ at $\hat{u}$ are

$$\delta M(\hat{y}, \hat{u}) = 0$$
$$\hat{u} - cy \leq 0 \quad \text{at} \quad t_f$$
$$\mu \geq 0, \qquad \underbrace{\mu(\hat{u} - c\hat{y})}_{} \quad \text{at} \quad t_f$$
$$\text{complementary slackness condition}$$

subject to the following additional preconditions:

1. The Gâteaux differential $dK(\hat{y}, u; \delta u)$ is weakly continuous near $\hat{u}$ for $t = t_f$. This means that the partial derivative K_u at $\hat{y}$ is continuous.

2. The constraint qualification — If the constraint is active, there exists a control variation δu for which $\delta K(\hat{y}, \hat{u}; \delta u) \neq 0$ for $t = t_f$. It means that $K_u \neq 0$ at $(\hat{y}, \hat{u})$ for $t = t_f$.

□

Example 4.7

Let us replace the point inequality constraint in the above example by

$$K \equiv u - cy \leq 0, \quad \text{for } all \text{ } t \text{ in } [0, t_f]$$

In this case, with the values of t defined as

$$t_0 \equiv 0, \quad t_1 \equiv t_0 + \Delta t, \quad t_2 \equiv t_1 + \Delta t, \quad \ldots, \quad t_{n-1} \equiv t_{n-2} + \Delta t, \quad t_n \equiv t_f$$

with Δt tending to zero, the inequality constraint can be rendered in terms of the following inequalities:

$$\underbrace{u(t_i) - cy(t_i)}_{K_i} \leq 0; \quad i = 0, 1, \ldots, (n \to \infty)$$

Hence from the serial application of the John Multiplier Theorem, [similar to that in Section 4.3.3.1 (p. 100)], the final augmented functional is given by

$$M(\hat{y}, u) = J(\hat{y}, u) + \sum_{i=0}^{n \to \infty} \mu_i K_i(\hat{y}_i, u_i) = J + \int_0^{t_f} \mu(u - c\hat{y}) \, dt$$

where the multiplier μ is a function of t and the necessary conditions for the minimum are

$$\delta M(\hat{y}, \hat{u}) = \delta J(\hat{y}, \hat{u}) + \int_0^{t_f} \mu \delta K(\hat{y}, \hat{u}) \, dt = 0$$

$$\left. \begin{array}{l} \hat{u} - c\hat{y} \leq 0 \\ \mu \geq 0, \qquad \underbrace{\mu(\hat{u} - c\hat{y}) = 0}_{\text{complementary slackness condition}} \end{array} \right\} \quad \text{for all } t \text{ in } [0, t_f]$$

The additional preconditions of Example 4.6 now apply to each t in the interval $[0, t_f]$. □

4.5.1 Generalized John Multiplier Theorem

The approach in the preceding section may be followed to arrive at the following the John Multiplier Theorem for several inequality constraints:

The necessary conditions for the minimum of $J(\mathbf{y}, \mathbf{u})$ subject to $\mathbf{K}(\mathbf{y}, \mathbf{u}) \leq \mathbf{k}$ at $\hat{\mathbf{u}}(t)$ are

$$\delta M(\hat{\mathbf{y}}, \hat{\mathbf{u}}) = \delta J(\hat{\mathbf{y}}, \hat{\mathbf{u}}) + \boldsymbol{\mu}^\top \delta \mathbf{K}(\hat{\mathbf{y}}, \hat{\mathbf{u}}) = 0$$
$$\mathbf{K}(\hat{\mathbf{y}}, \hat{\mathbf{u}}) \leq \mathbf{k}$$
$$\boldsymbol{\mu} \geq \mathbf{0}, \qquad \underbrace{\boldsymbol{\mu}^\top [\mathbf{K}(\hat{\mathbf{y}}, \hat{\mathbf{u}}) - \mathbf{k}]}_{\text{complementary slackness condition}} = 0$$

where $M \equiv J + \boldsymbol{\mu}^\top \mathbf{K}$. The the following preconditions must be satisfied at the minimum:

1. The Gâteaux differentials of J and $\mathbf{K}$ are weakly continuous near $\hat{\mathbf{u}}$.

2. The constraint qualification — Whenever l_a of the total l constraints are active, there exists a set of control variations $(\delta \mathbf{u}_1, \delta \mathbf{u}_2, \ldots, \delta \mathbf{u}_{l_a})$ for which

$$\eta_0 = \begin{vmatrix} \delta K_1(\hat{\mathbf{z}}; \delta \mathbf{u}_1) & \delta K_1(\hat{\mathbf{z}}; \delta \mathbf{u}_2) & \cdots & \delta K_1(\hat{\mathbf{z}}; \delta \mathbf{u}_{l_a}) \\ \delta K_2(\hat{\mathbf{z}}; \delta \mathbf{u}_1) & \delta K_2(\hat{\mathbf{z}}; \delta \mathbf{u}_2) & \cdots & \delta K_2(\hat{\mathbf{z}}; \delta \mathbf{u}_{l_a}) \\ \vdots & \vdots & \vdots & \vdots \\ \delta K_{l_a}(\hat{\mathbf{z}}; \delta \mathbf{u}_1) & \delta K_{l_a}(\hat{\mathbf{z}}; \delta \mathbf{u}_2) & \cdots & \delta K_{l_a}(\hat{\mathbf{z}}; \delta \mathbf{u}_{l_a}) \end{vmatrix} \neq 0$$

where $\hat{\mathbf{z}} = (\hat{\mathbf{y}}, \hat{\mathbf{u}})$ and each $\delta K_i(\hat{\mathbf{z}}; \delta \mathbf{u}_j)$ corresponds to an active constraint $K_i(\mathbf{y}, \mathbf{u}) = k_i$.

Example 4.8

Let us restrict the second control

$$u_{\min} \leq u_2 \leq u_{\max}$$

in the problem of Example 4.5 (p. 107). Thus, the problem is to minimize

$$I = \int_0^{t_f} \sum_{i=1}^2 \left[(x_i - x_i^s)^2 + (u_i - u_i^s)^2 \right] dt$$

subject to

$$\dot{x}_1 = \frac{u_1}{V}(x_f - x_1) - k x_1 x_2, \quad x_1(0) = x_{1,0}$$
$$\dot{x}_2 = \frac{u_2 - u_1 x_2}{V}, \qquad\qquad x_2(0) = x_{2,0}$$

and the following algebraic equality and inequality constraints:

$$\frac{x_2}{x_1} = a\left(\frac{1}{b^{u_2-c}} - 1\right)$$

$$-u_2 + u_{\min} \le 0$$

$$u_2 - u_{\max} \le 0$$

The equivalent problem is to minimize J given by Equation (4.15) on p. 105 subject to the initial conditions and the algebraic constraints. It is assumed that the preconditions for the minimum of J (see p. 105) are satisfied. The augmented functional for this modified problem is given by

$$M = J + \int_0^{t_f} \left\{ \mu_1 \underbrace{\left[\frac{x_2}{x_1} - a\left(\frac{1}{b^{u_2-c}} - 1\right)\right]}_{K_1} + \mu_2 \underbrace{(-u + u_{\min})}_{K_2} + \mu_3 \underbrace{(u - u_{\max})}_{K_3} \right\} dt$$

where μ_1, μ_2, and μ_3 are the time dependent Lagrange multipliers.

From the Generalized John Multiplier theorem, the necessary conditions for the minimum are $\delta M = 0$ and

$$-u_2 \le u_{\min} \qquad \mu_2 \ge 0 \qquad \mu_2(-u_2 + u_{\min}) = 0$$

$$u_2 \le u_{\max} \qquad \mu_3 \ge 0 \qquad \mu_3(u_2 - u_{\max}) = 0$$

throughout the time interval $[0, t_f]$ subject to the following additional preconditions:

1. The Gâteaux differentials of K_1, K_2, and K_3 are weakly continuous at $\hat{\mathbf{x}}$ near $\hat{\mathbf{u}}$. In other words, the partial derivatives $\partial K_i / \partial u_j$ $(i, j = 1, 2)$ at $\hat{\mathbf{x}}$ are continuous.

2. The constraint qualification —

    2.a The first constraint is always active, for which $K_1 = 0$. If only the first constraint is active, then there exists a control variation $\delta \mathbf{u}$ for which $\delta K_1(\hat{\mathbf{x}}, \hat{\mathbf{u}}; \delta \mathbf{u}) \ne 0$. It means that

    $$\hat{K}_{1\hat{u}} = \left[\frac{\partial K_1}{\partial u_1} \quad \frac{\partial K_1}{\partial u_2}\right]_{\hat{\mathbf{x}},\hat{\mathbf{u}}}^{\top} \ne \begin{bmatrix} 0 & 0 \end{bmatrix}^{\top}$$

    2.b At any time, if the first and any of the remaining two constraints are active, then the corresponding vectors of partial derivatives, $\hat{K}_{1\hat{u}}$ and $(\hat{K}_{2\hat{u}}$ or $\hat{K}_{3\hat{u}})$, are linearly independent.

    Note that the last two constraints can never be active together.

    ⬜

4.5.2 Remark on Numerical Solutions

When solving an inequality-constrained optimal control problem numerically, it is impossible to determine which constraints are active. The reason is one cannot obtain a μ exactly equal to zero. This difficulty is surmounted by considering a constraint to be active if the corresponding $\mu \leq \alpha$ where α is a small positive number such as 10^{-3} or less, depending on the problem. Slack variables may be used to convert inequalities into equalities and utilize the Lagrange Multiplier Rule.

Alternatively, increasing penalties may be applied on constraint violations during repeated applications of any computational algorithm used for unconstrained problems. We will use the latter approach in Chapter 7 to solve optimal control problems constrained by (in)equalities.

Note

When using Lagrange multipliers in the rest of the book, we will skip mentioning the preconditions assuming that they are satisfied.

4.A Inverse Function Theorem

Let $\mathbf{f}$ be a continuously differentiable function of $\mathbf{x}$, both having the same dimension n greater than zero. If the derivative $\mathbf{f}'(\mathbf{x})$ is non-zero at $\mathbf{x} = \mathbf{x}_0$ for which $\mathbf{y}_0 = \mathbf{f}(\mathbf{x}_0)$, then there exists a continuous inverse function $\mathbf{f}^{\text{inv}}(\mathbf{y})$, which maps an open set Y containing $\mathbf{y}_0$ to an open set X containing $\mathbf{x}_0$.

Note that an open set has each member completely surrounded by members of the same set (see Section 9.3, p. 268). Moreover, the inverse function is differentiable, the proof of which can be found in Rudin (1976).

Remark

Under the given conditions, the theorem assures that $\mathbf{y} = \mathbf{f}(\mathbf{x})$, which is a set of n equations

$$y_i = f_i(x_1, x_2, \ldots, x_n), \quad i = 1, 2, \ldots, n$$

can be uniquely solved as $\mathbf{x} = \mathbf{f}^{\text{inv}}(\mathbf{y})$, i. e.,

$$x_i = f_i^{\text{inv}}(y_1, y_2, \ldots, y_n), \quad i = 1, 2, \ldots, n$$

in sufficiently small neighborhoods of $\mathbf{x}_0$ and $\mathbf{y}_0$.

Outline of Proof

Based on the given function **f**, we will develop an auxiliary function **g**. We will show it to be a contraction, which is associated to a unique fixed point. This property will then lead to the existence of the inverse function $\mathbf{f}^{\mathrm{inv}}$. We start with the description of a contraction and its fixed point.

Contraction

A **contraction** is defined to be a function ϕ that maps a region X to itself such that for all $\mathbf{x}_i$ and $\mathbf{x}_j$ in X

$$\|\phi(\mathbf{x}_i) - \phi(\mathbf{x}_j)\| \le c\|\mathbf{x}_i - \mathbf{x}_j\|, \quad c < 1 \tag{4.22}$$

In the next three steps, we will show that there is a unique **fixed point x** in X such that $\mathbf{x} = \phi(\mathbf{x})$.

Step 1   Let us select an arbitrary $\mathbf{x}_0$ in X and obtain the series

$$\mathbf{x}_{i+1} = \phi(\mathbf{x}_i); \quad i = 0, 1, 2, \ldots \tag{4.23}$$

Then for $i > 0$

$$\|\mathbf{x}_{i+1} - \mathbf{x}_i\| \;=\; \|\phi_i - \phi_{i-1}\| \;\le\; c\|\mathbf{x}_i - \mathbf{x}_{i-1}\|$$

where $\phi_i \equiv \phi(\mathbf{x}_i)$ and the above inequality follows from the definition of a contraction, i. e., Inequality (4.22). Its recursive application yields

$$\begin{aligned}
\|\mathbf{x}_{i+1} - \mathbf{x}_i\| \;&\le\; c\|\mathbf{x}_i - \mathbf{x}_{i-1}\| \;\le\; c^2\|\mathbf{x}_{i-1} - \mathbf{x}_{i-2}\| \;\le\; \cdots \\
&\le\; c^i\|\mathbf{x}_1 - \mathbf{x}_0\|
\end{aligned} \tag{4.24}$$

Step 2   For $j > i$, we can write

$$\|\mathbf{x}_j - \mathbf{x}_i\| = \underbrace{\|(\mathbf{x}_{i+1} - \mathbf{x}_i) + (\mathbf{x}_{i+2} - \mathbf{x}_{i+1}) + \cdots + (\mathbf{x}_{j-1} - \mathbf{x}_{j-2}) + (\mathbf{x}_j - \mathbf{x}_{j-1})\|}_{\|r\|}$$

Applying the triangle inequality (Section 2.2.2, p. 26) on the right-hand side,

$$\|r\| \le \|\mathbf{x}_{i+1} - \mathbf{x}_i\| + \|\mathbf{x}_{i+2} - \mathbf{x}_{i+1}\| + \cdots + \|\mathbf{x}_{j-1} - \mathbf{x}_{j-2}\| + \|\mathbf{x}_j - \mathbf{x}_{j-1}\|$$

Combining the last two results,

$$\|\mathbf{x}_i - \mathbf{x}_j\| \le \|\mathbf{x}_{i+1} - \mathbf{x}_i\| + \|\mathbf{x}_{i+2} - \mathbf{x}_{i+1}\| + \cdots + \|\mathbf{x}_{j-1} - \mathbf{x}_{j-2}\| + \|\mathbf{x}_j - \mathbf{x}_{j-1}\|$$

Step 3 Using Inequality (4.24) for each right-hand side term above, we get

$$\|\mathbf{x}_i - \mathbf{x}_j\| \leq \underbrace{(c^i + c^{i+1} + \cdots + c^{j-2} + c^{j-1})}_{P}\|\mathbf{x}_1 - \mathbf{x}_0\|$$

Because c is a positive fraction, the coefficient on the right-hand side

$$P \leq P + c^j + c^{j+1} + \cdots = c^i(1 + c + c^2 + \ldots) = \frac{c^i}{1-c}$$

because of which

$$\|\mathbf{x}_i - \mathbf{x}_j\| \leq \frac{c^i}{1-c}\|\mathbf{x}_1 - \mathbf{x}_0\|$$

Fixed Point

In the above inequality, with $c < 1$ and i tending to infinity, $\|\mathbf{x}_i - \mathbf{x}_j\|$ tends to zero and so does $\|\phi_i - \phi_j\|$ due to Inequality (4.22). In other words, $\mathbf{x}_i$ and ϕ_i tend respectively to some $\mathbf{x}$ and $\phi[= \phi(\mathbf{x})]$ in the region X. Hence, Equation (4.23), i. e.,

$$\lim_{i \to \infty} \mathbf{x}_{i+1} = \lim_{i \to \infty} \phi_i$$

is equivalent to

$$\mathbf{x} = \phi(\mathbf{x})$$

where $\mathbf{x}$ is the fixed point.

Uniqueness of the Fixed Point

The fixed point is unique because otherwise if there is another fixed point, say, $\bar{\mathbf{x}} = \phi(\bar{\mathbf{x}})$, then $\|\phi(\mathbf{x}) - \phi(\bar{\mathbf{x}})\| = \|\mathbf{x} - \bar{\mathbf{x}}\|$, which contradicts Inequality (4.22).

Having shown that a contraction has a unique fixed point, we consider the givens of the Inverse Function Theorem. Based on $\mathbf{f}(\mathbf{x})$, we propose an auxiliary function and prove it to be a contraction.

Auxiliary Function

Consider the auxiliary function

$$\mathbf{g}(\mathbf{x}) \equiv \mathbf{x} - \frac{\mathbf{f}(\mathbf{x}) - \mathbf{y}}{\mathbf{D}} \tag{4.25}$$

where $\mathbf{y}$ is an n-dimensional vector, $\mathbf{D} \equiv \mathbf{f}'(\mathbf{x}_0)$ and $\mathbf{x}$ is in the vicinity of $\mathbf{x}_0$ such that $\|\mathbf{x} - \mathbf{x}_0\| < \delta$ for some $\delta > 0$. In the next four steps, we show that $\mathbf{g}(\mathbf{x})$ is a contraction.

Step 1   The continuity of $\mathbf{f}'$ at $\mathbf{x}_0$ implies that there exists an open set X in which

$$\|\mathbf{x} - \mathbf{x}_0\| < \delta, \quad \delta > 0 \tag{4.26}$$

such that

$$\|\mathbf{f}'(\mathbf{x}) - \mathbf{D}\| < \epsilon_1, \quad \epsilon_1 > 0 \tag{4.27}$$

Differentiating $\mathbf{g}(\mathbf{x})$ with respect to $\mathbf{x}$, we get

$$\mathbf{g}'(\mathbf{x}) = \mathbf{I} - \mathbf{D}^{-1}\mathbf{f}'(\mathbf{x}) = \mathbf{D}^{-1}[\mathbf{D} - \mathbf{f}'(\mathbf{x})]$$

where $\mathbf{I}$ is the identity matrix. Taking the norm on both sides of the above equation and applying Inequality (4.27),

$$\|\mathbf{g}'(\mathbf{x})\| = \|\mathbf{D}^{-1}\| \, \|\mathbf{D} - \mathbf{f}'(\mathbf{x})\|$$
$$< \|\mathbf{D}^{-1}\| \epsilon_1$$

With the choice $\epsilon_1 \equiv 0.5/\|\mathbf{D}^{-1}\|$, we have

$$\|\mathbf{g}'(\mathbf{x})\| < 0.5 \tag{4.28}$$

Step 2   Considering $\mathbf{x}$ as

$$\mathbf{x}(s) = (1-s)\mathbf{x}_0 + s\mathbf{x}_1, \quad 0 \le s \le 1 \tag{4.29}$$

we obtain $\mathbf{g}(s) = \mathbf{g}[\mathbf{x}(s)]$ so that

$$\frac{d\mathbf{g}}{ds} = \mathbf{g}'(\mathbf{x})\frac{d\mathbf{x}}{ds} = \mathbf{g}'(\mathbf{x})(\mathbf{x}_1 - \mathbf{x}_0)$$

Taking the norm on both sides of the above equation, we get

$$\left\|\frac{d\mathbf{g}}{ds}\right\| = \|\mathbf{g}'(\mathbf{x})(\mathbf{x}_1 - \mathbf{x}_0)\|$$

Applying the Operator Inequality (Section 9.24, p. 281) on the right-hand side of the above equation,

$$\|\mathbf{g}'(\mathbf{x})(\mathbf{x}_1 - \mathbf{x}_0)\| \le \|\mathbf{g}'(\mathbf{x})\| \|(\mathbf{x}_1 - \mathbf{x}_0)\|$$

Combining the last two results, we get

$$\left\|\frac{d\mathbf{g}}{ds}\right\| \le \|\mathbf{g}'(\mathbf{x})\| \|\mathbf{x}_1 - \mathbf{x}_0\|$$

Using Inequality (4.28) in the above result, we obtain

$$\left\|\frac{d\mathbf{g}}{ds}\right\| < 0.5\|\mathbf{x}_1 - \mathbf{x}_0\| \tag{4.30}$$

Step 3 Let $h(s) \equiv [\mathbf{g}(1) - \mathbf{g}(0)]^\top \mathbf{g}(s)$. Then for $0 \leq s \leq 1$, the Mean Value Theorem for derivatives (Section 9.14.1, p. 276) yields

$$h(1) - h(0) = \left[\frac{dh}{ds}\right]_s (1 - 0) = [\mathbf{g}(1) - \mathbf{g}(0)]^\top \left[\frac{d\mathbf{g}}{ds}\right]_s$$

Also, from the definition of $h(s)$,

$$h(1) - h(0) = [\mathbf{g}(1) - \mathbf{g}(0)]^\top [\mathbf{g}(1) - \mathbf{g}(0)] = \|\mathbf{g}(1) - \mathbf{g}(0)\|^2$$

From the last two equations,

$$\|\mathbf{g}(1) - \mathbf{g}(0)\|^2 = [\mathbf{g}(1) - \mathbf{g}(0)]^\top \left[\frac{d\mathbf{g}}{ds}\right]_s \tag{4.31}$$

Applying the Cauchy–Schwarz Inequality (Section 9.23, p. 281) to the right-hand side of the above equation, we get

$$[\mathbf{g}(1) - \mathbf{g}(0)]^\top \left[\frac{d\mathbf{g}}{ds}\right]_s \leq \|\mathbf{g}(1) - \mathbf{g}(0)\| \left\|\left[\frac{d\mathbf{g}}{ds}\right]_s\right\|$$

Using the above inequality in Equation (4.31), we get

$$\|\mathbf{g}(1) - \mathbf{g}(0)\|^2 \leq \|\mathbf{g}(1) - \mathbf{g}(0)\| \left\|\left[\frac{d\mathbf{g}}{ds}\right]_s\right\|$$

or

$$\|\mathbf{g}(1) - \mathbf{g}(0)\| \leq \left\|\left[\frac{d\mathbf{g}}{ds}\right]_s\right\|$$

Step 4 From Equation (4.29)

$$\mathbf{g}(1) = \mathbf{g}[\mathbf{x}(1)] = \mathbf{g}(\mathbf{x}_1) \quad \text{and} \quad \mathbf{g}(0) = \mathbf{g}[\mathbf{x}(0)] = \mathbf{g}(\mathbf{x}_0)$$

Therefore, the last inequality can be written as

$$\|\mathbf{g}(\mathbf{x}_1) - \mathbf{g}(\mathbf{x}_0)\| \leq \left\|\left[\frac{d\mathbf{g}}{ds}\right]_s\right\|$$

Applying Inequality (4.30) in the above result, we get

$$\|\mathbf{g}(\mathbf{x}_1) - \mathbf{g}(\mathbf{x}_0)\| < 0.5\|\mathbf{x}_1 - \mathbf{x}_0\|$$

Hence, $\mathbf{g}(\mathbf{x})$ is a contraction as defined by Inequality (4.22). Being a contraction, it has a unique fixed point.

Existence of the Inverse Function

We will now use $\mathbf{g}(\mathbf{x})$ to show that $\mathbf{f}(\mathbf{x})$ is one-to-one (injective) as well as onto (surjective) over regions X and Y, which are open sets. Note that this statement is equivalent to the fact that there exists a continuous inverse function $\mathbf{f}^{\text{inv}}(\mathbf{y})$ that maps Y to X.

Injection of f(x)

Consider $g(x)$ defined by Equation (4.25) in the open set described by Inequality (4.26). Since $g(x)$ is a contraction, it is associated with a unique fixed point given by $x = g(x)$ or

$$x = x - \frac{f(x) - y}{D}$$

Since the determinant of D is not zero, further simplification yields

$$f(x) = y$$

which holds for *exactly one* x in X. Hence, $f(x)$ is injective in X.

Surjection of f(x)

Let Y be the collection of points $y = f(x)$ for all points x in the open set X. Then obviously $f(x)$ is a surjection from X to Y.

That Y is an open set follows from the continuity of $f(x)$. It means that each point $y_i = f(x_i)$ in Y lies in an open set surrounded by neighboring points $y = f(x)$. In other words, the open set $\|y - y_i\| < \epsilon$ corresponds to an open set $\|x - x_i\| < \delta$ in X where both ϵ and δ are some positive real numbers. Thus, Y, being a collection of open sets, is an open set.

Bibliography

M.S. Bazaraa, H.D. Sherali, and C.M. Shetty. *Nonlinear Programming Theory and Algorithms*, Chapter 4. Wiley-Interscience, New Jersey, 3rd edition, 2006.

F.B. Hildebrand. *Methods of Applied Mathematics*, Chapter 2, pages 139–142. Dover Publications Inc., New York, 2nd edition, 1992.

F. John. Extremum problems with inequalities as subsidiary conditions. In J. Moser, editor, *Fritz John Collected Papers*, Volume 2, pages 543–560. Birkhäuser, Boston, 1985.

O.L. Mangasarian and S. Fromovitz. The Fritz John necessary optimality conditions in the presence of equality and inequality constraints. *J. Math. Anal. Appl.*, 17:37–47, 1967.

W. Rudin. *Principles of Mathematical Analysis*, Chapter 9, pages 221–223. McGraw-Hill, Inc., New York, 1976.

D.R. Smith. *Variational Methods in Optimization*, Chapter 3. Dover Publications Inc., New York, 1998.

A. Takayama. *Mathematical Economics*, Chapter 8, pages 646–667. Cambridge University Press, Cambridge, UK, 1985.

Exercises

4.1 Show that the costate variables in the optimal control problem

a. are continuous with respect to time, and

b. have piecewise continuous time derivatives

4.2 Show that the augmented functional formed by adjoining even the initial condition of the state equation leads to the same necessary conditions for the optimum.

4.3 Expand the determinant η_0 in Example 4.4 (p. 104).

4.4 Simplify the constraint qualifications in Section 4.3.3.3 (p. 105) and Section 4.5.1 (p. 113).

4.5 Derive the necessary conditions for the minimum of the batch distillation problem described in Section 1.3.1 (p. 5) without the purity specification. List all involved assumptions.

4.6 Repeat Problem 4.5 including the purity specification.

4.7 Find the necessary conditions for the maximum in Example 2.10 (p. 45) in presence of the following inequality constraints:

$$T \leq T_{\max}, \quad 0 \leq t \leq t_f$$
$$x(t_f) \leq x_f$$

State all assumptions involved.

Chapter 5

Pontryagin's Minimum Principle

One of the most profound results of applied mathematics, Pontryagin's minimum principle provides the necessary conditions for the minimum of an optimal control problem. The elegance of the principle lies in the simplicity of its application to a vast variety of optimal control problems. Boltyanskii et al. (1956) developed the principle originally as a maximum principle requiring the Hamiltonian to be maximized at the minimum.

In this chapter, we will present the proof of the minimum principle. The minimum principle uses a positive multiplier for the objective functional in the Hamiltonian formulation.* With this provision, the minimum principle concludes that the minimum of the problem requires minimization of the Hamiltonian in an optimal control problem whose minimum needs to be determined.

Some readers may first want to get the essence of the minimum principle and go cursorily over the derivation in Section 5.4. This section may be skipped during the initial reading.

5.1 Application

Before delving into the proof, let us take the simplest optimal control problem and examine the application of Pontryagin's minimum principle. We will realize that we already have been applying the minimum principle to our optimal control problems.

Consider the minimization of

$$I = \int_0^{t_f} F(y, u)\, \mathrm{d}t \tag{3.4}$$

$$\text{subject to} \quad G(y, \dot{y}, u) \equiv -\dot{y} + g(y, u) = 0 \tag{3.5}$$

---

* A negative multiplier, on the other hand, leads to the maximum principle, which stipulates that the Hamiltonian be *maximized* in a minimum problem.

with the initial condition

$$y(0) = y_0 \tag{3.6}$$

This problem is the same as the one in Section 3.2 (p. 58). The Hamiltonian for this problem is defined as

$$H(y, \lambda, u) = F(y, u) + \lambda g(y, u)$$

According to Pontryagin's minimum principle, if $\hat{u}$ is optimal, then the corresponding Hamiltonian $H(\hat{y}, \hat{\lambda}, \hat{u})$ at each time instant is minimum over all admissible choices for u. Thus, if $u = \hat{u} + \delta u$ is any admissible control, then

$$\boxed{H(\hat{y}, \hat{\lambda}, u) \geq H(\hat{y}, \hat{\lambda}, \hat{u}), \quad 0 \leq t \leq t_f} \tag{5.1}$$

Let us understand the import of the above result in light of the optimal control analysis we have done so far. The augmented functional for this problem is

$$J = \int_0^{t_f} \left[F(y, u) + \lambda G(y, \dot{y}, u) \right] dt \tag{3.7}$$

whose variation is given by (see Section 3.2.2, p. 59)

$$\delta J = \int_0^{t_f} (\delta F + \lambda \delta G) \, dt + \int_0^{t_f} G \delta \lambda \, dt$$

$$= \int_0^{t_f} (F_y + \lambda G_y + \dot{\lambda}) \delta y \, dt - \left[\lambda \delta y \right]_0^{t_f} + \int_0^{t_f} (F_u + \lambda G_u) \delta u \, dt + \int_0^{t_f} G \delta \lambda \, dt$$

$$= \int_0^{t_f} (H_y + \dot{\lambda}) \delta y \, dt - \left[\lambda \delta y \right]_0^{t_f} + \int_0^{t_f} H_u \delta u \, dt + \int_0^{t_f} (-\dot{y} + H_\lambda \delta \lambda) \, dt$$

If, for a given control $\hat{u}(t)$, we obtain
  1. the state $\hat{y}(t)$ that satisfies $\dot{y} = H_\lambda$ or $G = 0$ with $\hat{y}(0)$ fixed to y_0, and
  2. the corresponding costate $\hat{\lambda}(t)$ that satisfies $\dot{\lambda} = -H_y$ with $\hat{\lambda}(t_f)$ fixed to 0,

then we are left with

$$\delta J = \int_0^{t_f} H_u(\hat{y}, \hat{\lambda}, \hat{u}) \delta u \, dt$$

where, since $G = 0$ is satisfied,

$$\delta J = J(\hat{y}, \hat{\lambda}, \underbrace{\hat{u} + \delta u}_{u}) - J(\hat{y}, \hat{\lambda}, \hat{u}) = I(\hat{y}, u) - I(\hat{y}, \hat{u})$$

for sufficiently small variation δu in $\hat{u}$.

Now, for sufficiently δu at any time, we can express the Hamiltonian at $(\hat{y}, \hat{\lambda}, \hat{u} + \delta u)$ using the first order Taylor expansion as

$$H(\hat{y}, \hat{\lambda}, \underbrace{\hat{u} + \delta u}_{u}) = H(\hat{y}, \hat{\lambda}, \hat{u}) + H_u(\hat{y}, \hat{\lambda}, \hat{u})\delta u$$

where u is the perturbed control $(\hat{u} + \delta u)$. From the last two equations, we obtain

$$I(\hat{y}, u) - I(\hat{y}, \hat{u}) = \int_0^{t_f} H_u(\hat{y}, \hat{\lambda}, \hat{u})\delta u\,dt = \int_0^{t_f} [H(\hat{y}, \hat{\lambda}, u) - H(\hat{y}, \hat{\lambda}, \hat{u})]\,dt \quad (5.2)$$

Depending on the control u, the following cases arise for the above equation:

Case 1 There is no constraint on u. Then δu can be positive or negative at any time instant. Now a non-zero $H_u(\hat{y}, \hat{\lambda}, \hat{u})$ at any time allows the possibility of $I(\hat{y}, u) - I(\hat{y}, \hat{u}) < 0$. Therefore, $H_u(\hat{y}, \hat{\lambda}, \hat{u})$ must be zero to ensure the minimum of I. This necessary condition, as observed from Equation (5.2), is equivalent to

$$H(\hat{y}, \hat{\lambda}, u) = H(\hat{y}, \hat{\lambda}, \hat{u}), \quad 0 \le t \le t_f$$

which is included in the Pontryagin's minimum principle, i.e., Equation (5.1).

Case 2 The control u is in between (but not at) its specified upper and lower limits, u_{max} and u_{min}, respectively. Each time when that happens, the necessary condition for the minimum of I is

$$H(\hat{y}, \hat{\lambda}, u) = H(\hat{y}, \hat{\lambda}, \hat{u}),$$

for the same reason as in the previous case.

Case 3 The control is constrained by $u \le u_{max}$. If at any time $\hat{u} = u_{max}$, then δu can only be negative. A positive δu will make the control exceed u_{max} and therefore be inadmissible. Hence in this case, $H_u(\hat{y}, \hat{\lambda}, \hat{u})$ must be zero or negative to ensure the minimum of I, i.e., $I(\hat{y}, u) - I(\hat{y}, \hat{u}) \ge 0$. This necessary condition, as observed from Equation (5.2), is equivalent to

$$H(\hat{y}, \hat{\lambda}, u) \ge H(\hat{y}, \hat{\lambda}, \hat{u}),$$

as asserted by Pontryagin's minimum principle, i.e., Equation (5.1).

Case 4 The control is constrained by $u \ge u_{min}$. If at any time $\hat{u} = u_{min}$, then δu can only be positive. A negative δu will make the control less than u_{min} and therefore be inadmissible. Hence in this case, $H_u(\hat{y}, \hat{\lambda}, \hat{u})$ must be zero or positive to ensure the minimum of I, i.e., $I(\hat{y}, u) - I(\hat{y}, \hat{u}) \ge 0$. Thus, the necessary condition is the same as in the previous case.

Case 5 The integrand F in Equation (3.4) is a function of $|u|$. Now, when $u = 0$ the partial derivative of F_u does not exist, and we cannot apply the stationarity condition, $H_u = 0$. However, Pontryagin's minimum principle does not require the partial derivatives F_u and g_u to exist. According to the principle,

$$H(\hat{y}, \hat{\lambda}, u) \geq H(\hat{y}, \hat{\lambda}, \hat{u})$$

is the necessary condition for the minimum of I in this case as well.

Freedom from having the partial derivatives of F and g with respect to u means that the principle is also applicable to the case when only a finite number of controls are available.

The above cases show that Pontryagin's minimum principle provides an overarching necessary condition for the minimum. Appreciating this fact, we present a general optimal control problem involving a wide class of controls for which we will derive Pontryagin's minimum principle.

5.2 Problem Statement

It is desired to minimize the objective functional

$$I = \int_0^{t_f} F[y_1(t), y_2(t), \ldots, y_n(t), \, u_1(t), u_2(t), \ldots, u_m(t)] \, \mathrm{d}t$$

where t or time is the independent variable, $u_i(t)$s are the controls, and $y_i(t)$s are the state variables governed by the following ordinary differential equations:

$$\frac{\mathrm{d}y_i}{\mathrm{d}t} = g_i[y_1(t), y_2(t), \ldots, y_n(t), \, u_1(t), u_2(t), \ldots, u_m(t)]; \quad i = 1, 2, \ldots, n$$

called the state equations. These equations are autonomous because no g_i depends on the independent variable t. The objective functional I is also autonomous since F does not depend explicitly on t.

The initial conditions for the state equations are

$$y_i(0) = y_{i,0}; \quad i = 1, 2, \ldots, n$$

While the initial time is fixed at zero, the final time t_f is not specified or fixed.

5.2.1 Class of Controls

For the above problem, we consider a general class of piecewise constrained controls that are typically encountered in practice. The controls in this class have the following characteristics:

1. The controls u_is are *piecewise continuous* with respect to t. Figure 3.4 (p. 77) showed one such control, which is made of three continuous curves. When two such curves meet, there is a *jump discontinuity*, e. g., at time t_1, as shown in the figure. On either side of t_1, the control is provided by the curve on that side. At $t \geq t_1$ we take the control value from the right-hand side curve.

   Note that the number of jump continuities is finite.* In other words, each curve spans a non-zero time duration. Obviously, a single continuous curve for the control is also a member of this class of controls.

2. The controls are not discontinuous at the initial and final times.

   Thus, the control vector $\mathbf{u}(t)$ approaches $\mathbf{u}(0)$ as t tends to 0 from the right-hand side. Similarly, $\mathbf{u}(t)$ approaches $\mathbf{u}(t_f)$ as t tends to t_f from the left-hand side.

3. At all times, each control takes values from a bounded set of values. For example, a control u_i may be specified to take values from the constrained set $\{0 < u_i \leq 5\}$.

5.2.2 New State Variable

We also introduce $g_0 \equiv F$ and a new state variable y_0 defined by

$$\frac{dy_0}{dt} \equiv g_0[y_1(t), y_2(t), \ldots, y_n(t), u_1(t), u_2(t), \ldots, u_m(t)]; \quad y_0(0) = 0\theta$$

Thus, $y_0(t_f)$ is equal to I, as can be verified by integrating the above differential equation. Next, we define the Hamiltonian,

$$H \equiv \lambda_0(t)g_0 + \lambda_1(t)g_1 + \cdots + \lambda_n(t)g_n$$

where λ_is are the time dependent Lagrange multipliers or costate variables. This is the same definition as Equation (3.34) on p. 68 with $\lambda_0 = 1$.

5.2.3 Notation

For the sake of convenience, we adopt the following notation:

1. To reduce clutter, we omit (t) unless needed for clarity. The time-dependence of y_is, u_is, and λ_is is taken for granted.

2. The vectors $\mathbf{y}$, $\boldsymbol{\lambda}$, and $\mathbf{g}$ are, respectively,

   $$\begin{bmatrix} y_0 & y_1 & \cdots & y_n \end{bmatrix}^\top, \quad \begin{bmatrix} \lambda_0 & \lambda_1 & \cdots & \lambda_n \end{bmatrix}^\top, \quad \text{and}$$
   $$\begin{bmatrix} g_0 & g_1 & \cdots & g_n \end{bmatrix}^\top$$

---

* Otherwise, with infinite jump discontinuities, the control would never be continuous.

3. The i-th component of **g** is the function

$$g_i(\mathbf{y}, \mathbf{u}) \equiv g_i(y_1, y_2, \ldots, y_n, u_1, u_2, \ldots, u_m)$$

It is implicitly understood that any g_i does not depend on y_0.

4. The costate at the minimum is denoted by $\boldsymbol{\lambda}$ without the hat ^.

Using the above notation, the state equations can be written as

$$\dot{\mathbf{y}} \equiv \frac{d\mathbf{y}}{dt} = \mathbf{g}(\mathbf{y}, \mathbf{u}) \tag{5.3}$$

and the Hamiltonian can be expressed as

$$H(\mathbf{y}, \boldsymbol{\lambda}, \mathbf{u}) = \boldsymbol{\lambda}^\top \mathbf{g}(\mathbf{y}, \mathbf{u}) \tag{5.4}$$

From the last two equations, we get the identity

$$\dot{\mathbf{y}} = \frac{\partial H}{\partial \boldsymbol{\lambda}} \quad \text{or} \quad \dot{y}_i \equiv \frac{dy_i}{dt} = \frac{\partial H}{\partial \lambda_i}, \quad i = 0, 1, \ldots, n \tag{5.5}$$

5.3 Pontryagin's Minimum Principle

Pontryagin's minimum principle is a statement of necessary conditions for the control to be optimal. The main conclusion of this principle is that the optimal control minimizes the Hamiltonian at each point in the time interval whether or not the control there is continuous.

5.3.1 Assumptions

The principle is based on the following assumptions:

1. The controls belong to the class of piecewise continuous controls defined in Section 5.2.1.

2. The functions

$$g_i(\mathbf{y}, \mathbf{u}) \quad \text{and} \quad \frac{\partial g_i(\mathbf{y}, \mathbf{u})}{\partial y_j} \quad \text{for } i, j = 0, 1, \ldots, n$$

are continuous in the space of functions

$$(y_1, y_2, \ldots, y_n, \mathbf{u})$$

where **u** takes values from the m-dimensional constrained space including all boundary points.

For example, if $1 \leq u_1 < 2$, then the functions are required to be continuous at $u_1 = 2$ even though u_1 cannot be equal to 2.

5.3.2   Statement

If $\hat{\mathbf{u}}$ is the optimal control resulting in the optimal state $\hat{\mathbf{y}}$ over the time interval $[0, \hat{t}_f]$ in which I is minimum, then there exists a non-zero vector $\boldsymbol{\lambda}$ such that the following conclusions hold:

1. The corresponding Hamiltonian

$$\hat{H} \equiv H(\hat{\mathbf{y}}, \boldsymbol{\lambda}, \hat{\mathbf{u}})$$

   is minimum over the set of all admissible controls at each point in $[0, \hat{t}_f]$ whether or not the controls are continuous at the point. To be specific,

$$\hat{H} \leq H(\hat{\mathbf{y}}, \boldsymbol{\lambda}, \mathbf{u})$$

   at each t in $[0, \hat{t}_f]$.

2. Both $\hat{H}$ and λ_0 are time invariant, i.e., constant over $[0, \hat{t}_f]$.

3. If the final time is not fixed, then $\hat{H} = 0$ at the final time $\hat{t}_f$.

Remarks

1. The first conclusion means that throughout the time interval, the optimal control $\hat{\mathbf{u}}$ minimizes the Hamiltonian $H(\hat{\mathbf{y}}, \boldsymbol{\lambda}, \mathbf{u})$.

2. The last two conclusions together imply that if the final time is not fixed, then

$$\hat{H} = 0$$

   throughout the time interval.

5.4   Derivation of Pontryagin's Minimum Principle

Very versatile in applications, Pontryagin's minimum principle is among the most profound and difficult results to derive. Figure 5.1 presents the outline of the derivation of the principle.

Basically, we intend to perturb the optimal control along the control axis as well as change the final time to examine the effect on the Hamiltonian.

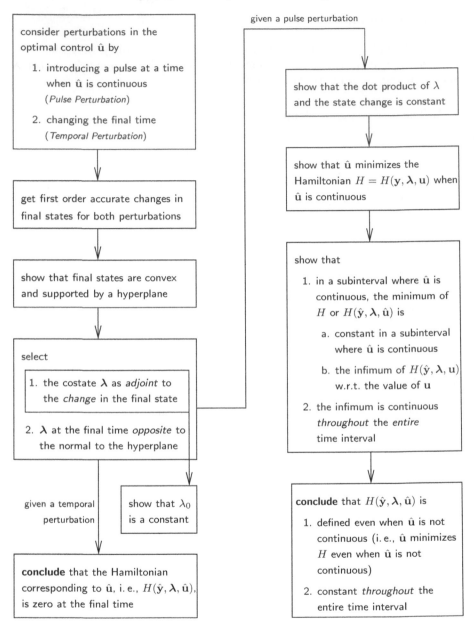

Figure 5.1 Outline of the proof of Pontryagin's minimum principle

Figure 5.2 shows these two types of perturbations. While the first type involves a **pulse perturbation** in a finite subinterval, the second one has

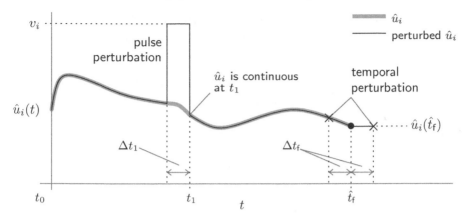

Figure 5.2 Two types of perturbations in $\hat{u}_i$, the i-th component of the optimal control $\hat{\mathbf{u}}$

temporal perturbation in the final time. These perturbations change the optimal state generated by the optimal control.

The time intervals of both perturbations are small enough to ensure the accuracy of the first-order approximation of the resulting new state. We will now examine these perturbations in the optimal control and determine the resulting states.

5.4.1 Pulse Perturbation of Optimal Control

Consider a subinterval

$$[t_1 - \Delta t_1, t_1), \quad \text{i.e.,} \quad t_1 - \Delta t_1 \leq t < t_1$$

where the optimal control $\hat{\mathbf{u}}(t)$ is continuous at t_1. In this subinterval, we consider a perturbed control $\mathbf{u}(t)$ to be constant, i.e., a vector $\mathbf{v}$ of some constants within the specified control limits. Elsewhere in the time interval $[0, \hat{t}_f]$, $\mathbf{u}$ is the same as the optimal control $\hat{\mathbf{u}}$. Figure 5.2 shows a pulse perturbation in the i-th component of $\hat{\mathbf{u}}$.

In the next three steps, we determine the final state $\mathbf{y}(\hat{t}_f)$ due to a pulse perturbation in $\hat{\mathbf{u}}$ at time t_1.

Step 1 Integrating the state equation over the subinterval $[t_1 - \Delta t_1, t_1)$ for the controls $\mathbf{u}$ and $\hat{\mathbf{u}}$, we get, respectively,

$$[\mathbf{y}]_{t_1 - \Delta t_1}^{t_1} = \int_{t_1 - \Delta t_1}^{t_1} \mathbf{g}(\mathbf{y}, \mathbf{u}) \, dt \quad \text{and} \quad [\hat{\mathbf{y}}]_{t_1 - \Delta t_1}^{t_1} = \int_{t_1 - \Delta t_1}^{t_1} \mathbf{g}(\hat{\mathbf{y}}, \hat{\mathbf{u}}) \, dt$$

Since there is no perturbation prior to $(t_1 - \Delta t_1)$, $\mathbf{y}$ at this time is the same as $\hat{\mathbf{y}}$. Hence, the difference between the above two equations gives

$$\mathbf{y}(t_1) - \hat{\mathbf{y}}(t_1) = \int_{t_1-\Delta t_1}^{t_1} [\mathbf{g}(\mathbf{y}, \mathbf{u}) - \mathbf{g}(\hat{\mathbf{y}}, \hat{\mathbf{u}})]\, dt \qquad (5.6)$$

Step 2 Being a continuous function of $(y_0, y_1, \ldots, y_n)$ as per Assumption 1 (Section 5.3.1, p. 128), $\mathbf{g}(\mathbf{y}, \mathbf{v})$ approaches $\mathbf{g}(\hat{\mathbf{y}}, \mathbf{v})$ as $\mathbf{y}$ tends to $\hat{\mathbf{y}}$ in the subinterval $[t_1 - \Delta t_1, t_1)$. Thus, we can write

$$\frac{|g_i(\mathbf{y}, \mathbf{v}) - g_i(\hat{\mathbf{y}}, \mathbf{v})|}{\|\mathbf{y} - \hat{\mathbf{y}}\|} < c_i; \quad i = 0, 1, \ldots, n$$

where c_is are some positive constants,* and $\mathbf{y}$ is sufficiently close to $\hat{\mathbf{y}}$. Similarly, since $\mathbf{y}$ approaches $\hat{\mathbf{y}}$ as the perturbation interval Δt_1 tends to zero,

$$\frac{\|\mathbf{y} - \hat{\mathbf{y}}\|}{\Delta t_1} < c_0$$

where c_0 is some positive constant, and Δt_1 is sufficiently small. Multiplying together the last two inequalities,

$$\lim_{\Delta t_1 \to 0} |g_i(\mathbf{y}, \mathbf{v}) - g_i(\hat{\mathbf{y}}, \mathbf{v})| < d_i \Delta t_1; \quad i = 0, 1, \ldots, n$$

where each $d_i \equiv c_0 c_i$ is a non-negative constant. The above inequality is symbolically expressed as

$$g_i(\mathbf{y}, \mathbf{v}) - g_i(\hat{\mathbf{y}}, \mathbf{v}) = O(\Delta t_1); \quad i = 0, 1, \ldots, n$$

using the big-O notation explained in Section 9.5.1 (p. 269). Here $O(\Delta t_1)$ denotes the maximum of the absolute error $|g_i(\mathbf{y}, \mathbf{v}) - g_i(\hat{\mathbf{y}}, \mathbf{v})|$. This maximum is equal to some positive constant times Δt_1 when Δt_1 is sufficiently small. In vector form,

$$\mathbf{g}(\mathbf{y}, \mathbf{v}) = \mathbf{g}(\hat{\mathbf{y}}, \mathbf{v}) + \mathbf{O}(\Delta t_1)$$

Using similar reasoning as above, we obtain

$$\mathbf{g}[\hat{\mathbf{y}}(t), \mathbf{v}] = \mathbf{g}[\hat{\mathbf{y}}(t_1), \mathbf{v}] + \mathbf{O}(\Delta t_1)$$
$$\mathbf{g}[\hat{\mathbf{y}}(t), \hat{\mathbf{u}}(t)] = \mathbf{g}[\hat{\mathbf{y}}(t_1), \hat{\mathbf{u}}(t_1)] + \mathbf{O}(\Delta t_1)$$

where t lies in the subinterval $[t_1 - \Delta t_1, t_1)$. The last three equations combined with Equation (5.6) provide the new state due to the pulse perturbation at time t_1 as follows:

---

* Since the left-hand side of the inequality cannot be negative, c_is cannot be zero or negative.

$$\mathbf{y}(t_1) = \hat{\mathbf{y}}(t_1) + \int_{t_1-\Delta t_1}^{t_1} \Big\{ \mathbf{g}[\hat{\mathbf{y}}(t_1), \mathbf{v}] - \mathbf{g}[\hat{\mathbf{y}}(t_1), \hat{\mathbf{u}}(t_1)] + \mathbf{O}(\Delta t_1) \Big\} \, dt$$

$$= \hat{\mathbf{y}}(t_1) + \underbrace{\Big\{ \mathbf{g}[\hat{\mathbf{y}}(t_1), \mathbf{v}] - \mathbf{g}[\hat{\mathbf{y}}(t_1), \hat{\mathbf{u}}(t_1)] \Big\} \Delta t_1}_{\text{first order term}} + \underbrace{\mathbf{O}(\Delta t_1^2)}_{\text{second order term}} \qquad (5.7)$$

We consider Δt_1 to be sufficiently small so that the second order term vanishes. With this provision, the change of state at t_1 is

$$\Delta \mathbf{y}(t_1) \equiv \mathbf{y}(t_1) - \hat{\mathbf{y}}(t_1) = \Big\{ \mathbf{g}[\hat{\mathbf{y}}(t_1), \mathbf{v}] - \mathbf{g}[\hat{\mathbf{y}}(t_1), \hat{\mathbf{u}}(t_1)] \Big\} \Delta t_1 \qquad (5.8)$$

Step 3 From time t_1 to the final time $\hat{t}_f$, the control is optimal, i. e., $\hat{\mathbf{u}}$, and the state equation is $\dot{\mathbf{y}} = \mathbf{g}(\mathbf{y}, \hat{\mathbf{u}})$. Replacing $\mathbf{y}$ by $(\hat{\mathbf{y}} + \Delta \mathbf{y})$ and using the first order Taylor expansion for sufficiently small $\Delta \mathbf{y}$, the state equation can be written as

$$\frac{d}{dt}(\hat{\mathbf{y}} + \Delta \mathbf{y}) = \mathbf{g}(\hat{\mathbf{y}} + \Delta \mathbf{y}, \hat{\mathbf{u}}) = \mathbf{g}(\hat{\mathbf{y}}, \hat{\mathbf{u}}) + \mathbf{g}_{\hat{\mathbf{y}}} \Delta \mathbf{y}, \qquad t_1 \le t \le \hat{t}_f \qquad (5.9)$$

where $\mathbf{g}_{\hat{\mathbf{y}}}$ is $\mathbf{g}_{\mathbf{y}}$ evaluated at $(\hat{\mathbf{y}}, \hat{\mathbf{u}})$. Since $d\hat{\mathbf{y}}/dt$ is $\mathbf{g}(\hat{\mathbf{y}}, \hat{\mathbf{u}})$, Equation (5.9) simplifies to

$$\frac{d}{dt} \Delta \mathbf{y} = \mathbf{g}_{\hat{\mathbf{y}}} \Delta \mathbf{y}, \qquad t_1 \le t \le \hat{t}_f \qquad (5.10)$$

which is a homogeneous linear differential equation with the initial condition given by Equation (5.8). Integrating the above equation, we get for $t > t_1$

$$\Delta \mathbf{y}(t) = \int^t \frac{d}{dt} \Delta \mathbf{y} \, dt = \boldsymbol{\Psi}(t) \boldsymbol{\Psi}^{-1}(t_1) \Delta \mathbf{y}(t_1) \qquad (5.11)$$

where $\boldsymbol{\Psi}(t)$ is an $(n+1) \times (n+1)$ fundamental matrix whose columns are the linearly independent solutions of the differential equation, Equation (5.10) (see Section 9.26, p. 283). Observe that the above equation along with Equation (5.8) reveals that $\Delta \mathbf{y}$ is directly proportional to Δt_1. We ensure the requisite size of $\Delta \mathbf{y}$ in Equation (5.9) by having Δt_1 suitably small.

Now the new state at time $t > t_1$, utilizing Equation (5.11), is given by

$$\mathbf{y}(t) = \int^t \frac{d\mathbf{y}}{dt} \, dt = \int^t \frac{d\hat{\mathbf{y}}}{dt} \, dt + \int^t \frac{d}{dt} \Delta \mathbf{y} \, dt = \hat{\mathbf{y}}(t) + \boldsymbol{\Psi}(t) \boldsymbol{\Psi}^{-1}(t_1) \Delta \mathbf{y}(t_1)$$

The above equation, upon substituting for $\Delta \mathbf{y}(t_1)$ given by Equation (5.8) and replacing t by $\hat{t}_f$, yields the final state in the presence of the pulse perturbation,

$$\mathbf{y}(\hat{t}_f) = \hat{\mathbf{y}}(\hat{t}_f) + \boldsymbol{\Psi}(\hat{t}_f) \boldsymbol{\Psi}^{-1}(t_1) \Big\{ \mathbf{g}[\hat{\mathbf{y}}(t_1), \mathbf{v}] - \mathbf{g}[\hat{\mathbf{y}}(t_1), \hat{\mathbf{u}}(t_1)] \Big\} \Delta t_1$$

which is first order accurate with respect to Δt_1, the interval of pulse perturbation. We take Δt_1 to be small enough for the above equation to be true.

5.4.2 Temporal Perturbation of Optimal Control

As shown in Figure 5.2, we now consider a truncation or extension of the optimal control $\hat{\mathbf{u}}$ at the final time $\hat{t}_f$. If Δt_f is negative, then the control remains $\hat{\mathbf{u}}$ in the truncated interval $[0, \hat{t}_f - |\Delta t_f|]$. But if Δt_f is positive, then the control remains $\hat{\mathbf{u}}$ in the original interval $[0, \hat{t}_f]$ and stays at the optimal value $\hat{\mathbf{u}}(\hat{t}_f)$ for $t > \hat{t}_f$ in the added subinterval $[\hat{t}_f, \hat{t}_f + \Delta t_f]$.

Integrating the state equations from $\hat{t}_f$ to $\hat{t}_f + \Delta t_f$, we get

$$\mathbf{y}(\hat{t}_f + \Delta t_f) = \mathbf{y}(\hat{t}_f) + \int_{\hat{t}_f}^{\hat{t}_f + \Delta t_f} \mathbf{g}[\mathbf{y}(t), \hat{\mathbf{u}}(\hat{t}_f)]\, dt$$

For sufficiently small Δt_f, the integral in the above equation is given by its first order Taylor expansion, $\mathbf{g}[\mathbf{y}(\hat{t}_f), \hat{\mathbf{u}}(\hat{t}_f)]\Delta t_f$. Thus,

$$\mathbf{y}(\hat{t}_f + \Delta t_f) = \mathbf{y}(\hat{t}_f) + \mathbf{g}[\mathbf{y}(\hat{t}_f), \hat{\mathbf{u}}(\hat{t}_f)]\Delta t_f \tag{5.12}$$

is the first order approximation of the final state due to the temporal perturbation. Similar to Δt_1 above, we take Δt_f to be small enough for the above equation to be true.

5.4.3 Effect on Final State

Consider a collection of final states generated by the optimal control altered by all possible combinations of pulse and temporal perturbations. This collection also includes the optimal final state generated by the optimal control without perturbations. Appendix 5.A (p. 145) shows that this collection is a *convex set*. As shown in Figure 5.3, a convex set contains all points that lie on a straight line joining *any* two points of the set.

In this set, let $\mathbf{y}_f$ be a point representing a final state, which is either

1. $\mathbf{y}(\hat{t}_f)$ due to pulse perturbation, or

2. $\mathbf{y}(\hat{t}_f + \Delta t_f)$ due to temporal perturbation.

Then, since the collection is convex, the final state given by

$$\mathbf{z}_f = \alpha \mathbf{y}_f + (1 - \alpha)\tilde{\mathbf{y}}_f \quad \text{for} \quad 0 \le \alpha \le 1$$

also belongs to the set where $\tilde{\mathbf{y}}_f$ is another member of the set, i.e., another final state.

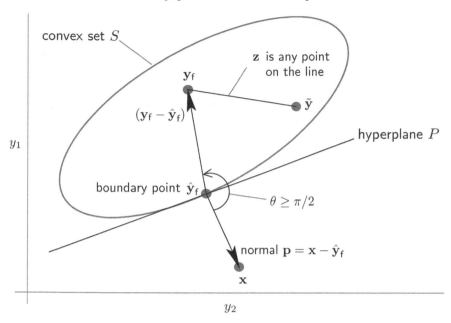

Figure 5.3 A convex set S of final states with a supporting hyperplane P. A hyperplane is a tangent line in two dimensions

Supporting Hyperplane

Now for a convex set, there exists a supporting hyperplane at any boundary point of the set (see Appendix 5.B, p. 149). We apply this result to our convex set of final states, in which the final optimal state

$$\hat{\mathbf{y}}_{\mathrm{f}} \equiv \begin{bmatrix} \hat{y}_0(t_{\mathrm{f}}) & \hat{y}_1(t_{\mathrm{f}}) & \cdots & \hat{y}_n(t_{\mathrm{f}}) \end{bmatrix}^{\top}$$

is a boundary point since it cannot be an interior point. Otherwise, $\hat{\mathbf{y}}_{\mathrm{f}}$ would have been surrounded in all directions by other points of the set, including those with $y_0(\hat{t}_{\mathrm{f}})$ components less than $\hat{y}_0(\hat{t}_{\mathrm{f}})$. This situation is contradictory since no point in the set can have the $y_0(\hat{t}_{\mathrm{f}})$ component — the objective function value I — less than $\hat{y}_0(\hat{t}_{\mathrm{f}})$, which is the minimum I.

In light of the above facts, there exists at $\hat{\mathbf{y}}_{\mathrm{f}}$ a hyperplane supporting the convex set of final states. Let

$$\mathbf{p} \equiv \begin{bmatrix} p_0 & p_1 & \cdots & p_n \end{bmatrix}^{\top}$$

be the normal to the hyperplane such that no point of $\mathbf{p}$ is in the set, as shown in Figure 5.3. Hence, for any point $\mathbf{y}_{\mathrm{f}}$ in the set, the angle between the vector $(\mathbf{y}_{\mathrm{f}} - \hat{\mathbf{y}}_{\mathrm{f}})$ and the normal $\mathbf{p}$ is greater than or equal to $\pi/2$. Derived in Appendix 5.B, this result is mathematically expressed by the inequality

$$\mathbf{p}^{\top}(\mathbf{y}_{\mathrm{f}} - \hat{\mathbf{y}}_{\mathrm{f}}) \leq 0$$

5.4.4   Choice of Final Costate

Let us choose $\boldsymbol{\lambda}(\hat{t}_f) = -\mathbf{p}$. Then the above inequality becomes

$$\boldsymbol{\lambda}(\hat{t}_f)^\top (\mathbf{y}_f - \hat{\mathbf{y}}_f) \geq 0 \tag{5.13}$$

Simply put, we choose the final costate to be a vector opposite to the normal $\mathbf{p}$ and pointing toward the convex set of final states (see Figure 5.3).* The first component of the costate, λ_0, is a constant, as we show next.

Constancy of λ_0

Consider the set of linear first order differential equations

$$\dot{\boldsymbol{\lambda}} \equiv \frac{d\boldsymbol{\lambda}}{dt} = -\mathbf{g}_{\hat{\mathbf{y}}}^\top \boldsymbol{\lambda}, \qquad t_1 \leq t \leq \hat{t}_f \tag{5.14}$$

which is called the adjoint of Equation (5.10). Note that $\mathbf{g}_{\hat{\mathbf{y}}}$ is $\mathbf{g}_{\mathbf{y}}$ evaluated at $(\hat{\mathbf{y}}, \hat{\mathbf{u}})$. Expanding this set of adjoint equations,

$$\begin{bmatrix} \dot{\lambda}_0 \\ \dot{\lambda}_1 \\ \vdots \\ \dot{\lambda}_n \end{bmatrix} = -\begin{bmatrix} \dfrac{\partial g_0}{\partial y_0} & \dfrac{\partial g_1}{\partial y_0} & \dfrac{\partial g_2}{\partial y_0} & \cdots & \dfrac{\partial g_n}{\partial y_0} \\[2mm] \dfrac{\partial g_0}{\partial y_1} & \dfrac{\partial g_1}{\partial y_1} & \dfrac{\partial g_2}{\partial y_1} & \cdots & \dfrac{\partial g_n}{\partial y_1} \\[2mm] \vdots & \vdots & \vdots & \vdots & \vdots \\[2mm] \dfrac{\partial g_0}{\partial y_n} & \dfrac{\partial g_1}{\partial y_n} & \dfrac{\partial g_2}{\partial y_n} & \cdots & \dfrac{\partial g_n}{\partial y_n} \end{bmatrix}_{(\hat{\mathbf{y}}, \hat{\mathbf{u}})} \begin{bmatrix} \lambda_0 \\ \lambda_1 \\ \vdots \\ \lambda_n \end{bmatrix} = -\begin{bmatrix} \dfrac{\partial H}{\partial y_0} \\[2mm] \dfrac{\partial H}{\partial y_1} \\[2mm] \vdots \\[2mm] \dfrac{\partial H}{\partial y_n} \end{bmatrix}_{(\hat{\mathbf{y}}, \hat{\mathbf{u}})}$$

where the last equality follows from the definition of the Hamiltonian. We rewrite this equality as

$$\dot{\boldsymbol{\lambda}} = -\frac{\partial H}{\partial \mathbf{y}}\bigg|_{(\hat{\mathbf{y}}, \hat{\mathbf{u}})} \quad \text{or} \quad \dot{\lambda}_i \equiv \frac{d\lambda_i}{dt} = -\frac{\partial H}{\partial y_i}\bigg|_{(\hat{\mathbf{y}}, \hat{\mathbf{u}})}, \quad i = 0, 1, \ldots, n \tag{5.15}$$

Since H does not depend explicitly on y_0, the first adjoint equation is $\dot{\lambda}_0 = 0$, thereby implying that λ_0 is a constant.

Specification of λ_0

If λ_0 is zero, then the Hamiltonian would be independent of the objective functional. Consequently, the optimal control problem would have nothing to do with the objective functional. Since an objective functional is essentially

---

* In the original derivation of the maximum principle, Pontryagin (1986) chose the final costate along the normal $\mathbf{p}$, i.e., away from the set of final states.

relevant to an engineering optimal control problem, we implicitly assume that λ_0 is not zero. It could be zero in a rare problem where the solution is independent of the objective functional, i.e., the latter is inconsequential. See Kamien and Schwartz (1991) for an example.

In practice, therefore, we expect to get solutions with any non-zero λ_0 and skip the exercise to prove that λ_0 is surely a non-zero constant. Nonetheless, it still needs to be specified. We make the simple choice, $\lambda_0 = 1$, and introduce it in Inequality (5.13) by dividing both sides by $-p_0$ to obtain

$$\underbrace{\begin{bmatrix} 1 & p_1/p_0 & \cdots & p_n/p_0 \end{bmatrix}}_{\boldsymbol{\lambda}^\top(\hat{t}_\mathrm{f})}(\mathbf{y}_\mathrm{f} - \hat{\mathbf{y}}_\mathrm{f}) \geq 0$$

Considering $\mathbf{y}_\mathrm{f} \equiv \mathbf{y}(\hat{t}_\mathrm{f})$, i.e., the final state due to a pulse perturbation in the optimal control, we can rewrite the above inequality as

$$\boldsymbol{\lambda}^\top(\mathbf{y} - \hat{\mathbf{y}}) \geq 0, \quad \text{at} \quad t = \hat{t}_\mathrm{f} \tag{5.16}$$

5.4.5 Minimality of the Hamiltonian

Next, we will show that the dot product $\boldsymbol{\lambda}^\top(\mathbf{y} - \hat{\mathbf{y}})$ in Inequality (5.16) is constant so that the inequality also holds at t_1, the time of pulse perturbation. This outcome will reveal that at any time when the optimal control is continuous, the corresponding Hamiltonian is minimum, and any control perturbation does not decrease the Hamiltonian further. Finally, we will extend this minimality to times when the optimal control is not continuous.

Constancy of the dot product $\boldsymbol{\lambda}^\top(\mathbf{y} - \hat{\mathbf{y}})$ in $[t_1, \hat{t}_\mathrm{f}]$

Let $\Delta\mathbf{y} \equiv (\mathbf{y} - \hat{\mathbf{y}})$. Then utilizing Equations (5.10) and (5.14), the time derivative of the dot product $\boldsymbol{\lambda}^\top(\mathbf{y} - \hat{\mathbf{y}})$ is

$$\frac{\mathrm{d}}{\mathrm{d}t}(\boldsymbol{\lambda}^\top\Delta\mathbf{y}) = \frac{\mathrm{d}\boldsymbol{\lambda}^\top}{\mathrm{d}t}\Delta\mathbf{y} + \boldsymbol{\lambda}^\top\frac{\mathrm{d}\Delta\mathbf{y}}{\mathrm{d}t} = (-\mathbf{g}_{\hat{\mathbf{y}}}^\top\boldsymbol{\lambda})^\top\Delta\mathbf{y} + \boldsymbol{\lambda}^\top(\mathbf{g}_{\hat{\mathbf{y}}}\Delta\mathbf{y})$$

$$= -\boldsymbol{\lambda}^\top\mathbf{g}_{\hat{\mathbf{y}}}\Delta\mathbf{y} + \boldsymbol{\lambda}^\top\mathbf{g}_{\hat{\mathbf{y}}}\Delta\mathbf{y} = 0$$

Hence, the dot product $\boldsymbol{\lambda}^\top(\mathbf{y} - \hat{\mathbf{y}})$ is constant so that Inequality (5.16) holds throughout the time interval $[t_1, \hat{t}_\mathrm{f}]$. In particular,

$$\boldsymbol{\lambda}^\top(t_1)[\mathbf{y}(t_1) - \hat{\mathbf{y}}(t_1)] \geq 0$$

which yields

$$\boldsymbol{\lambda}^\top(t_1)\Big\{\mathbf{g}[\hat{\mathbf{y}}(t_1), \mathbf{v}] - \mathbf{g}[\hat{\mathbf{y}}(t_1), \hat{\mathbf{u}}(t_1)]\Big\} \geq 0$$

after substituting for $[\mathbf{y}(t_1) - \hat{\mathbf{y}}(t_1)]$ from [Equation (5.8), p. 133] and canceling out the positive time duration Δt_1. In terms of the Hamiltonian as expressed by [Equation (5.4), p. 128], the above inequality becomes

$$H[\hat{\mathbf{y}}(t_1), \boldsymbol{\lambda}, \mathbf{v}] \geq H[\hat{\mathbf{y}}(t_1), \boldsymbol{\lambda}, \hat{\mathbf{u}}(t_1)]$$

where $\mathbf{v}$ is any admissible control value at t_1, including $\hat{\mathbf{u}}(t_1)$. Hence, the above inequality shows that at t_1, the Hamiltonian achieves the minimum with the optimal control function.

Note that t_1 is any arbitrary time instant when the optimal control is continuous. Hence the optimal control, whenever continuous, minimizes the Hamiltonian.

5.4.5.1 Minimality Even When $\hat{u}(t)$ Is Discontinuous

Belonging to the class defined in Section 5.2.1 (p. 126), the optimal control $\hat{\mathbf{u}}$ could be discontinuous in the interval $[0, t_f]$. Therefore, we need to show that even when $\hat{\mathbf{u}}$ is discontinuous, it minimizes the Hamiltonian. We will derive this result in the following three steps:

1. We will show that the optimal Hamiltonian $\hat{H}(t) \equiv \boldsymbol{\lambda}^{\mathsf{T}}(t)\mathbf{g}[\hat{\mathbf{y}}(t), \hat{\mathbf{u}}(t)]$ is

   a. constant in a subinterval in which the optimal control $\hat{\mathbf{u}}(t)$ is continuous, and

   b. equal to a certain function $h(t)$ — an infimum of H with respect to the value of control — in the subinterval.

2. We will show that the function $h(t)$ is continuous *throughout* the interval $[0, \hat{t}_f]$.

3. We will utilize the above steps to show that $\hat{H}(t)$ is

   a. defined, and the Hamiltonian $H(t)$ is minimized by $\hat{\mathbf{u}}$ even when $\hat{\mathbf{u}}$ is discontinuous, and

   b. the same constant throughout the time interval $[0, \hat{t}_f]$.

Step 1a   Given a closed subinterval $[t_i, t_j]$ where $\hat{\mathbf{u}}(t)$ is continuous, we wish to show that $\hat{H}(t)$ is constant.

Continuity of the Hamiltonian

Observe that the Hamiltonian defined as

$$H(t) \equiv \boldsymbol{\lambda}^{\mathsf{T}}(t)\mathbf{g}[\mathbf{y}(t), \mathbf{u}(t)]$$

is a continuous function of t in the subinterval $[t_i, t_j]$. The reasons are as follows:

1. Both $\mathbf{g}(\mathbf{y}, \mathbf{u})$ and $\mathbf{g}_\mathbf{y}(\mathbf{y}, \mathbf{u})$ are continuous in the space $(y_1, y_2, \ldots, y_n, \mathbf{u})$ as assumed in Section 5.3.1 (p. 128).

2. As specified, $\hat{\mathbf{u}}$ is a continuous function of t in the subinterval $[t_i, t_j]$.

3. Due to the above two reasons, $\mathbf{y}$ and $\boldsymbol{\lambda}$, being respective solutions of Equation (5.3) on p. 128 and Equation (5.14) on p. 136, are continuous functions of t.

As a consequence, H is a continuous function of t in the subinterval $[t_i, t_j]$. Next, we examine a function defined as

$$\hat{H}(t) \equiv H[\hat{\mathbf{y}}(t), \boldsymbol{\lambda}(t), \hat{\mathbf{u}}(t)] \quad \text{and} \quad \hat{H}(t, t_c) \equiv H[\hat{\mathbf{y}}(t), \boldsymbol{\lambda}(t), \hat{\mathbf{u}}(t_c)]$$

for some fixed arbitrary time t_c in the subinterval $[t_i, t_j]$.

Properties of $\hat{H}(t, t_c)$ and its Time Derivative

As a particular case of the continuous Hamiltonian, $\hat{H}(t, t_c)$ is a continuous function of t in the subinterval $[t_i, t_j]$. Moreover, observe that

$$\frac{d\hat{H}(t, t_c)}{dt} = \sum_{i=0}^{n} \left\{ \frac{\partial H[\hat{\mathbf{y}}(t), \boldsymbol{\lambda}(t), \hat{\mathbf{u}}(t_c)]}{\partial y_i} \frac{d\hat{y}_i}{dt} + \frac{\partial H[\hat{\mathbf{y}}(t), \boldsymbol{\lambda}(t), \hat{\mathbf{u}}(t_c)]}{\partial \lambda_i} \frac{d\lambda_i}{dt} \right\}$$

$$= \sum_{i=0}^{n} \sum_{j=0}^{n} \left(\lambda_j \frac{\partial g_j[\hat{\mathbf{y}}(t), \hat{\mathbf{u}}(t_c)]}{\partial y_i} \frac{d\hat{y}_i}{dt} + g_j[\hat{\mathbf{y}}(t), \hat{\mathbf{u}}(t_c)] \frac{d\lambda_i}{dt} \right) \quad (5.17)$$

is also a continuous function of t in the interval $[t_i, t_j]$ for the same reasons given previously for the continuity of the Hamiltonian.

At $t = t_c$, the time variable in $\hat{H}(t, t_c)$ is t_c throughout, and we can apply Equation (5.5) on p. 128 and Equation (5.15) on p. 136 with $\mathbf{y} = \hat{\mathbf{y}}$ and $\mathbf{u} = \hat{\mathbf{u}}$ to obtain

$$\left[\frac{\partial \hat{H}(t, t_c)}{\partial y_i} \right]_{t=t_c} = -\frac{d\lambda_i}{dt} \quad \text{and} \quad \left[\frac{\partial \hat{H}(t, t_c)}{\partial \lambda_i} \right]_{t=t_c} = \frac{d\hat{y}_i}{dt}, \quad i = 0, 1, \ldots, n$$

The above two equations, when substituted in Equation (5.17), yield

$$\left[\frac{dH(t, t_c)}{dt} \right]_{t=t_c} = 0$$

Since the derivative $d/dt[\hat{H}(t, t_c)]$ is continuous, for each $\epsilon > 0$ there exists a $\delta > 0$ such that $|t - t_c| < \delta$ implies

$$\left| \frac{d\hat{H}(t, t_c)}{dt} - \left[\frac{d\hat{H}(t, t_c)}{dt} \right]_{t=t_c} \right| < \epsilon$$

where the second term in the above inequality is zero from the last equation. Hence, the above inequality simplifies to

$$-\epsilon < \frac{d\hat{H}(t, t_c)}{dt} < \epsilon \quad (5.18)$$

Now we are ready to utilize the properties of $\hat{H}(t, t_c)$ and $d/dt[\hat{H}(t, t_c)]$ in conjunction with the minimality of the Hamiltonian.

Minimality of the Hamiltonian

Being continuous in the subinterval $[t_i, t_j]$, $\hat{\mathbf{u}}$ minimizes the Hamiltonian so that

$$\underbrace{H[\hat{\mathbf{y}}(t), \boldsymbol{\lambda}(t), \hat{\mathbf{u}}(t)]}_{\hat{H}(t)} \;\leq\; \underbrace{H[\hat{\mathbf{y}}(t), \boldsymbol{\lambda}(t), \hat{\mathbf{u}}(t_c)]}_{\hat{H}(t,t_c)}$$

Subtracting $\hat{H}(t_c)$ from both the sides of the above inequality,

$$\hat{H}(t) - \hat{H}(t_c) \;\leq\; \hat{H}(t, t_c) - \hat{H}(t_c) \tag{5.19}$$

Depending on t we have the following two cases:

1. $t > t_c$ for which we will prove $\hat{H}(t_j) \leq \hat{H}(t_i)$, and

2. $t < t_c$ for which we will prove $\hat{H}(t_j) \geq \hat{H}(t_i)$.

These cases, when proved, will together imply that $\hat{H}(t_j) = \hat{H}(t_i)$ because t could be on either side of t_c.

Case 1   Here $t > t_c$. Since $\hat{H}(t, t_c)$ is a continuous function of t, the Mean Value Theorem for derivatives (Section 9.14.1, p. 276) yields

$$\hat{H}(t, t_c) - \underbrace{\hat{H}(t_c, t_c)}_{\text{or } H(t_c)} = \left[\frac{\mathrm{d}\hat{H}(t, t_c)}{\mathrm{d}t} \right]_{t=t_d} (t - t_c) \tag{5.20}$$

for some t_d in between t and t_c. The above equation, when combined with the right-hand inequality of Inequality set (5.18), yields

$$\hat{H}(t, t_c) - \hat{H}(t_c) < \epsilon(t - t_c)$$

Comparing the above inequality with Inequality (5.19), we get

$$\hat{H}(t) - \hat{H}(t_c) < \epsilon(t - t_c) \tag{5.21}$$

where both t and t_c lie in the closed subinterval $[t_i, t_j]$.

Now consider the auxiliary function

$$H_a(t) \equiv \hat{H}(t) - \frac{\hat{H}(t_j) - \hat{H}(t_i)}{t_j - t_i}(t - t_i)$$

which is continuous since both terms on the right-hand side of the above equation are continuous functions of t in the subinterval. As a consequence, $H_a(t)$ must have a minimum at some time, since a continuous function in a closed interval achieves a minimum as well as a maximum according to the Weierstrass Theorem (Section 9.18, p. 278). Let the minimum of $H_a(t)$ be at $t_{\min}$ so that

$$H_a(t_{\min}) \leq H_a(t)$$

in the subinterval. Expanding the above inequality, we get

$$\frac{\hat{H}(t_j) - \hat{H}(t_i)}{t_j - t_i} \leq \frac{\hat{H}(t) - \hat{H}(t_{min})}{t - t_{min}} \tag{5.22}$$

Now Inequality (5.21), which is also satisfied for $t_c = t_{min}$, provides

$$\frac{\hat{H}(t) - \hat{H}(t_{min})}{t - t_{min}} < \epsilon \tag{5.23}$$

From Inequalities (5.22) and (5.23),

$$\frac{\hat{H}(t_j) - \hat{H}(t_i)}{t_j - t_i} < \epsilon \quad \text{or} \quad \hat{H}(t_j) - \hat{H}(t_i) < (t_j - t_i)\epsilon$$

Since both $(t_j - t_i)$ and ϵ are positive real numbers that can be made arbitrarily small,

$$\hat{H}(t_j) - \hat{H}(t_i) \leq 0 \quad \text{or} \quad \hat{H}(t_j) \leq \hat{H}(t_i) \tag{5.24}$$

Case 2 In this case $t < t_c$. The Mean Value theorem for derivatives provides [compare with Equation (5.20)]

$$\hat{H}(t_c) - \hat{H}(t, t_c) = \left[\frac{\mathrm{d}\hat{H}(t, t_c)}{\mathrm{d}t}\right]_{t=t_d} (t_c - t)$$

Combining the above equation with the left-hand inequality of Inequality set (5.18) and Inequality (5.19) yields

$$\hat{H}(t_c) - \hat{H}(t) > -\epsilon(t_c - t) \tag{5.25}$$

The auxiliary function $H_a(t)$ must also have a maximum at some time, say, t_{max}, so that

$$H_a(t_{max}) \geq H_a(t)$$

in the subinterval. Expanding this inequality, we get

$$\frac{\hat{H}(t_j) - \hat{H}(t_i)}{t_j - t_i} \geq \frac{\hat{H}(t) - \hat{H}(t_{max})}{t - t_{max}} \tag{5.26}$$

Now Inequality (5.25), which is also satisfied for $t_c = t_{max}$, provides

$$\frac{\hat{H}(t) - \hat{H}(t_{max})}{t - t_{max}} > -\epsilon \tag{5.27}$$

Altogether, Inequalities (5.26) and (5.27) yield

$$\frac{\hat{H}(t_j) - \hat{H}(t_i)}{t_j - t_i} > -\epsilon \quad \text{or} \quad \hat{H}(t_j) - \hat{H}(t_i) > -(t_j - t_i)\epsilon$$

Since both $(t_j - t_i)$ and ϵ are positive real numbers that can be made arbitrarily small,

$$\hat{H}(t_j) - \hat{H}(t_i) \geq 0 \quad \text{or} \quad \hat{H}(t_j) \geq \hat{H}(t_i) \tag{5.28}$$

Constancy of $\hat{H}$ in $[t_i, t_j]$

Since t is arbitrary within the closed subinterval $[t_i, t_j]$, the intersection of Inequalities (5.24) and (5.28) yields $\hat{H}(t_i) = \hat{H}(t_j)$. Hence if t_k is an arbitrary time in $[t_i, t_j]$, then

$$\hat{H}(t_i) = \hat{H}(t_k) = \hat{H}(t_j)$$

By necessity therefore, $\hat{H}$ is constant throughout the subinterval where the optimal control is continuous.

Step 1b   Consider the function $h(t)$, which is the infimum or the greatest lower bound (Section 9.4, p. 269) of the Hamiltonian with respect to the value of the control in the set U of admissible control values. Hence, $h(t)$ is the minimum value of the Hamiltonian at time t, if the minimum exists for some control in U. Otherwise, $h(t)$ is the greatest value of the Hamiltonian, which is *less* than all Hamiltonian values obtainable from U. In either case, we specify the infimum to occur for $\mathbf{u} = \mathbf{w}$ at any time t so that

$$h(t) \;=\; \inf_{\mathbf{u} \in U} H[\hat{\mathbf{y}}(t), \boldsymbol{\lambda}(t), \mathbf{u}(t)] \;=\; H[\hat{\mathbf{y}}(t), \boldsymbol{\lambda}(t), \mathbf{w}(t)] \tag{5.29}$$

Note that $\mathbf{w}(t)$ is simply an m-dimensional coordinate of control that provides the infimum of the Hamiltonian at time t. We now show that whenever $\hat{\mathbf{u}}$ is continuous,

$$h(t) = H[\hat{\mathbf{y}}(t), \boldsymbol{\lambda}(t), \hat{\mathbf{u}}(t)]$$

Let $\hat{\mathbf{u}}$ be continuous at time t_i. Then

$$H[\hat{\mathbf{y}}(t_i), \boldsymbol{\lambda}(t_i), \hat{\mathbf{u}}(t_i)] \geq h(t_i) \tag{5.30}$$

by virtue of Equation (5.29). Now consider an admissible control function $\mathbf{u}[(t, \mathbf{w}(t_i)]$, which is $\hat{\mathbf{u}}(t)$ having a pulse perturbation $\mathbf{w}$ at time t_i when $\hat{\mathbf{u}}$ is continuous. At such a time instant, we have already shown that the Hamiltonian is minimum when the control is optimal. Thus,

$$H[\hat{\mathbf{y}}(t_i), \boldsymbol{\lambda}(t_i), \hat{\mathbf{u}}(t_i)] \;\leq\; H\{[\hat{\mathbf{y}}(t_i), \boldsymbol{\lambda}(t_i), \mathbf{u}[(t_i, \mathbf{w}(t_i)]\}$$

Since $\mathbf{u}[(t_i, \mathbf{w}(t_i)] = \mathbf{w}(t_i)$, the right-hand side of the above inequality is $h(t_i)$ as per Equation (5.29). Therefore,

$$H[\hat{\mathbf{y}}(t_i), \boldsymbol{\lambda}(t_i), \hat{\mathbf{u}}(t_i)] \;\leq\; h(t_i) \tag{5.31}$$

The intersection of the above inequality with Inequality (5.30) yields

$$h(t) = H[\hat{\mathbf{y}}(t), \boldsymbol{\lambda}(t), \hat{\mathbf{u}}(t)] \quad \text{or} \quad \hat{H}(t)$$

at any time t when $\hat{\mathbf{u}}$ is continuous.

Step 2   We need to show that $h(t)$ is continuous in the time interval $[0, \hat{t}_f]$. Thus, for each $\epsilon > 0$ there exists a $\delta > 0$ such that $|s - t| < \delta$ implies

$$|h(s) - h(t)| < \epsilon$$

In terms of the function denoted by

$$\mathcal{H}[t, \mathbf{u}(t)] \equiv H[\hat{\mathbf{y}}(t), \boldsymbol{\lambda}(t), \mathbf{u}(t)] \tag{5.32}$$

we can write

$$h(t) = H[\hat{\mathbf{y}}(t), \boldsymbol{\lambda}(t), \mathbf{w}(t)] = \mathcal{H}[t, \mathbf{w}(t)] \tag{5.33}$$

Because of the continuity of H (see p. 138), $\mathcal{H}$ is continuous in the space of time and control. Given a time instant s, for each $\epsilon > 0$ there exists a $\delta_1 > 0$ such that the norm of the variable vector $[s - t, \; \mathbf{u}(s) - \mathbf{w}(t)]^\top$ less than δ_1 implies

$$\left| \mathcal{H}[s, \mathbf{u}(s)] - \mathcal{H}[t, \mathbf{w}(t)] \right| = \left| \mathcal{H}[s, \mathbf{u}(s)] - h(t) \right| < \epsilon$$

The above inequality for $\mathbf{u}(s) = \mathbf{w}(t)$ particularly yields

$$\mathcal{H}[s, \mathbf{w}(t)] < h(t) + \epsilon$$

From the definition

$$h(s) = \inf_{\mathbf{u} \in U} H[\hat{\mathbf{y}}(s), \boldsymbol{\lambda}(s), \mathbf{u}(s)] = \mathcal{H}[s, \mathbf{w}(s)] \leq \mathcal{H}[s, \mathbf{w}(t)]$$

Combining the last two inequalities, we obtain

$$h(s) < h(t) + \epsilon \tag{5.34}$$

which establishes the upper semi-continuity (see Section 9.2.1, p. 268) of h at t. To complete the proof of the continuity of $h(t)$, we need to prove the lower semi-continuity of h at t, i.e.,

$$h(s) > h(t) - \epsilon \quad \text{for} \quad |s - t| < \delta_2 > 0 \tag{5.35}$$

We prove it by contradiction. Suppose there is no $\delta_2 > 0$ for which the above set of inequalities is true. It means that in any time interval $|s - t| < \delta_2$, we have

$$h(s) = \mathcal{H}[s, \mathbf{w}(s)] \leq h(t) - \epsilon$$

The above inequality for an infinite sequence of progressively smaller intervals, $|s_n - t| < 1/n; \quad n = 1, 2, \ldots$, can be written as

$$h(s_n) = \mathcal{H}[s_n, \mathbf{w}_n] \leq h(t) - \epsilon$$

where s_n and $\mathbf{w}_n$ are the respective values of s and $\mathbf{w}$ in the n-th interval. Belonging to the set of admissible control values, the sequence of $\mathbf{w}_n$ is bounded.

The Bolzano–Weierstrass Theorem (Section 9.17, p. 277) assures that this sequence contains a subsequence that is convergent on some value, say, $\mathbf{w}_0$.* As $\mathbf{w}_n$ approaches $\mathbf{w}_0$, s_n approaches t, thereby implying

$$\mathcal{H}[t, \mathbf{w}_0] \leq h(t) - \epsilon$$

which contradicts $h(t)$ being the infimum. Hence, Inequality (5.35) is true. Combining it with Inequality (5.34) for δ as the lower of δ_1 and δ_2, we obtain

$$|h(s) - h(t)| < \epsilon \quad \text{for} \quad |s - t| < \delta$$

thereby proving the continuity of $h(t)$ in the time interval $[0, \hat{t}_f]$.

Finally, we prove that the minimality of $\hat{H}$ extends to even those times when the optimal control is discontinuous. Furthermore, $\hat{H}$ is constant throughout time interval $[0, \hat{t}_f]$.

Step 3a Let $\hat{\mathbf{u}}$ be discontinuous at a time t_i somewhere in the open time interval $(0, \hat{t}_f)$, as shown in Figure 5.4. Since $h(t)$ is continuous throughout

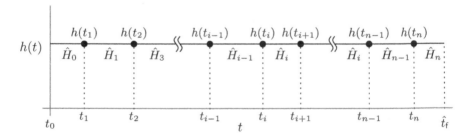

Figure 5.4 The function $h(t)$. Solid dots correspond to times when $\hat{\mathbf{u}}$ is not continuous

the interval, and equal to $\hat{H}(t)$ whenever $\hat{\mathbf{u}}(t)$ is continuous,

$$\lim_{t \to t_i-} h(t) = h(t_i) = \lim_{t \to t_i+} h(t)$$

Note that the number of t_is is finite, and $h(t)$ is equal to $\hat{H}(t)$ when $\hat{\mathbf{u}}(t)$ is continuous in a sufficiently small neighborhood of t_i, as shown earlier in Step 1b. Hence, the above set of equations becomes

$$\lim_{t \to t_i-} \hat{H}(t) = h(t_i) = \lim_{t \to t_i+} \hat{H}(t)$$

which shows that $\hat{H}(t)$ is defined at t_i to be equal to $h(t_i)$. Thus, $\hat{\mathbf{u}}$, even when discontinuous, minimizes the Hamiltonian.

---

* If $\mathbf{w}$ were a continuous function of t, then $\mathbf{w}_0 = \mathbf{w}(t)$. However, we cannot impose continuity on $\mathbf{w}$ with respect to t, hence the need for this theorem.

Step 3b Now $\hat{H}(t)$ is a constant, not necessarily the same, in the left as well as the right part of the neighborhood corresponding to $t < t_i$ and $t > t_i$. Let the respective constants be $\hat{H}_{i-1}$ and $\hat{H}_i$, as shown in Figure 5.4. Then the above set of equations becomes

$$\hat{H}_{i-1} = h(t_i) = \hat{H}_i$$

which shows that $\hat{H}(t)$ is constant over the entire neighborhood, including t_i. This reasoning, when applied to all such t_is $(i = 1, 2, \ldots, n)$ where

$$0 < t_1 < t_2 < \cdots < t_{n-1} < t_n < \hat{t}_f$$

yields

$$\hat{H}_0 = h(t_1) = \hat{H}_1 = h(t_2) = \hat{H}_3 = \cdots = \hat{H}_{n-1} = h(t_n) = \hat{H}_n$$

as shown in the figure. As a consequence, the optimal Hamiltonian is constant throughout the time interval.

5.4.6 Zero Hamiltonian at Free Final Time

If the final time is not fixed, then from Equation (5.12) on p. 134, the final state is

$$\mathbf{y}_f \equiv \mathbf{y}(\hat{t}_f + \Delta t_f) = \mathbf{y}(\hat{t}_f) + \underbrace{\mathbf{g}[\mathbf{y}(\hat{t}_f), \hat{\mathbf{u}}(\hat{t}_f)]}_{\hat{\mathbf{g}}}\Delta t_f$$

Hence the change in the final state from the optimal time $\hat{t}_f$ is

$$\mathbf{y}_f - \mathbf{y}(\hat{t}_f) = \hat{\mathbf{g}}\Delta t_f$$

which, when substituted into Inequality (5.13), yields

$$(\boldsymbol{\lambda}^\top \hat{\mathbf{g}})\Delta t_f \geq 0$$

Since Δt_f is arbitrary, it is necessary that

$$\boldsymbol{\lambda}^\top \hat{\mathbf{g}} = \hat{H}(\hat{t}_f) = 0$$

5.A Convexity of Final States

Consider the following control function

$$\mathbf{u}(t) = \begin{cases} \hat{\mathbf{u}}(t), & 0 \leq t < t_1 - \Delta t_1 \\ v_1, & t_1 - \Delta t_1 \leq t < t_1 \quad (\text{subinterval } I_1) \\ \hat{\mathbf{u}}(t), & t_1 \leq t < t_2 - \Delta t_2 \\ v_2, & t_2 - \Delta t_2 \leq t < t_2 \quad (\text{subinterval } I_2) \\ \hat{\mathbf{u}}(t), & t \geq t_2 \end{cases}$$

This function has pulse perturbations $\mathbf{v}_1$ and $\mathbf{v}_2$ in the two disjunct subintervals I_1 and I_2, respectively. Elsewhere in $[0, \hat{t}_f]$, the function is the same as the optimal control. Derived earlier in Section 5.4.1 (p. 131), the state change at time t_1 is

$$\Delta \mathbf{y}(t_1) = \Big\{ \mathbf{g}[\hat{\mathbf{y}}(t_1), \mathbf{v}] - \mathbf{g}[\hat{\mathbf{y}}(t_1), \hat{\mathbf{u}}(t_1)] \Big\} \Delta t_1 \qquad (5.8)$$

This state change, as per Equation (5.11) on p. 133, evolves to

$$\Delta \mathbf{y}(t_2 - \Delta t_2) = \mathbf{\Psi}(t_2 - \Delta t_2) \mathbf{\Psi}^{-1}(t_1) \Delta \mathbf{y}(t_1) \qquad (5.36)$$

at time $(t_2 - \Delta t_2)$, the onset of the second subinterval I_2. Let us simplify the above equation. From Equation (5.11), $\mathbf{\Psi}(t)$ depends on $\Delta \mathbf{y}(t)$ — a continuous function of t whose derivative exists as defined by Equation (5.10). Thus, $\mathbf{\Psi}(t)$ is a continuous function of t in the subinterval $[t_1, t_2 - \Delta t_2]$. Hence, $\mathbf{\Psi}(t_2 - \Delta t_2)$ approaches $\mathbf{\Psi}(t_2)$ as Δt_2 tends to zero. Using the big-O notation (see p. 269), we have the matrix equation

$$\mathbf{\Psi}(t_2 - \Delta t_2) = \mathbf{\Psi}(t_2) + \mathbf{O}(\Delta t_2)$$

which, when used in Equation (5.36), yields

$$\begin{aligned}
\Delta \mathbf{y}(t_2 - \Delta t_2) &= \mathbf{\Psi}(t_2) \mathbf{\Psi}^{-1}(t_1) \Delta \mathbf{y}(t_1) + \mathbf{O}(\Delta t_2) \mathbf{\Psi}^{-1}(t_1) \Delta \mathbf{y}(t_1) \\
&= \mathbf{\Psi}(t_2) \mathbf{\Psi}^{-1}(t_1) \Delta \mathbf{y}(t_1) + \underbrace{\mathbf{O}(\Delta t_1 \Delta t_2)}_{\text{second order term}}
\end{aligned}$$

For sufficiently small Δt_1 and Δt_2, the second order term vanishes, and the state change at $(t_2 - \Delta t_2)$ — the onset of I_2 — simplifies to

$$\Delta \mathbf{y}(t_2 - \Delta t_2) = \mathbf{\Psi}(t_2) \mathbf{\Psi}^{-1}(t_1) \Delta \mathbf{y}(t_1) \qquad (5.37)$$

To determine the state change at the end of I_2, we repeat the steps of Section 5.4.1, which is concerned with a single pulse perturbation corresponding to the first subinterval I_1.

Step 1 Thus, in place of Equation (5.6) on p. 132, we get for the second subinterval I_2

$$\mathbf{y}(t_2) - \hat{\mathbf{y}}(t_2) - \underbrace{[\mathbf{y}(t_2 - \Delta t_2) - \hat{\mathbf{y}}(t_2 - \Delta t_2)]}_{\Delta \mathbf{y}(t_2 - \Delta t_2)} = \int_{t_2 - \Delta t_2}^{t_2} [\mathbf{g}(\mathbf{y}, \mathbf{u}) - \mathbf{g}(\hat{\mathbf{y}}, \hat{\mathbf{u}})] \, dt$$

Step 2 Doing the analysis that led to Equation (5.7) on p. 133, we get

$$\int_{t_2 - \Delta t_2}^{t_2} [\mathbf{g}(\mathbf{y}, \mathbf{u}) - \mathbf{g}(\hat{\mathbf{y}}, \hat{\mathbf{u}})] \, dt = \Big\{ \mathbf{g}[\hat{\mathbf{y}}(t_2), \mathbf{v}_2] - \mathbf{g}[\hat{\mathbf{y}}(t_2), \hat{\mathbf{u}}(t_2)] \Big\} \Delta t_2 + \mathbf{O}(\Delta t_2^2)$$

Hence, for sufficiently small Δt_2, the above equation when substituted in the result of Step 1, yields [compare with Equation (5.8), p. 133]

$$\Delta \mathbf{y}(t_2) \equiv \mathbf{y}(t_2) - \hat{\mathbf{y}}(t_2) = \underbrace{\left\{ \mathbf{g}[\hat{\mathbf{y}}(t_2), \mathbf{v}_2] - \mathbf{g}[\hat{\mathbf{y}}(t_2), \hat{\mathbf{u}}(t_2)] \right\} \Delta t_2}_{\delta \mathbf{y}(t_2)} + \Delta \mathbf{y}(t_2 - \Delta t_2)$$

In the above equation, $\delta \mathbf{y}(t_2)$ indicates the state change at t_2 if there were no previous perturbation. Thus, $\Delta \mathbf{y}(t_2)$ is $\delta \mathbf{y}(t_2)$ if $\Delta \mathbf{y}(t_2 - \Delta t_2)$ is zero. Obviously, for the first perturbation, $\Delta \mathbf{y}(t_1)$ is $\delta \mathbf{y}(t_1)$ so that Equation (5.37) can be written as

$$\Delta \mathbf{y}(t_2 - \Delta t_2) = \mathbf{\Psi}(t_2) \mathbf{\Psi}^{-1}(t_1)\, \delta \mathbf{y}(t_1)$$

Combining the last two equations, we obtain

$$\Delta \mathbf{y}(t_2) = \delta \mathbf{y}(t_2) + \mathbf{\Psi}(t_2) \mathbf{\Psi}^{-1}(t_1)\, \delta \mathbf{y}(t_1)$$

Final State

Now from t_2 to $\hat{t}_\mathrm{f}$ during which the control is optimal, the evolution of $\Delta \mathbf{y}$ is governed by Equation (5.11) on p. 133 so that the state change at the final time is

$$\Delta \mathbf{y}(\hat{t}_\mathrm{f}) \equiv \mathbf{y}(\hat{t}_\mathrm{f}) - \hat{\mathbf{y}}(\hat{t}_\mathrm{f}) = \mathbf{\Psi}(\hat{t}_\mathrm{f}) \mathbf{\Psi}^{-1}(t_2) \Delta \mathbf{y}(t_2)$$

$$= \mathbf{\Psi}(\hat{t}_\mathrm{f}) \mathbf{\Psi}^{-1}(t_2) \delta \mathbf{y}(t_2) + \mathbf{\Psi}(\hat{t}_\mathrm{f}) \mathbf{\Psi}^{-1}(t_1) \delta \mathbf{y}(t_1)$$

In general, if the optimal control has pulse perturbations in n disjunct intervals I_i $(i = 1, 2, \ldots, n)$, then the change in final state is

$$\Delta \mathbf{y}(\hat{t}_\mathrm{f}) = \sum_{i=1}^{n} \mathbf{\Phi}(t_i) \delta \mathbf{y}(t_i)$$

where $\mathbf{\Phi}(t_i) \equiv \mathbf{\Psi}(\hat{t}_\mathrm{f}) \mathbf{\Psi}^{-1}(t_i)$. Now if the final time changes by Δt_f, it would bring about an additional final state change

$$\mathbf{y}(\hat{t}_\mathrm{f} + \Delta t_\mathrm{f}) - \mathbf{y}(\hat{t}_\mathrm{f}) = \mathbf{g}[\mathbf{y}(\hat{t}_\mathrm{f}), \hat{\mathbf{u}}(\hat{t}_\mathrm{f})] \Delta t_\mathrm{f}$$

given by Equation (5.12). Thus, the final state resulting from pulse as well as temporal perturbations to the optimal control is given by

$$\mathbf{y}(\hat{t}_\mathrm{f}) = \hat{\mathbf{y}}(\hat{t}_\mathrm{f}) + \sum_{i=1}^{n} \mathbf{\Phi}(t_i) \delta \mathbf{y}(t_i) + \mathbf{g}[\hat{\mathbf{y}}(\hat{t}_\mathrm{f}), \hat{\mathbf{u}}(\hat{t}_\mathrm{f})]\, \Delta t_\mathrm{f}$$

$$= \hat{\mathbf{y}}(\hat{t}_\mathrm{f}) + \sum_{i=1}^{n} \mathbf{\Phi}(t_i) \underbrace{\left\{ \mathbf{g}[\hat{\mathbf{y}}(t_i), \mathbf{v}_i] - \mathbf{g}[\hat{\mathbf{y}}(t_i), \hat{\mathbf{u}}(t_i)] \right\} \Delta t_i}_{\mathbf{a}_i} + \mathbf{g}[\hat{\mathbf{y}}(\hat{t}_\mathrm{f}), \hat{\mathbf{u}}(\hat{t}_\mathrm{f})]\, \Delta t_\mathrm{f}$$

The optimal control thus perturbed is an admissible control. Let us consider a collection of all such controls with

1. arbitrarily fixed but admissible pulse perturbations $\mathbf{v}_1, \mathbf{v}_2, \ldots, \mathbf{v}_n$ at a finite number of time instants, $t_1, t_2, \ldots, t_n$;[*]

2. corresponding intervals $\Delta t_1, \Delta t_2, \ldots \Delta t_n$ — each of variable size, including zero; and

3. intervals of temporal perturbations at $\hat{t}_f$, which also have variable sizes including zero.

In terms of $\mathbf{a}_i$ introduced in the last equation, the final state is then

$$\mathbf{y}(\hat{t}_f) = \hat{\mathbf{y}}(\hat{t}_f) + \sum_{i=1}^{n} \mathbf{a}_i \Delta t_i + \mathbf{g}[\hat{\mathbf{y}}(\hat{t}_f), \hat{\mathbf{u}}(\hat{t}_f)] \, \Delta t_f \qquad (5.38)$$

Containing all combinations of the pulse and temporal perturbations, the collection of admissible controls includes the optimal control, which has all Δt_is and Δt_f of zero size. Each member of this collection results in a final state. The final state is optimal when the control is the optimal control.

Convexity

We now show that the set of all final states given by Equation (5.38) is convex. Thus, given any two final states $\mathbf{y}(\hat{t}_f)$ and $\tilde{\mathbf{y}}(\hat{t}_f)$ of this set, their combination

$$\mathbf{z}(\hat{t}_f) = \alpha \mathbf{y}(\hat{t}_f) + (1 - \alpha)\tilde{\mathbf{y}}(\hat{t}_f), \quad 0 \le \alpha \le 1$$

also belongs to this set. It means that we need to show that $\mathbf{z}(\hat{t}_f)$ is generated by a member of the collection of admissible controls.

Let $\mathbf{y}(\hat{t}_f)$ and $\tilde{\mathbf{y}}(\hat{t}_f)$ be generated by respective admissible controls $\mathbf{u}(t)$ and $\tilde{\mathbf{u}}(t)$. The control $\mathbf{u}(t)$ has perturbation intervals

$$\Delta t_1, \ \Delta t_2, \ \ldots, \ \Delta t_n$$

and a temporal perturbation Δt_f. On the other hand, the control $\tilde{\mathbf{u}}(t)$ has perturbation intervals

$$\Delta \tilde{t}_1, \ \Delta \tilde{t}_2, \ \ldots, \ \Delta \tilde{t}_n$$

and a temporal perturbation $\Delta \tilde{t}_f$. Then

$$\mathbf{y}(\hat{t}_f) = \hat{\mathbf{y}}(\hat{t}_f) + \sum_{i=1}^{n} \mathbf{a}_i \Delta t_i \ + \ \mathbf{g}[\hat{\mathbf{y}}(\hat{t}_f), \hat{\mathbf{u}}(\hat{t}_f)] \, \Delta t_f$$

$$\tilde{\mathbf{y}}(\hat{t}_f) = \hat{\mathbf{y}}(\hat{t}_f) + \sum_{i=1}^{n} \mathbf{a}_i \Delta \tilde{t}_i \ + \ \mathbf{g}[\hat{\mathbf{y}}(\hat{t}_f), \hat{\mathbf{u}}(\hat{t}_f)] \, \Delta \tilde{t}_f$$

---

[*] If n is infinite, then the control would be discontinuous everywhere in $[0, \hat{t}_f]$ and thus not admissible.

and

$$\mathbf{z}(\hat{t}_f) = \hat{\mathbf{y}}(\hat{t}_f) + \alpha \mathbf{y}(\hat{t}_f) + (1-\alpha)\tilde{\mathbf{y}}(\hat{t}_f)$$

$$= \hat{\mathbf{y}}(\hat{t}_f) + \sum_{i=1}^{n} \mathbf{a}_i \underbrace{\left[\alpha \Delta t_i + (1-\alpha)\Delta \tilde{t}_i\right]}_{\Delta \check{t}_i} + \mathbf{g}[\hat{\mathbf{y}}(\hat{t}_f), \hat{\mathbf{u}}(\hat{t}_f)] \underbrace{\left[\alpha \Delta t_f + (1-\alpha)\Delta \tilde{t}_f\right]}_{\Delta \check{t}_f}$$

Thus, $\mathbf{z}(\hat{t}_f)$ is the result of a control $\check{\mathbf{u}}(t)$ from the collection of admissible controls with non-negative time intervals $\Delta \check{t}_i$s and $\Delta \check{t}_f$.* Therefore, the set of final states is convex.

5.B   Supporting Hyperplane of a Convex Set

A hyperplane is a generalization of a plane in a coordinate system of a number of dimensions. Any point $\mathbf{y}$ in a hyperplane satisfies a set of linear equations

$$\sum_{i=1}^{n} p_i y_i = \alpha, \quad i = 1, 2, \ldots, n \quad \text{or} \quad \mathbf{p}^\top \mathbf{y} = \alpha$$

where α is some constant. For a specific point $\hat{\mathbf{y}}$ on the hyperplane, we have

$$\mathbf{p}^\top \hat{\mathbf{y}} = \alpha$$

The difference between the last two equations yields

$$\mathbf{p}^\top (\mathbf{y} - \hat{\mathbf{y}}) = 0$$

which is the equation of the hyperplane relative to $\hat{\mathbf{y}}$. The vector $\mathbf{p}$ is called the normal to the hyperplane at $\hat{\mathbf{y}}$.

Consider a set of points, S. If and only if all points of S lie on one side of P, then P is called a **supporting hyperplane** of S. For our purpose, we consider S to lie on the side opposite to the normal to P at a given point. For a convex set of two-dimensional points, Figure 5.3 (p. 135) shows a hyperplane P supporting a closed set S at a point $\hat{\mathbf{y}}$. Belonging to both P and S, the point lies on the boundary of S.

We will now show that if S is a convex set, then at each of its boundary point we have a supporting hyperplane, or equivalently $\mathbf{p}^\top(\mathbf{y} - \hat{\mathbf{y}}) \le 0$ for all $\mathbf{y}$ in S. In the following two steps we show that:

---

* They are non-negative since they depend on non-negative Δt_is, $\Delta \tilde{t}_i$s, Δt_f, $\Delta \tilde{t}_f$, and α.

1. In a closed convex set S in R^n, there exists a unique point $\hat{\mathbf{y}}$, which is closest to a given point $\mathbf{x}$ outside the set S.

2. If $\hat{\mathbf{y}}$ in the set S is closest to $\mathbf{x}$, then for all $\mathbf{y}$ in S the dot product $(\mathbf{x} - \hat{\mathbf{y}})^\top (\mathbf{y} - \hat{\mathbf{y}}) \leq 0$ and vice versa.

3. Given any boundary point $\hat{\mathbf{y}}$ of the set S, there exists a non-zero vector $\mathbf{p}$ such that the dot product $\mathbf{p}^\top (\mathbf{y} - \hat{\mathbf{y}}) \leq 0$ for all $\mathbf{y}$ in the S. This result means that a hyperplane supports the set S at each of its boundary point.

Step 1   Consider within S a point $\mathbf{y}_0$ and all points $\mathbf{y}_i$ ($i = 1, 2, \ldots$) equidistant from or closer to $\mathbf{x}$ than $\mathbf{y}_0$. Then finding $\hat{\mathbf{y}}$, a point of S closest to $\mathbf{x}$, is equivalent to finding the minimum element of the set

$$\{\|\mathbf{x} - \mathbf{y}_i\|; \quad i = 0, 1, 2, \ldots\}$$

Now the above set is closed and bounded (by 0 and $\|\mathbf{x} - \mathbf{y}_0\|$). In this set, consider the continuous function $f(\mathbf{y}_i) = \|\mathbf{x} - \mathbf{y}_i\|$. According to the Weierstrass Theorem (Section 9.18, p. 278), there exists a minimum of f, say, $f(\hat{\mathbf{y}})$, corresponding to the smallest element of the set, $\|\mathbf{x} - \hat{\mathbf{y}}\|$. Thus, there exists a point $\hat{\mathbf{y}}$ in S that is closest to $\mathbf{x}$.

Next, we show that $\hat{\mathbf{y}}$ is unique. Suppose there exists a point $\mathbf{y}_a$ equidistant from $\mathbf{x}$. Let

$$d \equiv \|\mathbf{x} - \hat{\mathbf{y}}\| = \|\mathbf{x} - \mathbf{y}_a\| \tag{5.39}$$

Since S is convex, it contains $0.5(\hat{\mathbf{y}} + \mathbf{y}_a)$. From the triangle inequality (Section 2.2.2, p. 26), the distance

$$\|\mathbf{x} - 0.5(\hat{\mathbf{y}} + \mathbf{y}_a)\| = 0.5\|(\mathbf{x} - \hat{\mathbf{y}}) + (\mathbf{x} - \mathbf{y}_a)\| \leq 0.5\underbrace{\|(\mathbf{x} - \hat{\mathbf{y}})\|}_{d} + 0.5\underbrace{\|(\mathbf{x} - \mathbf{y}_a)\|}_{d}$$

Strict inequality in the above relation is contradictory since it would imply that $0.5(\hat{\mathbf{y}} + \mathbf{y}_a)$ is closer to $\mathbf{x}$ than the minimum distance d. Therefore,

$$\|(\mathbf{x} - \hat{\mathbf{y}}) + (\mathbf{x} - \mathbf{y}_a)\| = \|(\mathbf{x} - \hat{\mathbf{y}})\| + \|(\mathbf{x} - \mathbf{y}_a)\|$$

which means that the vectors $(\mathbf{x} - \hat{\mathbf{y}})$ and $(\mathbf{x} - \mathbf{y}_a)$ are collinear.* Thus,

$$(\mathbf{x} - \hat{\mathbf{y}}) = \gamma(\mathbf{x} - \mathbf{y}_a)$$

where γ is some scalar. Taking the norm on both sides of the above equation and comparing it with Equation (5.39), we get

$$\|\gamma\| = |\gamma| = 1$$

---

* Let $\|\mathbf{a} + \mathbf{b}\| = \|\mathbf{a}\| + \|\mathbf{b}\|$. Squaring both sides, $\sum_{i=1}^n (a_i + b_i)^2 = \sum_{i=1}^n (a_i^2 + b_i^2) + 2\|\mathbf{a}\|\|\mathbf{b}\|$. Upon simplifying, $\sum_{i=1}^n a_i b_i = \mathbf{a}^\top \mathbf{b} = \|\mathbf{a}\|\|\mathbf{b}\|$. Now $\mathbf{a}^\top \mathbf{b} = \|\mathbf{a}\|\|\mathbf{b}\| \cos\theta$ where θ is the angle between the two vectors. In this case where $\|\mathbf{a} + \mathbf{b}\| = \|\mathbf{a}\| + \|\mathbf{b}\|$, we have $\theta = 0°$.

If $\gamma = -1$, then $\mathbf{x} = (\hat{\mathbf{y}} + \mathbf{y}_a)/2$, which belongs to the convex set S. This is contradictory since $\mathbf{x}$ lies outside S. Thus, $\gamma \neq -1$. Hence, $\gamma = 1$, thereby implying that $\hat{\mathbf{y}} = \mathbf{y}_a$. In other words, $\hat{\mathbf{y}}$ is a unique point of S for which the distance $\|\mathbf{x} - \hat{\mathbf{y}}\|$ is minimum.

Step 2 Next, we show that if $\hat{\mathbf{y}}$ is closest to $\mathbf{x}$, then for any point $\mathbf{y}$ in S,

$$(\mathbf{x} - \hat{\mathbf{y}})^\top (\mathbf{y} - \hat{\mathbf{y}}) \leq 0 \tag{5.40}$$

Consider a point $\mathbf{y}_b = \hat{\mathbf{y}} + \alpha(\mathbf{y} - \hat{\mathbf{y}})$, which lies in the convex set S for $0 \leq \alpha \leq 1$. Since $\hat{\mathbf{y}}$ is closest to $\mathbf{x}$, we have $\|\mathbf{x} - \hat{\mathbf{y}}\| \leq \|\mathbf{x} - \mathbf{y}_b\|$ or equivalently,

$$\|\mathbf{x} - \hat{\mathbf{y}}\|^2 \leq \|\mathbf{x} - \mathbf{y}_b\|^2 = \|(\mathbf{x} - \hat{\mathbf{y}}) - \alpha(\mathbf{y} - \hat{\mathbf{y}})\|^2 \tag{5.41}$$

Using the Parallelogram Identity (Section 9.21, p. 280),

$$\|(\mathbf{x} - \hat{\mathbf{y}}) - \alpha(\mathbf{y} - \hat{\mathbf{y}})\|^2 = \|\mathbf{x} - \hat{\mathbf{y}}\|^2 + \alpha^2 \|\mathbf{y} - \hat{\mathbf{y}}\|^2 - 2\alpha(\mathbf{x} - \hat{\mathbf{y}})^\top (\mathbf{y} - \hat{\mathbf{y}})$$

Using the above equation in Inequality (5.41), we get

$$(\mathbf{x} - \hat{\mathbf{y}})^\top (\mathbf{y} - \hat{\mathbf{y}}) \leq (\alpha/2)\|\mathbf{y} - \hat{\mathbf{y}}\|^2, \quad 0 \leq \alpha \leq 1$$

In the above inequality, taking the right-hand limit of α to zero, we get the desired result,

$$(\mathbf{x} - \hat{\mathbf{y}})^\top (\mathbf{y} - \hat{\mathbf{y}}) \leq 0$$

Finally, we show that if $(\mathbf{x} - \hat{\mathbf{y}})^\top (\mathbf{y} - \hat{\mathbf{y}}) \leq 0$, then $\hat{\mathbf{y}}$ is closest to $\mathbf{x}$ for any point $\mathbf{y}$ in S. Consider

$$\|\mathbf{x} - \mathbf{y}\|^2 = \|(\mathbf{x} - \hat{\mathbf{y}}) - (\mathbf{y} - \hat{\mathbf{y}})\|^2 = \|\mathbf{x} - \hat{\mathbf{y}}\|^2 + \|\mathbf{y} - \hat{\mathbf{y}}\|^2 - 2(\mathbf{x} - \hat{\mathbf{y}})^\top (\mathbf{y} - \hat{\mathbf{y}})$$

where the right-hand side follows from the Parallelogram Identity. Since we are given

$$(\mathbf{x} - \hat{\mathbf{y}})^\top (\mathbf{y} - \hat{\mathbf{y}}) \leq 0$$

the last term of the above equation is positive or zero. Hence,

$$\|\mathbf{x} - \mathbf{y}\|^2 \geq \|\mathbf{x} - \hat{\mathbf{y}}\|^2 \quad \text{or} \quad \|\mathbf{x} - \mathbf{y}\| \geq \|\mathbf{x} - \hat{\mathbf{y}}\|$$

which implies that $\hat{\mathbf{y}}$ is closest to $\mathbf{x}$.

Step 3 In terms of a unit vector $\mathbf{p} \equiv (\mathbf{x} - \hat{\mathbf{y}})/\|\mathbf{x} - \hat{\mathbf{y}}\|$, Inequality (5.40) becomes

$$\mathbf{p}^\top \mathbf{y} \leq \underbrace{\mathbf{p}^\top \hat{\mathbf{y}}}_{\alpha}$$

where we denote the right-hand side, a dot product, as a scalar α. But $\alpha < \mathbf{p}^\top \mathbf{x}$ since

$$\mathbf{p}^\top \mathbf{x} - \alpha = (\mathbf{x} - \hat{\mathbf{y}})^\top (\mathbf{x} - \hat{\mathbf{y}}) = \|\mathbf{x} - \hat{\mathbf{y}}\|^2 > 0$$

Consequently, $\mathbf{p}^\top\mathbf{y} < \mathbf{p}^\top\mathbf{x}$. In other words, given a point $\mathbf{x}_i$ outside the convex set S, there is a unique point $\hat{\mathbf{y}}_i$ in S, corresponding to the unit vector $\mathbf{p}_i = (\mathbf{x}_i - \hat{\mathbf{y}}_i)/\|\mathbf{x}_i - \hat{\mathbf{y}}_i\|$, such that

$$\mathbf{p}_i^\top\mathbf{y} < \mathbf{p}_i^\top\mathbf{x}_i$$

for each $\mathbf{y}$ in S.

Now consider an infinite sequence of $\mathbf{x}_i$ tending to a point $\hat{\mathbf{y}}$ on the boundary of S. The corresponding sequence $\mathbf{p}_i$ of unit vectors is bounded. According to the Bolzano–Weierstrass Theorem (Section 9.17, p. 277), this sequence has a subsequence converging to a limit, say, $\mathbf{p}$ whose norm is also unity. Hence, as i tends to infinity, the above inequality becomes

$$\mathbf{p}^\top\mathbf{y} < \mathbf{p}^\top\hat{\mathbf{y}} \quad \text{or} \quad \mathbf{p}^\top(\mathbf{y} - \hat{\mathbf{y}}) < 0$$

Including the case $\mathbf{y} = \hat{\mathbf{y}}$, we finally have

$$\mathbf{p}^\top(\mathbf{y} - \hat{\mathbf{y}}) \leq 0$$

Bibliography

M. Athans and P.L. Falb. *Optimal Control — An Introduction to the Theory and Its Applications*, Chapter 5. Dover Publications Inc., New York, 2007.

M.S. Bazaraa, H.D. Sherali, and C.M. Shetty. *Nonlinear Programming Theory and Algorithms*, Chapter 2. Wiley-Interscience, New Jersey, 3rd edition, 2006.

V.G. Boltyanskii, R.V. Gamkrelidze, and L.S. Pontryagin. Towards a theory of optimal processes. *Reports Acad. Sci. USSR*, 110(1), 1956.

M.I. Kamien and N.L. Schwartz. *Dynamic Optimization*, Chapter 6, page 149. Elsevier, Amsterdam, 1991.

E.R. Pinch. *Optimal Control and the Calculus of Variations*, Chapter 6. Oxford University Press Inc., New York, 1993.

L.S. Pontryagin. *The Mathematical Theory of Optimal Processes*, Chapter II. Gordon and Breach Science Publishers, New York, 1986.

Chapter 6

Different Types of Optimal Control Problems

In this chapter, we engage with different types of optimal control problems and derive the necessary conditions for the minimum. The constraints in the problems are handled using the Lagrange Multiplier Rule and the John Multiplier Theorem. Since derivatives are available in most of the engineering problems, we assume the functions involved are sufficiently differentiable and the constraint qualifications explained in Chapter 4 are satisfied. Simply put, we take for granted the existence of a set of Lagrange multipliers in the augmented functional incorporating all constraints of an optimal control problem.

6.1   Free Final Time

We first consider optimal control problems in which the final time is free, i. e., not specified or fixed. Thus, in addition to finding optimal control functions, we need to determine the optimal final time as well in these problems.

6.1.1   Free Final State

This is the optimal control problem of Section 5.2 (p. 126) in which both the final time and the final state are unspecified or free. Using vectors, the objective is to minimize the functional

$$I = \int\limits_{0}^{t_f} F[\mathbf{y}(t), \mathbf{u}(t)]\, dt \tag{6.1}$$

subject to the autonomous ordinary differential equations

$$\dot{\mathbf{y}} = \mathbf{g}[\mathbf{y}(t), \mathbf{u}(t)] \tag{6.2}$$

with the initial conditions

$$\mathbf{y}(0) = \mathbf{y}_0 \tag{6.3}$$

Based on the Lagrange Multiplier Rule (see Section 4.3.3.2, p. 103), the above problem is equivalent to minimizing the augmented functional

$$J = \int\limits_0^{t_f} \left[F + \boldsymbol{\lambda}^\top (-\dot{\mathbf{y}} + \mathbf{g}) \right] \mathrm{d}t \tag{6.4}$$

subject to the initial conditions. The vectors $\mathbf{y}$, $\dot{\mathbf{y}}$, $\mathbf{g}$, and $\boldsymbol{\lambda}$ are each of dimension n with the first element indexed 1. For example,

$$\boldsymbol{\lambda} = \begin{bmatrix} \lambda_1 & \lambda_2 & \dots & \lambda_n \end{bmatrix}^\top$$

Variation of J

We begin with the definition of the Hamiltonian

$$H \equiv F + \boldsymbol{\lambda}^\top \mathbf{g}$$

which is the same as Equation (5.4) on p. 128 with $\lambda_0 = 1$ and $g_0 \equiv F$. In terms of H, the augmented functional becomes

$$J = \int\limits_0^{t_f} (H - \boldsymbol{\lambda}^\top \dot{\mathbf{y}}) \, \mathrm{d}t$$

The variation of J is given by

$$\delta J = \int\limits_0^{t_f} \left[\sum_{i=1}^n \left(\frac{\partial H}{\partial y_i} \delta y_i + \frac{\partial H}{\partial \lambda_i} \delta \lambda_i + \frac{\partial H}{\partial u_i} \delta u_i - \lambda_i \delta \dot{y}_i - \dot{y}_i \delta \lambda_i \right) \right] \mathrm{d}t$$

$$+ \left[H - \boldsymbol{\lambda}^\top \dot{\mathbf{y}} \right]_{t_f} \delta t_f$$

$$= \int\limits_0^{t_f} \left(H_{\mathbf{y}}^\top \delta \mathbf{y} + H_{\boldsymbol{\lambda}}^\top \delta \boldsymbol{\lambda} + H_{\mathbf{u}}^\top \delta \mathbf{u} - \boldsymbol{\lambda}^\top \delta \dot{\mathbf{y}} - \dot{\mathbf{y}}^\top \delta \boldsymbol{\lambda} \right) \mathrm{d}t + \underbrace{\left[H - \boldsymbol{\lambda}^\top \dot{\mathbf{y}} \right]_{t_f} \delta t_f}_{\partial J / \partial t_f}$$

where the last term above is the partial derivative of J with respect to t_f resulting from the application of the Leibniz Integral Rule (see Section 9.10, p. 273). Applying integration by parts to the integral of $\boldsymbol{\lambda}^\top \delta \dot{\mathbf{y}}$, we get

$$\int\limits_0^{t_f} \boldsymbol{\lambda}^\top \delta \dot{\mathbf{y}} \, \mathrm{d}t = \left[\boldsymbol{\lambda}^\top \delta \mathbf{y} \right]_0^{t_f} - \int\limits_0^{t_f} \dot{\boldsymbol{\lambda}}^\top \delta \mathbf{y} \, \mathrm{d}t = \left[\boldsymbol{\lambda}^\top \delta \mathbf{y} \right]_{t_f} - \int\limits_0^{t_f} \dot{\boldsymbol{\lambda}}^\top \delta \mathbf{y} \, \mathrm{d}t$$

where in the last step we have made use of the fact that $\delta \mathbf{y}(0)$ must be zero since $\mathbf{y}(0)$ cannot vary — it is fixed at $\mathbf{y}_0$ by the initial condition. Substituting the above equation in the expression for δJ, we obtain

$$\delta J = \int_0^{t_f} \left[(H_y + \dot{\boldsymbol{\lambda}})^\top \delta \mathbf{y} + (H_\lambda - \dot{\mathbf{y}})^\top \delta \boldsymbol{\lambda} + H_u^\top \delta \mathbf{u} \right] dt - \boldsymbol{\lambda}^\top (t_f)\, \delta \mathbf{y}(t_f)$$
$$+ \left[H - \boldsymbol{\lambda}^\top \dot{\mathbf{y}} \right]_{t_f} \delta t_f \tag{6.5}$$

Note that $\delta \mathbf{y}(t_f)$ is the variation in the optimal state $\hat{\mathbf{y}}$ at time t_f, i.e.,

$$\delta \mathbf{y}(t_f) = \mathbf{y}(t_f) - \hat{\mathbf{y}}(t_f) \tag{6.6}$$

This variation is different from the variation in the *final* state

$$\delta \mathbf{y}_f \equiv \mathbf{y}(t_f + \delta t_f) - \hat{\mathbf{y}}(t_f) \tag{6.7}$$

which also incorporates a change δt_f in t_f. Figure 6.1 illustrates both of these variations for a single optimal state $\hat{y}$. To incorporate the specifications of free final state and free final time, we need to introduce $\delta \mathbf{y}_f$ in Equation (6.5). For this purpose, we obtain a first-order approximation of $\delta \mathbf{y}_f$ in terms of $\delta \mathbf{y}(t_f)$ as follows.

The variation $\delta \mathbf{y}_f$ can be expressed as

$$\delta \mathbf{y}_f = \mathbf{y}(t_f + \delta t_f) - \hat{\mathbf{y}}(t_f) = \mathbf{y}(t_f) + \dot{\mathbf{y}}(t_f)\delta t_f - \hat{\mathbf{y}}(t_f)$$
$$= \delta \mathbf{y}(t_f) + \dot{\mathbf{y}}(t_f)\delta t_f$$

where we have used the first order Taylor expansion for $\mathbf{y}(t_f + \delta t_f)$. Next, differentiating Equation (6.6) with respect to t, we obtain

$$\dot{\mathbf{y}}(t_f) = \dot{\hat{\mathbf{y}}}(t_f) + \delta \dot{\mathbf{y}}(t_f)$$

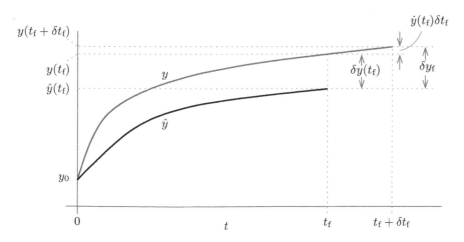

Figure 6.1 Relation between δy_f and $\delta y(t_f)$

After combining the last two equations and discarding the second-order term $\delta\dot{\mathbf{y}}(t_f)\delta t_f$, we get

$$\delta\mathbf{y}_f = \delta\mathbf{y}(t_f) + \dot{\mathbf{y}}(t_f)\delta t_f \quad \text{or} \quad \delta\mathbf{y}(t_f) = \delta\mathbf{y}_f - \dot{\mathbf{y}}(t_f)\delta t_f \qquad (6.8)$$

Finally, substituting the above equation in Equation (6.5), we obtain at the minimum of J [where $\delta J = 0$, $\mathbf{u} = \hat{\mathbf{u}}$, $\mathbf{y} = \hat{\mathbf{y}}$, and, in particular, $\dot{\mathbf{y}}(t_f) = \dot{\hat{\mathbf{y}}}(t_f)$]

$$\underbrace{\int_0^{t_f} \left[(H_\mathbf{y} + \dot{\boldsymbol{\lambda}})^\top \delta\mathbf{y} + (H_{\boldsymbol{\lambda}} - \dot{\mathbf{y}})^\top \delta\boldsymbol{\lambda} + H_\mathbf{u}^\top \delta\mathbf{u} \right] dt - \boldsymbol{\lambda}^\top(t_f)\,\delta\mathbf{y}_f + H(t_f)\delta t_f}_{\delta J} \;=\; 0$$

$$(6.9)$$

Necessary Conditions for the Minimum

Hence, at the minimum of J, and equivalently of I, the following equations must be satisfied:

$$\dot{\boldsymbol{\lambda}} = -H_\mathbf{y} \qquad\qquad \dot{\mathbf{y}} = H_{\boldsymbol{\lambda}} = \mathbf{g} \qquad\qquad H_\mathbf{u} = \mathbf{0}$$
$$\boldsymbol{\lambda}(t_f) = \mathbf{0} \qquad\qquad H(t_f) = 0$$

along with the initial conditions, $\mathbf{y}(0) = \mathbf{y}_0$. The control for which the above conditions are satisfied is called the optimal control $\hat{\mathbf{u}}$. This result is valid when the following conditions are met with:

1. the functions F, $\mathbf{g}$, $H_\mathbf{y}$, and $H_\mathbf{u}$ exist and are continuous with respect to $\mathbf{y}$ and $\mathbf{u}$.

2. the optimal control is

   2.a either unconstrained, i.e., a $\hat{u}_i$ can take any value

   2.b or in the interior of a specified set containing all admissible control values

Condition 2a is applicable to the present problem, which does not impose any constraints on the control.

It might happen that the first condition is not satisfied, e.g., when the objective functional depends on $|\mathbf{u}|$ or the set of admissible controls contains discrete elements. To obtain necessary conditions for the minimum in that situation, we can rely on Pontryagin's minimum principle, which does not depend on the existence of $H_\mathbf{u}$ or its continuity (see Section 5.3.1, p. 128). According to the principle, the optimal control $\hat{\mathbf{u}}$ minimizes H, i.e.,

$$H(\hat{\mathbf{y}}, \boldsymbol{\lambda}, \mathbf{u}) \geq H(\hat{\mathbf{y}}, \boldsymbol{\lambda}, \hat{\mathbf{u}})$$

at each time in the interval $[0, t_f]$ where $\hat{\mathbf{y}}$ and $\boldsymbol{\lambda}$ are the state and costate at the minimum, respectively.

Example 6.1

For the batch reactor problem in Example 2.10 (p. 45), find the necessary conditions for the minimum of

$$I = -\int_0^{t_f} ckx^a \, dt, \quad k = k_o \exp\left(-\frac{E}{RT}\right)$$

subject to the satisfaction of the state equation

$$\dot{x} = -akx^a \tag{6.10}$$

with $T(t)$ as the control, $x(0) = x_0$ as the initial condition, and t_f as the unspecified final time.

In this problem, x and T are the state (y) and control (u), respectively. The Hamiltonian for this problem is

$$H = -ckx^a - \lambda akx^a$$

Thus, for I to be minimum the necessary conditions are as follows:

$$\underbrace{\dot{\lambda} = akx^{a-1}(c + a\lambda)}_{-H_x} \qquad \underbrace{\dot{x} = -akx^a}_{H_\lambda} \qquad \underbrace{-\frac{kE}{RT^2}x^a(c + a\lambda) = 0}_{H_T}$$

$$\lambda(t_f) = 0 \qquad\qquad x(0) = x_0 \qquad\qquad \underbrace{-\left[ckx^a + \lambda akx^a\right]_{t_f}}_{H(t_f)} = 0$$

□

6.1.2   Fixed Final State

Consider the optimal control problem of Section 6.1.1 (p. 153). If the final state is fixed, say, at $y(t_f) = y_f$, then the variation δy_f must be zero. Consequently, Equation (6.9) simplifies to

$$\delta J = \int_0^{t_f} \left[(H_y + \dot{\lambda})^\top \delta y + (H_\lambda - \dot{y})^\top \delta\lambda + H_u^\top \delta u\right] dt + H(t_f)\delta t_f = 0$$

and the necessary conditions for the minimum of J, and equivalently of I, are

$$\dot{\lambda} = -H_y, \quad \dot{y} = H_\lambda = g, \quad H_u = 0, \quad \text{and} \quad H(t_f) = 0$$

along with the initial and final conditions

$$y(0) = y_0 \quad \text{and} \quad y(t_f) = y_f.$$

Example 6.2
Let the final state in Example 6.1 be specified as $x(t_f) = x_f$. In this case, the necessary conditions for the minimum are the same except $\lambda(t_f) = 0$, which is replaced with $x(t_f) = x_f$.   ⬜

6.1.3   Final State on Hypersurfaces

In this case, the objective is to minimize the functional

$$I = \int_0^{t_f} F[\mathbf{y}(t), \mathbf{u}(t)]\,dt$$

subject to

$$\dot{\mathbf{y}} = \mathbf{g}[\mathbf{y}(t), \mathbf{u}(t)], \quad \mathbf{y}(0) = \mathbf{y}_0$$

as well as

$$\mathbf{q}[\mathbf{y}(t_f)] = \mathbf{0}$$

where $\mathbf{q} = \begin{bmatrix} q_1 & q_2 & \cdots & q_l \end{bmatrix}^\top$ and each $q_i = 0$ is an equation of a hypersurface in the space of coordinates

$$(y_1, y_2, \ldots, y_n).$$

Thus, the final state is constrained to lie on these hypersurfaces. Given the initial conditions, the optimal control problem is to minimize the augmented functional J given by Equation (6.4) subject to the hypersurface constraint $\mathbf{q}[\mathbf{y}(t_f)] = \mathbf{0}$. According to the Lagrange Multiplier Rule, this problem is equivalent to that of minimizing the further-augmented objective functional

$$M = J + \boldsymbol{\mu}^\top \mathbf{q} = \int_0^{t_f} \left[F + \boldsymbol{\lambda}^\top(-\dot{\mathbf{y}} + \mathbf{g}) \right] dt + \boldsymbol{\mu}^\top \mathbf{q} \qquad (6.11)$$

where

$$\boldsymbol{\mu} = \begin{bmatrix} \mu_1 & \mu_2 & \cdots & \mu_n \end{bmatrix}^\top$$

is the vector of additional Lagrange multipliers corresponding to the vector of constraint hypersurfaces.

The variation of M is

$$\delta M = \delta J + \delta\boldsymbol{\mu}^\top \mathbf{q} + \boldsymbol{\mu}^\top \delta\mathbf{q}$$

Because $\mathbf{q}$ depends on $\mathbf{y}(t_f)$, which in turn can additionally vary with t_f,

$$\delta\mathbf{q} = \mathbf{q}_{\mathbf{y}(t_f)}\delta\mathbf{y}(t_f) + \mathbf{q}_{\mathbf{y}(t_f)}\dot{\mathbf{y}}(t_f)\delta t_f = \mathbf{q}_{\mathbf{y}(t_f)}\delta\mathbf{y}_f \qquad (6.12)$$

at the minimum. In the above equation, $q_{y(t_f)}$ is the matrix of partial derivatives of q with respect to $y(t_f)$, and the right-hand side is the result of substituting $\delta y(t_f)$ given by Equation (6.8).

Substituting Equations (6.9) and (6.12) in the expression for δM, we obtain

$$
\delta M = \int_0^{t_f} \left[(H_y + \dot{\lambda})^\top \delta y + (H_\lambda - \dot{y})^\top \delta \lambda + H_u^\top \delta u \right] dt + H(t_f)\delta t_f
$$
$$
+ \delta \mu^\top q - [\lambda^\top(t_f) - \mu^\top q_{y(t_f)}] \delta y_f \tag{6.13}
$$

Since M has to be zero at the minimum, the necessary conditions for the minimum of M, and equivalently of J and I, are

$$
\dot{\lambda} = -H_y, \quad \dot{y} = H_\lambda, \quad H_u = 0, \quad H(t_f) = 0, \quad q = 0, \quad \text{and} \quad \lambda^\top(t_f) = \mu^\top q_{y(t_f)}
$$

along with the initial conditions $y(0) = y_0$.

Example 6.3

Consider the consecutive reaction A $\longrightarrow$ B $\rightleftharpoons$ C carried out in a batch reactor. The mole fractions y_1, y_2, and y_3 of species A, B, and C are respectively governed by

$$
\begin{aligned}
\dot{y}_1 &= -a_1 y_1^2 e^{E_1/T}, & y_1(0) &= 1 \\
\dot{y}_2 &= a_1 y_1^2 e^{E_1/T} - a_2 y_2 e^{E_2/T}, & y_2(0) &= 0 \\
\dot{y}_3 &= a_2 y_2 e^{E_2/T} - a_3 y_3 e^{E_3/T}, & y_3(0) &= 0
\end{aligned}
$$

where a_i and E_i $(i = 1, 2, 3)$ are constants. It is desired to find the temperature policy $T(t)$ that maximizes the product mole fraction y_3 at the final time t_f, which is not fixed. Moreover, y_3 must satisfy the two selectivity constraints

$$
y_3 = b_1 y_1 \qquad \text{and} \qquad y_3 = b_2 y_2 \qquad \text{at} \quad t = t_f
$$

where b_1 and b_2 are some constants.

In this problem, T is the control (u). The equivalent objective is to minimize the functional

$$
I = -y_3(t_f) = -\int_0^{y_3(t_f)} dy_3 = -\int_0^{t_f} \dot{y}_3 \, dt = \int_0^{t_f} \left(-a_2 y_2 e^{E_2/T} + a_3 y_3 e^{E_3/T} \right) dt
$$

for which we have used the state equation for the species C in the last step. The above objecive is subject to the satisfaction of all state equations and the selectivity constraints. The Hamiltonian is then given by

$$
H = -a_2 y_2 e^{E_2/T} + a_3 y_3 e^{E_3/T} - \lambda_1 a_1 y_1^2 e^{E_1/T} + \lambda_2 \left(a_1 y_1^2 e^{E_1/T} - a_2 y_2 e^{E_2/T} \right)
$$
$$
+ \lambda_3 \left(a_2 y_2 e^{E_2/T} - a_3 y_3 e^{E_3/T} \right)
$$

The necessary conditions for the minimum of I are

$$
\begin{bmatrix} \dot{y}_1 \\ \dot{y}_2 \\ \dot{y}_3 \end{bmatrix} = \underbrace{\begin{bmatrix} -a_1 y_1^2 e^{E_1/T} \\ a_1 y_1^2 e^{E_1/T} - a_2 y_2 e^{E_2/T} \\ a_2 y_2 e^{E_2/T} - a_3 y_3 e^{E_3/T} \end{bmatrix}}_{H_\lambda}, \quad \begin{bmatrix} y_1 \\ y_2 \\ y_3 \end{bmatrix}_{t=0} = \begin{bmatrix} 1 \\ 0 \\ 0 \end{bmatrix}
$$

$$
\begin{bmatrix} q_1 \\ q_2 \end{bmatrix} = \begin{bmatrix} y_3 - b_1 y_1 \\ y_3 - b_2 y_2 \end{bmatrix}_{t_f} = \begin{bmatrix} 0 \\ 0 \end{bmatrix}
$$

$$
\begin{bmatrix} \dot{\lambda}_1 \\ \dot{\lambda}_2 \\ \dot{\lambda}_3 \end{bmatrix} = \underbrace{\begin{bmatrix} 2(-\lambda_1 + \lambda_2)a_1 y_1 e^{E_1/T} \\ (-1 - \lambda_2 + \lambda_3)a_2 e^{E_2/T} \\ (1 - \lambda_3)a_3 e^{E_3/T} \end{bmatrix}}_{-H_y}
$$

$$
\begin{bmatrix} \lambda_1 \\ \lambda_2 \\ \lambda_3 \end{bmatrix}_{t_f}^\top = \begin{bmatrix} \mu_1 \\ \mu_2 \end{bmatrix}^\top \underbrace{\begin{bmatrix} \dfrac{\partial q_1}{\partial y_1} & \dfrac{\partial q_1}{\partial y_2} & \dfrac{\partial q_1}{\partial y_3} \\ \dfrac{\partial q_2}{\partial y_1} & \dfrac{\partial q_2}{\partial y_2} & \dfrac{\partial q_2}{\partial y_3} \end{bmatrix}_{t_f}}_{q_{y(t_f)}} = \begin{bmatrix} \mu_1 \\ \mu_2 \end{bmatrix}^\top \begin{bmatrix} -b_1 & 0 & 1 \\ 0 & -b_2 & 1 \end{bmatrix}
$$

$$
= \begin{bmatrix} -\mu_1 b_1 & -\mu_2 b_2 & \mu_1 + \mu_2 \end{bmatrix}
$$

$$
H_T = -\frac{1}{T^2}\Big[(-\lambda_1 + \lambda_2)a_1 y_1^2 E_1 e^{E_1/T} + (-1 - \lambda_2 + \lambda_3)a_2 y_2 E_2 e^{E_2/T}
$$
$$
+ (1 - \lambda_3)a_3 y_3 E_3 e^{E_3/T}\Big] = 0
$$

$$
H(t_f) = \Big[(-\lambda_1 + \lambda_2)a_1 y_1^2 e^{E_1/T} + (-1 - \lambda_2 + \lambda_3)a_2 y_2 e^{E_2/T}
$$
$$
+ (1 - \lambda_3)a_3 y_3 e^{E_3/T}\Big]_{t_f} = 0
$$

□

6.2 Fixed Final Time

In this section, we take up problems in which the final time is specified or fixed. The control functions are the only optimization parameters.

6.2.1 Free Final State

Consider again the optimal control problem of Section 6.1.1 (p. 153). If the final time is fixed, then its variation δt_{f} must be zero in Equation (6.9) on p. 156. Moreover, $\delta \mathbf{y}_{\mathrm{f}} = \delta \mathbf{y}(t_{\mathrm{f}})$ from Equation (6.8). Consequently, Equation (6.9) simplifies to

$$\delta J = \int_0^{t_{\mathrm{f}}} \left[(H_{\mathbf{y}} + \dot{\boldsymbol{\lambda}})^{\top} \delta \mathbf{y} + (H_{\boldsymbol{\lambda}} - \dot{\mathbf{y}})^{\top} \delta \boldsymbol{\lambda} + H_{\mathbf{u}}^{\top} \delta \mathbf{u} \right] dt - \boldsymbol{\lambda}^{\top}(t_{\mathrm{f}}) \, \delta \mathbf{y}(t_{\mathrm{f}}) = 0$$

(6.14)

and the necessary conditions for the minimum of J, and equivalently of I, are

$$\dot{\boldsymbol{\lambda}} = -H_{\mathbf{y}}, \quad \dot{\mathbf{y}} = H_{\boldsymbol{\lambda}} = \mathbf{g}, \quad H_{\mathbf{u}} = \mathbf{0}, \quad \text{and} \quad \boldsymbol{\lambda}(t_{\mathrm{f}}) = \mathbf{0}$$

along with the initial conditions $\mathbf{y}(0) = \mathbf{y}_0$.

Example 6.4
Let the final time be fixed in Example 6.1 (p. 157). Then the necessary conditions for the minimum are as follows:

$$\dot{\lambda} = akx^{a-1}(c + \lambda) \qquad \dot{x} = -akx^a \qquad -\frac{kE}{RT^2}x^a(c + a\lambda) = 0$$

$$\lambda(t_{\mathrm{f}}) = 0 \qquad x(0) = x_0$$

□

6.2.2 Fixed Final State

In this case, we have the optimal control problem of the last section, but with the final state fixed. Thus, Equation (6.14) becomes

$$\delta J = \int_0^{t_{\mathrm{f}}} \left[(H_{\mathbf{y}} + \dot{\boldsymbol{\lambda}})^{\top} \delta \mathbf{y} + (H_{\boldsymbol{\lambda}} - \dot{\mathbf{y}})^{\top} \delta \boldsymbol{\lambda} + H_{\mathbf{u}}^{\top} \delta \mathbf{u} \right] dt = 0$$

Thus, the necessary conditions for the minimum of J, and equivalently of I, are

$$\dot{\boldsymbol{\lambda}} = -H_{\mathbf{y}}, \quad \dot{\mathbf{y}} = H_{\boldsymbol{\lambda}} = \mathbf{g}, \quad \text{and} \quad H_{\mathbf{u}} = \mathbf{0}$$

along with the respective initial and final conditions

$$\mathbf{y}(0) = \mathbf{y}_0 \quad \text{and} \quad \mathbf{y}(t_f) = \mathbf{y}_f$$

Example 6.5
Let both final time and final state be fixed in Example 6.1 (p. 157). Thus, for fixed t_f and $x(t_f) = x_f$, the necessary conditions for the minimum are as follows:

$$\dot{\lambda} = akx^{a-1}(c + \lambda) \qquad \dot{x} = -akx^a \qquad -\frac{kE}{RT^2}x^a(c + a\lambda) = 0$$

$$x(t_f) = x_f \qquad\qquad x(0) = x_0$$

⬛

6.2.3 Final State on Hypersurfaces

In this case, we have the optimal control problem of Section 6.1.3 (p. 158), but with fixed final time. This problem is equivalent to that of minimizing the augmented functional

$$M = \int_0^{t_f} \left[F + \boldsymbol{\lambda}^\top (-\dot{\mathbf{y}} + \mathbf{g}) \right] \mathrm{d}t + \boldsymbol{\mu}^\top \mathbf{q} = J + \boldsymbol{\mu}^\top \mathbf{q} \tag{6.11}$$

whose variation from Equation (6.13) for fixed t_f, i. e., $\delta t_f = 0$, is

$$\delta M = \int_0^{t_f} \left[(H_\mathbf{y} + \dot{\boldsymbol{\lambda}})^\top \delta \mathbf{y} + (H_{\boldsymbol{\lambda}} - \dot{\mathbf{y}})^\top \delta \boldsymbol{\lambda} + H_\mathbf{u}^\top \delta \mathbf{u} \right] \mathrm{d}t + \delta \boldsymbol{\mu}^\top \mathbf{q}$$
$$- [\boldsymbol{\lambda}^\top(t_f) - \boldsymbol{\mu}^\top \mathbf{q}_{\mathbf{y}(t_f)}] \, \delta \mathbf{y}(t_f)$$

where $\mathbf{q}_{\mathbf{y}(t_f)}$ is the matrix of partial derivatives of $\mathbf{q}$ with respect to $\mathbf{y}(t_f)$. Since M has to be zero at the minimum, the necessary conditions for the minimum of I, and equivalently of J and I, are

$$\dot{\boldsymbol{\lambda}} = -H_\mathbf{y}, \quad \dot{\mathbf{y}} = H_{\boldsymbol{\lambda}} = \mathbf{g}, \quad H_\mathbf{u} = \mathbf{0}, \quad \mathbf{q} = \mathbf{0} \quad \text{and} \quad \boldsymbol{\lambda}^\top(t_f) - \boldsymbol{\mu}^\top \mathbf{q}_{\mathbf{y}(t_f)} = \mathbf{0}$$

along with the initial conditions $\mathbf{y}(0) = \mathbf{y}_0$.

Example 6.6
Let the final time in Example 6.3 (p. 159) be fixed at t_f. In this case, all necessary conditions of Example 6.3 hold except $H(t_f) = 0$.

⬛

6.3 Algebraic Constraints

This section deals with optimal control problems constrained by algebraic equalities and inequalities.

6.3.1 Algebraic Equality Constraints

Consider the objective to minimize the functional

$$I = \int_0^{t_f} F[\mathbf{y}(t), \mathbf{u}(t)]\, dt$$

subject to the state equations

$$\dot{\mathbf{y}} = \mathbf{g}[\mathbf{y}(t), \mathbf{u}(t)], \quad \mathbf{y}(0) = \mathbf{y}_0$$

and the equality constraints

$$f_i(\mathbf{y}, \mathbf{u}) = 0, \quad i = 1, 2, \ldots, l \quad \text{or} \quad \mathbf{f}(\mathbf{y}, \mathbf{u}) = \mathbf{0}$$

where it is provided that

1. the functions $\mathbf{f}$, $\mathbf{f_y}$, and $\mathbf{f_u}$ exist and are continuous with respect to $\mathbf{y}$ and $\mathbf{u}$.

2. The number of equality constraints l is less than the number of controls, i. e., m.

   The reason for the above provision is that if $l = m$, then the constraints would uniquely determine the control, and there would not be any optimal control problem remaining.

According to the Lagrange Multiplier Rule, the above problem is equivalent to the minimization of the augmented functional

$$M = \int_0^{t_f} \left[F + \boldsymbol{\lambda}^\top (-\dot{\mathbf{y}} + \mathbf{g}) + \boldsymbol{\mu}^\top \mathbf{f} \right] dt = \int_0^{t_f} (L - \boldsymbol{\lambda}^\top \dot{\mathbf{y}})\, dt \tag{6.15}$$

where

$$\boldsymbol{\mu} = \begin{bmatrix} \mu_1 & \mu_2 & \cdots & \mu_l \end{bmatrix}^\top$$

is the vector of time dependent Lagrange multipliers corresponding to the equality constraints, and L is the Lagrangian defined as

$$L \equiv H + \boldsymbol{\mu}^\top \mathbf{f} = F + \boldsymbol{\lambda}^\top \mathbf{g} + \boldsymbol{\mu}^\top \mathbf{f} \tag{6.16}$$

Considering that both final state and time are free, i. e., unspecified, the variation of M upon simplification is given by [compare with Equation (6.9) on p. 156]

$$\delta M = \int_0^{t_f} \left[(L_\mathbf{y} + \dot{\boldsymbol{\lambda}})^\top \delta \mathbf{y} + (L_{\boldsymbol{\lambda}} - \dot{\mathbf{y}})^\top \delta \boldsymbol{\lambda} + L_{\boldsymbol{\mu}}^\top \delta \boldsymbol{\mu} + L_\mathbf{u}^\top \delta \mathbf{u} \right] dt$$
$$- \boldsymbol{\lambda}^\top (t_f)\, \delta \mathbf{y}_f + L(t_f)\delta t_f \tag{6.17}$$

Since δM should be zero at the minimum, the following equations are necessary at the minimum of M, and equivalently of J and I:

$$\dot{\boldsymbol{\lambda}} = -L_\mathbf{y} \qquad \dot{\mathbf{y}} = L_{\boldsymbol{\lambda}} = \mathbf{g} \qquad L_{\boldsymbol{\mu}} = \mathbf{f} = \mathbf{0}$$
$$L_\mathbf{u} = \mathbf{0} \qquad \boldsymbol{\lambda}(t_f) = \mathbf{0} \qquad L(t_f) = 0$$

along with $\mathbf{y}(0) = \mathbf{y}_0$.

Example 6.7

Consider the isothermal operation of the CSTR in Example 3.5 (p. 69). The reactant and catalyst concentrations, namely, y_1 and y_2, are governed by the state equations

$$\dot{y}_1 = u_1(y_f - y_1) - ky_1y_2, \qquad y_1(0) = y_{1,0}$$
$$\dot{y}_2 = u_2 - u_1y_2, \qquad y_2(0) = y_{2,0}$$

where the controls u_1 and u_2 are, respectively, the volumetric flow rate of the CSTR and the catalyst mass flow rate, both per unit reactor volume. Let us say we want u_2 to be increasingly higher for smaller amounts of the catalyst-to-reactant ratio. For this purpose we enforce the following equality constraint at all times:

$$\frac{y_2}{y_1} = a\left(\frac{1}{b^{u_2 - c}} - 1 \right) \qquad \text{or} \qquad f \equiv \frac{y_2}{y_1} - a\left(\frac{1}{b^{u_2 - c}} - 1 \right) = 0$$

where a, b, and c are some suitable parameters. Subject to the above state equations and equality constraint, the objective is to minimize the deviation of the state of the CSTR with minimum control action, i. e., to minimize the functional

$$I = \int_0^{t_f} \sum_{i=1}^2 \left[(y_i - y_i^s)^2 + (u_i - u_i^s)^2 \right] dt$$

where the superscript s denotes the steady state value.

Based on Equation (6.16), the Lagrangian for this problem is

$$L = \sum_{i=1}^{2} \left[(y_i - y_i^s)^2 + (u_i - u_i^s)^2 \right] + \lambda_1 [u_1(y_f - y_1) - ky_1 y_2]$$

$$+ \lambda_2 [u_2 - u_1 y_2] + \mu \underbrace{\left[\frac{y_2}{y_1} - a \left(\frac{1}{b^{u_2 - c}} - 1 \right) \right]}_{f}$$

Then the necessary conditions for the minimum of I are as follows:

$$\begin{bmatrix} \dot{y}_1 \\ \\ \dot{y}_2 \end{bmatrix} = \underbrace{\begin{bmatrix} u_1(y_f - y_1) - ky_1 y_2 \\ \\ u_2 - u_1 y_2 \end{bmatrix}}_{L_\lambda}, \quad \begin{bmatrix} y_1 \\ \\ y_2 \end{bmatrix}_{t=0} = \begin{bmatrix} y_{1,0} \\ \\ y_{2,0} \end{bmatrix}$$

$$\begin{bmatrix} \dot{\lambda}_1 \\ \\ \dot{\lambda}_2 \end{bmatrix} = \underbrace{\begin{bmatrix} -2(y_1 - y_1^s) + \lambda_1(u_1 + ky_2) + \mu \frac{y_2}{y_1^2} \\ \\ -2(y_2 - y_2^s) + \lambda_1 ky_1 + \lambda_2 u_1 - \frac{\mu}{y_1} \end{bmatrix}}_{-L_y}, \quad \begin{bmatrix} \lambda_1 \\ \\ \lambda_2 \end{bmatrix}_{t_f}^\top = \begin{bmatrix} 0 \\ \\ 0 \end{bmatrix}$$

$$\begin{bmatrix} L_{u_1} \\ \\ L_{u_2} \end{bmatrix} = \begin{bmatrix} 2(u_1 - u_1^s) + \lambda_1(y_f - y_1) - \lambda_2 y_2 \\ \\ 2(u_2 - u_2^s) + \lambda_2 + \mu \frac{a \ln b}{b^{u_2 - c}} \end{bmatrix} = \begin{bmatrix} 0 \\ \\ 0 \end{bmatrix}$$

$$L_\mu = \underbrace{\frac{y_2}{y_1} - a \left(\frac{1}{b^{u_2 - c}} - 1 \right)}_{f} = 0$$

$$L(t_f) = \left\{ \sum_{i=1}^{2} \left[(y_i - y_i^s)^2 + (u_i - u_i^s)^2 \right] + \lambda_1 [u_1(y_f - y_1) - ky_1 y_2] \right.$$

$$\left. + \lambda_2 [(u_2 - u_1 y_2)] + \mu \left[\frac{y_2}{y_1} - a \left(\frac{1}{b^{u_2 - c}} - 1 \right) \right] \right\}_{t_f} = 0$$

□

6.3.2 Algebraic Inequality Constraints

Consider the optimal control problem in which the objective is to minimize the functional

$$I = \int_0^{t_f} F[\mathbf{y}(t), \mathbf{u}(t)] \, dt$$

subject to

$$\dot{\mathbf{y}} = \mathbf{g}[\mathbf{y}(t), \mathbf{u}(t)], \quad \mathbf{y}(0) = \mathbf{y}_0$$

and the inequality constraints

$$f_i(\mathbf{y}, \mathbf{u}, t) \leq 0, \quad i = 1, 2, \ldots, l \quad \text{or} \quad \mathbf{f}(\mathbf{y}, \mathbf{u}, t) \leq \mathbf{0}$$

where each f_i and its derivative with respect to $\mathbf{y}$ and $\mathbf{u}$ are continuous.

The augmented objective functional for this problem is M, which is given by Equation (6.15) on p. 163. From the John Multiplier Theorem (Section 4.5.1, p. 113), the following equations must be satisfied at the minimum of M, and equivalently of I:

$$\dot{\boldsymbol{\lambda}} = -L_\mathbf{y} \qquad \dot{\mathbf{y}} = L_\boldsymbol{\lambda} = \mathbf{g} \qquad L_\boldsymbol{\mu} = \mathbf{f} \leq \mathbf{0} \qquad L_\mathbf{u} = \mathbf{0}$$

$$\boldsymbol{\lambda}(t_f) = \mathbf{0} \qquad L(t_f) = 0 \qquad \boldsymbol{\mu} \geq \mathbf{0} \qquad \boldsymbol{\mu}^\mathsf{T} L_\boldsymbol{\mu} = \mathbf{0}$$

along with $\mathbf{y}(0) = \mathbf{y}_0$ where L is the Lagrangian given by Equation (6.16).

Remarks

The number of inequality constraints, l, can be greater than m, the number of controls. That the number of active constraints at any time does not exceed the m is assured by the constraint qualification. It requires that if p inequality constraints are active (i. e., $f_i = 0$ for $i = 1, 2, \ldots, p$), then p should be the rank of the matrix of partial derivatives of $\mathbf{f}$ with respect to $\mathbf{u}$.[*] Note that p is the number of linearly independent rows or columns of the matrix (see Section 4.3.2.1, p. 97).

Example 6.8

Let us consider Example 6.7 with its equality constraint replaced with the inequality

$$\frac{y_2}{y_1} - a\left(\frac{1}{b^{u_2 - c}} - 1\right) \leq 0$$

Then the necessary conditions for the minimum are

---

[*] This is one of the four constraint qualifications, any one of which must be satisfied to obtain the necessary conditions for the minimum in inequality constrained problems (Arrow et al., 1961). See Takayama (1985) for a thorough exposition.

- the ones in the previous example except $L_\mu = 0$, which is replaced with

$$L_\mu = \frac{y_2}{y_1} - a\left(\frac{1}{b^{u_2-c}} - 1\right) \le 0$$

and

- $\mu \ge 0$, and the complimentary slackness condition, i. e.,

$$\mu\left[\frac{y_2}{y_1} - a\left(\frac{1}{b^{u_2-c}} - 1\right)\right] = 0$$

$\square$

Example 6.9

Let us consider Example 6.3 (p. 159) without the selectivity constraints and impose the constraints $T_{\min} \le T \le T_{\max}$ on the temperature control. Thus, instead of the selectivity constraints, we have the two inequalities

$$-T + T_{\min} \le 0 \quad \text{and} \quad T - T_{\max} \le 0$$

For this modified problem, the Lagrangian is given by

$$L = H + \mu_1(-T + T_{\min}) + \mu_2(T - T_{\max})$$
$$= -a_2 y_2 e^{E_2/T} + a_3 y_3 e^{E_3/T} - \lambda_1 a_1 y_1^2 e^{E_1/T} + \lambda_2\left(a_1 y_1^2 e^{E_1/T} - a_2 y_2 e^{E_2/T}\right)$$
$$+ \lambda_3\left(a_2 y_2 e^{E_2/T} - a_3 y_3 e^{E_3/T}\right) + \mu_1(-T + T_{\min}) + \mu_2(T - T_{\max})$$

The necessary conditions for the minimum are

$$\begin{bmatrix} \dot{y}_1 \\ \dot{y}_2 \\ \dot{y}_3 \end{bmatrix} = \underbrace{\begin{bmatrix} -a_1 y_1^2 e^{E_1/T} \\ a_1 y_1^2 e^{E_1/T} - a_2 y_2 e^{E_2/T} \\ a_2 y_2 e^{E_2/T} - a_3 y_3 e^{E_3/T} \end{bmatrix}}_{L_\lambda}, \quad \begin{bmatrix} y_1 \\ y_2 \\ y_3 \end{bmatrix}_{t=0} = \begin{bmatrix} 1 \\ 0 \\ 0 \end{bmatrix}$$

$$\begin{bmatrix} \dot{\lambda}_1 \\ \dot{\lambda}_2 \\ \dot{\lambda}_3 \end{bmatrix} = \underbrace{\begin{bmatrix} 2(-\lambda_1 + \lambda_2)a_1 y_1 e^{E_1/T} \\ (-1 - \lambda_2 + \lambda_3)a_2 e^{E_2/T} \\ (1 - \lambda_3)a_3 e^{E_3/T} \end{bmatrix}}_{-L_y}, \quad \begin{bmatrix} \lambda_1 \\ \lambda_2 \\ \lambda_3 \end{bmatrix}_{t_f} = \begin{bmatrix} 0 \\ 0 \\ 0 \end{bmatrix}$$

$$L_T = -\frac{1}{T^2} \Big[(-\lambda_1 + \lambda_2) a_1 y_1^2 E_1 e^{E_1/T} + (-1 - \lambda_2 + \lambda_3) a_2 y_2 E_2 e^{E_2/T}$$

$$+ (1 - \lambda_3) a_3 y_3 E_3 e^{E_3/T} \Big] - \mu_1 + \mu_2 = 0$$

$$\begin{bmatrix} L_{\mu_1} \\ L_{\mu_2} \end{bmatrix} = \begin{bmatrix} -T + T_{\min} \\ T - T_{\max} \end{bmatrix} \leq \begin{bmatrix} 0 \\ 0 \end{bmatrix}$$

$$L(t_{\mathrm{f}}) = \Big[(-\lambda_1 + \lambda_2) a_1 y_1^2 e^{E_1/T} + (-1 - \lambda_2 + \lambda_3) a_2 y_2 e^{E_2/T}$$

$$+ (1 - \lambda_3) a_3 y_3 e^{E_3/T} + \mu_1 (-T + T_{\min}) + \mu_2 (T - T_{\max}) \Big]_{t_{\mathrm{f}}} = 0$$

$$\begin{bmatrix} \mu_1 \\ \mu_2 \end{bmatrix}_{t=0} \geq \begin{bmatrix} 0 \\ 0 \end{bmatrix} \quad \text{and} \quad \begin{bmatrix} \mu_1 \\ \mu_2 \end{bmatrix}^\top \underbrace{\begin{bmatrix} -T + T_{\min} \\ T - T_{\max} \end{bmatrix}}_{L_\mu} = \begin{bmatrix} 0 \\ 0 \end{bmatrix}$$

□

6.4 Integral Constraints

Integral constraints could be equality or inequality constraints. We first consider integral equality constraints in an optimal control problem with free state and free final time.

6.4.1 Integral Equality Constraints

Consider the objective to minimize

$$I = \int_0^{t_{\mathrm{f}}} F[\mathbf{y}(t), \mathbf{u}(t)] \, \mathrm{d}t$$

subject to

$$\dot{\mathbf{y}} = \mathbf{g}[\mathbf{y}(t), \mathbf{u}(t)], \quad \mathbf{y}(0) = \mathbf{y}_0$$

and the integral equality constraints

$$\int_0^{t_f} F_i(\mathbf{y}, \mathbf{u}, t)\, dt = k_i, \quad i = 1, 2, \ldots, l$$

where each F_i and its partial derivative with respect to $\mathbf{y}$ and $\mathbf{u}$ are continuous. According to the Lagrange Multiplier Rule, the above problem is equivalent to the minimization of the augmented functional

$$
\begin{aligned}
M &= \int_0^{t_f} \left[F + \boldsymbol{\lambda}^\top (-\dot{\mathbf{y}} + \mathbf{g}) \right] dt + \sum_{i=1}^{l} \mu_i \left(\int_0^{t_f} F_i\, dt - k_i \right) \\
&= \int_0^{t_f} \left(\underbrace{H + \boldsymbol{\mu}^\top \mathbf{F}}_{L} - \boldsymbol{\lambda}^\top \dot{\mathbf{y}} \right) dt - \boldsymbol{\mu}^\top \mathbf{k} = \int_0^{t_f} (L - \boldsymbol{\lambda}^\top \dot{\mathbf{y}})\, dt - \boldsymbol{\mu}^\top \mathbf{k} \quad (6.18)
\end{aligned}
$$

where

1. $\boldsymbol{\mu} = \begin{bmatrix} \mu_1 & \mu_2 & \cdots & \mu_l \end{bmatrix}^\top$ is the vector of *time invariant* Lagrange multipliers corresponding to the integral equality constraints;

2. $\mathbf{F}$ and $\mathbf{k}$ are respective vectors of F_is and k_is; and

3. L is the Lagrangian defined as

$$L \equiv H + \boldsymbol{\mu}^\top \mathbf{F} = F + \boldsymbol{\lambda}^\top \mathbf{g} + \boldsymbol{\mu}^\top \mathbf{F} \quad (6.19)$$

Note that the μ_is are undetermined constants in contrast to λ_is, which are time dependent.

The variation of M is given by [compare with Equation (6.17) on p. 164]

$$
\delta M = \int_0^{t_f} \left[(L_\mathbf{y} + \dot{\boldsymbol{\lambda}})^\top \delta\mathbf{y} + (L_\boldsymbol{\lambda} - \dot{\mathbf{y}})^\top \delta\boldsymbol{\lambda} + L_\mathbf{u}^\top \delta\mathbf{u} \right] dt - \boldsymbol{\lambda}^\top(t_f)\, \delta\mathbf{y}_f
$$

$$
+ L(t_f)\delta t_f + \delta\boldsymbol{\mu}^\top \left(\underbrace{\int_0^{t_f} L_\boldsymbol{\mu}\, dt}_{\mathbf{F}} - \mathbf{k} \right) \quad (6.20)
$$

Since δM should be zero at the minimum of M, the following equations must be satisfied at the minimum of M, and equivalently of I:

$$\dot{\boldsymbol{\lambda}} = -L_\mathbf{y} \qquad \dot{\mathbf{y}} = L_\boldsymbol{\lambda} = \mathbf{g} \qquad L_\mathbf{u} = 0$$

$$\boldsymbol{\lambda}(t_f) = 0 \qquad L(t_f) = 0 \qquad \int_0^{t_f} \mathbf{F}\, dt = \mathbf{k}$$

along with $\mathbf{y}(0) = \mathbf{y}_0$. Observe that $L_\boldsymbol{\mu}$ is the vector $\mathbf{F}$, and the last equation is the set of the integral constraints.

Remarks

The integral constraints are equivalent to having additional state equations

$$\frac{\mathrm{d}y_{n+i}}{\mathrm{d}t} = F_i, \quad y_{n+i}(0) = 0, \quad y_{n+i}(t_f) = k_i; \quad i = 1, 2, \ldots, l$$

with state variables y_{n+i}, $i = 1, 2, \ldots, l$. The reader can easily verify this equivalence by taking the integral on both sides of the above equations and applying the boundary conditions. Hence, an integral constraint is equivalent to a differential equation for an additional state variable with the initial and final conditions specified. Consequently, there is no limit on the number of integral constraints in an optimal control problem.

Example 6.10

Let us re-state the problem of batch distillation in Section 1.3.1 (p. 5). The objective is to minimize the functional

$$I = -\int_0^{t_f} u \, \mathrm{d}t$$

subject to the state equations

$$\dot{z}_1 = -u, \qquad\qquad z_1(0) = z_{1,0}$$
$$\dot{z}_2 = u[z_1 - y(z_1, z_2)], \qquad z_2(0) = z_{2,0}$$

and the purity specification

$$\int_0^{t_f} (y^* - y)u \, \mathrm{d}t = 0$$

The Lagrangian for this problem is given by

$$L = -u - \lambda_1 u + \lambda_2 u(z_1 - y) + \mu(y^* - y)u$$

Note that while λ_1 and λ_2 are time dependent, μ is a constant. The necessary conditions for the minimum are as follows:

$$\begin{bmatrix} \dot{z}_1 \\ \dot{z}_2 \end{bmatrix} = \underbrace{\begin{bmatrix} -u \\ u(z_1 - y) \end{bmatrix}}_{L_\lambda}, \quad \begin{bmatrix} z_1 \\ z_2 \end{bmatrix}_{t=0} = \begin{bmatrix} z_{1,0} \\ z_{2,0} \end{bmatrix}$$

$$\begin{bmatrix} \dot{\lambda}_1 \\[2mm] \dot{\lambda}_2 \end{bmatrix} = \underbrace{\begin{bmatrix} -\lambda_2 u\left(1 - \dfrac{\partial y}{\partial z_1}\right) + \mu u \dfrac{\partial y}{\partial z_1} \\[4mm] \lambda_2 u \dfrac{\partial y}{\partial z_2} + \mu u \dfrac{\partial y}{\partial z_2} \end{bmatrix}}_{-L_z}, \qquad \begin{bmatrix} \lambda_1 \\[2mm] \lambda_2 \end{bmatrix}_{t_f} = \begin{bmatrix} 0 \\[2mm] 0 \end{bmatrix}$$

$$L_u = -1 - \lambda_1 + \lambda_2(z_1 - y) + \mu(y^* - y) = 0$$

$$\int_0^{t_f} L_\mu \, dt = \int_0^{t_f} (y^* - y)u \, dt = 0$$

$$L(t_f) = \left[-u - \lambda_1 u + \lambda_2 u (z_1 - y) + \mu(y^* - y)u\right]_{t_f} = 0$$

□

6.4.2 Integral Inequality Constraints

Consider the optimal control problem in the last section, but with integral equality constraints replaced with the inequality constraints

$$\int_0^{t_f} F_i(\mathbf{y}, \mathbf{u}, t) \, dt \le k_i, \quad i = 1, 2, \dots, l$$

In this case, Equation (6.18) provides the augmented functional M whose variation is given by Equation (6.20). From the John Multiplier Theorem (Section 4.5.1, p. 113), the necessary conditions for the minimum are

$$\dot{\boldsymbol{\lambda}} = -L_\mathbf{y} \qquad \dot{\mathbf{y}} = L_{\boldsymbol{\lambda}} = \mathbf{g} \qquad L_\mathbf{u} = \mathbf{0}$$
$$\boldsymbol{\lambda}(t_f) = \mathbf{0} \qquad L(t_f) = 0 \qquad \boldsymbol{\mu} \ge \mathbf{0}$$
$$\boldsymbol{\mu}^\top \left(\int_0^{t_f} L_\mu \, dt - \mathbf{k} \right) = 0 \qquad \int_0^{t_f} L_\mu \, dt \le \mathbf{k}$$

The integral constraints are equivalent to the differential equations governing the additional state variables as follows:

$$\frac{dy_{n+i}}{dt} = F_i, \quad y_{n+i}(0) = 0, \quad y_{n+i}(t_f) \le k_i; \quad i = 1, 2, \dots, l$$

Example 6.11

Let us consider Example 6.10 with its equality constraint replaced with

$$\int_0^{t_f} (y^* - y)u \, dt \le 0$$

In other words, we want distillate purity greater than or equal to y^*. Then the necessary conditions for the minimum are

- those of Example 6.10 except $\int_0^{t_f} L_\mu \, dt = 0$, which is replaced with

$$\int_0^{t_f} L_\mu \, dt = \int_0^{t_f} (y^* - y)u \, dt \le 0$$

- as well as

$$\mu \ge 0 \qquad \text{and} \qquad \mu \int_0^{t_f} (y^* - y)u \, dt = 0$$

□

6.5 Interior Point Constraints

Consider the objective to minimize the functional

$$I = \int_0^{t_f} F[\mathbf{y}(t), \mathbf{u}(t)] \, dt$$

subject to the state equations

$$\dot{\mathbf{y}} = \mathbf{g}[\mathbf{y}(t), \mathbf{u}(t)]; \quad \mathbf{y}(0) = \mathbf{y}_0$$

and the interior point constraints

$$\mathbf{q}[\mathbf{y}(t_1), t_1] = \mathbf{0}$$

where

$$\mathbf{q} = \begin{bmatrix} q_1 & q_2 & \cdots & q_l \end{bmatrix}^{\top}$$

is the vector of functions depending on $\mathbf{y}(t_1)$ and t_1, which is an unspecified or free time in the interval $[0, t_f]$. With the help of Lagrange multipliers, the above problem is equivalent to that of minimizing the augmented functional

$$
M = \int_0^{t_f} \left[F + \boldsymbol{\lambda}^\top (-\dot{\mathbf{y}} + \mathbf{g}) \right] dt + \boldsymbol{\mu}^\top \mathbf{q}
$$

$$
= \int_0^{t_1} (H - \boldsymbol{\lambda}^\top \dot{\mathbf{y}}) \, dt + \int_{t_1}^{t_f} (H - \boldsymbol{\lambda}^\top \dot{\mathbf{y}}) \, dt + \boldsymbol{\mu}^\top \mathbf{q}
$$

where $H = F + \boldsymbol{\lambda}^\top \mathbf{g}$ and

$$
\boldsymbol{\mu} = \begin{bmatrix} \mu_1 & \mu_2 & \cdots & \mu_l \end{bmatrix}^\top
$$

is the vector of *time invariant* Lagrange multipliers corresponding to the interior point constraints. With the help of

$$
a \equiv H_\mathbf{y}^\top \delta \mathbf{y} + H_\lambda^\top \delta \boldsymbol{\lambda} + H_\mathbf{u}^\top \delta \mathbf{u}
$$

the variation of M is given by

$$
\delta M = \int_0^{t_{1-}} \left(a - \boldsymbol{\lambda}^\top \delta \dot{\mathbf{y}} - \dot{\mathbf{y}}^\top \delta \boldsymbol{\lambda} \right) dt + \left[H - \boldsymbol{\lambda}^\top \dot{\mathbf{y}} \right]_{t_{1-}} \delta t_1 - \left[H - \boldsymbol{\lambda}^\top \dot{\mathbf{y}} \right]_{t_{1+}} \delta t_1
$$

$$
+ \int_{t_{1+}}^{t_f} \left(a - \boldsymbol{\lambda}^\top \delta \dot{\mathbf{y}} - \dot{\mathbf{y}}^\top \delta \boldsymbol{\lambda} \right) dt + \left[H - \boldsymbol{\lambda}^\top \dot{\mathbf{y}} \right]_{t_f} \delta t_f
$$

$$
+ \mathbf{q}^\top \delta \boldsymbol{\mu} + \boldsymbol{\mu}^\top \underbrace{\left[\mathbf{q}_{\mathbf{y}(t_1)} \delta \mathbf{y}(t_1) + \mathbf{q}_{\mathbf{y}(t_1)} \dot{\mathbf{y}}(t_1) \delta t_1 + \mathbf{q}_{t_1} \delta t_1 \right]}_{\displaystyle = \mathbf{q}_{\mathbf{y}(t_1)} \delta \mathbf{y}_1 \text{ [compare with Equation (6.12) on p. 158]}}
$$

where

$$
t_{1-} \equiv t_1 - |\delta t_1|, \quad \text{and} \quad t_{1+} \equiv t_1 + |\delta t_1|, \quad \text{with} \quad \delta t_1 \to 0
$$

and $\delta \mathbf{y}_1$ is the variation in $\mathbf{y}$ for unspecified t_1, similar to $\mathbf{y}_f$ given by Equation (6.7) on p. 155.

Applying integration by parts to the time integrals of $\boldsymbol{\lambda}^\top \delta \dot{\mathbf{y}}$ and combining the integrals having the same integrand, we get

$$\delta M = \int_0^{t_f} (a + \dot{\boldsymbol{\lambda}}^\top \delta \mathbf{y} - \dot{\mathbf{y}}^\top \delta \boldsymbol{\lambda})\, dt - \left[\boldsymbol{\lambda}^\top \delta \mathbf{y}\right]_0^{t_{1-}} - \left[\boldsymbol{\lambda}^\top \delta \mathbf{y}\right]_{t_{1+}}^{t_f} + \left[H - \boldsymbol{\lambda}^\top \dot{\mathbf{y}}\right]_{t_{1-}} \delta t_1$$

$$- \left[H - \boldsymbol{\lambda}^\top \dot{\mathbf{y}}\right]_{t_{1+}} \delta t_1 + \left[H - \boldsymbol{\lambda}^\top \dot{\mathbf{y}}\right]_{t_f} \delta t_f$$

$$+ \mathbf{q}^\top \delta \boldsymbol{\mu} + \boldsymbol{\mu}^\top \left[\mathbf{q}_{\mathbf{y}(t_1)} \delta \mathbf{y}_1 + \mathbf{q}_{t_1} \delta t_1\right]$$

Expanding a and using Equation (6.8) on p. 156 as well as its equivalents for t_{1-} and t_{1+}, which are, respectively,

$$\delta \mathbf{y}(t_{1-}) = \delta \mathbf{y}_{t_1} - \dot{\mathbf{y}}(t_{1-})\delta t_1 \quad \text{and} \quad \delta \mathbf{y}(t_{1+}) = \delta \mathbf{y}_{t_1} - \dot{\mathbf{y}}(t_{1+})\delta t_1$$

we finally obtain at the minimum of M

$$\delta M = \int_0^{t_f} \left[(H_{\mathbf{y}} + \dot{\boldsymbol{\lambda}})^\top \delta \mathbf{y} + (H_{\boldsymbol{\lambda}} - \dot{\mathbf{y}})^\top \delta \boldsymbol{\lambda} + H_{\mathbf{u}}^\top \delta \mathbf{u}\right] dt$$

$$+ \mathbf{q}^\top \delta \boldsymbol{\mu} + \left[\boldsymbol{\mu}^\top \mathbf{q}_{\mathbf{y}(t_1)} + \boldsymbol{\lambda}(t_{1+}) - \boldsymbol{\lambda}(t_{1-})\right] \delta \mathbf{y}_{t_1}$$

$$+ \left[\boldsymbol{\mu}^\top \mathbf{q}_{t_1} + H(t_{1-}) - H(t_{1+})\right] \delta t_1 - \boldsymbol{\lambda}^\top(t_f)\, \delta \mathbf{y}_f + H(t_f)\delta t_f = 0$$

Thus, the necessary conditions for the minimum of M, and equivalently of I, are

$$\dot{\boldsymbol{\lambda}} = -H_{\mathbf{y}} \qquad \dot{\mathbf{y}} = H_{\boldsymbol{\lambda}} \qquad\qquad H_{\mathbf{u}} = \mathbf{0}$$

$$\mathbf{q} = \mathbf{0} \qquad \boldsymbol{\lambda}(t_{1-}) = \boldsymbol{\lambda}(t_{1+}) + \mathbf{q}_{\mathbf{y}(t_1)}^\top \boldsymbol{\mu} \qquad H(t_{1-}) = H(t_{1+}) - \boldsymbol{\mu}^\top \mathbf{q}_{t_1}$$

$$\boldsymbol{\lambda}(t_f) = \mathbf{0} \qquad H(t_f) = 0 \qquad\qquad \mathbf{y}(0) = \mathbf{y}_0$$

Example 6.12

Based on the chemotherapy example in Section 1.3.8 (p. 15), we pose a problem in which the ratios of immune-to-cancer and healthy-to-cancer cells must be equal to some desired values during the treatment period. Hence, it is desired to minimize the number of cancer cells (y_4) at the final time t_f as well as the use of the drug, i.e., the functional

$$I = y_4(t_f) + \int_0^{t_f} u\, dt = \int_0^{t_f} [\dot{y}_4 + u]\, dt$$

where u is the drug injection rate. The drug concentration y_1 and the number of immune, healthy, and cancer cells (y_2, y_3, and y_4) are governed by the state equations

$$\frac{dy_1}{dt} = u - \gamma_6 y_1$$

$$\frac{dy_2}{dt} = \dot{y}_{2,\text{in}} + r_2 \frac{y_2 y_4}{\beta_2 + y_4} - \gamma_3 y_2 y_4 - \gamma_4 y_2 - \alpha_2 y_2 \left(1 - e^{-y_1 \lambda_2}\right)$$

$$\frac{dy_3}{dt} = r_3 y_3 (1 - \beta_3 y_3) - \gamma_5 y_3 y_4 - \alpha_3 y_3 \left(1 - e^{-y_1 \lambda_3}\right)$$

$$\frac{dy_4}{dt} = r_1 y_4 (1 - \beta_1 y_4) - \gamma_1 y_3 y_4 - \gamma_2 y_2 y_4 - \alpha_1 y_4 \left(1 - e^{-y_1 \lambda_1}\right)$$

along with the initial conditions

$$y_i(0) = y_{i,0}, \quad i = 1, 2, 3, 4$$

and the interior equality constraints

$$\left. \frac{y_2}{y_4} \right|_{t_1} = a \quad \text{and} \quad \left. \frac{y_3}{y_4} \right|_{t_1} = b$$

where t_1 is an unspecified time in the interval $[0, t_f]$ and a and b are certain specified constants.

With the help of the state equation for y_4, the objective functional becomes

$$I = \int_0^{t_f} \left[r_1 y_4 (1 - \beta_1 y_4) - \gamma_1 y_4 y_3 - \gamma_2 y_2 y_4 - \alpha_1 y_4 \left(1 - e^{-y_1 \lambda_1}\right) + u \right] dt$$

which needs to be minimized subject to all state equations as well as the two interior equality constraints.

The Hamiltonian for this problem is given by

$$H = (1 + \lambda_4) \left[r_1 y_4 (1 - \beta_1 y_4) - \gamma_1 y_3 y_4 - \gamma_2 y_2 y_4 - \alpha_1 y_4 \left(1 - e^{-y_1 \lambda_1}\right) \right] + u$$

$$+ \lambda_1 [u - \gamma_6 y_1] + \lambda_2 \left[\dot{y}_{2,\text{in}} + r_2 \frac{y_2 y_4}{\beta_2 + y_4} - \gamma_3 y_2 y_4 - \gamma_4 y_2 - \right.$$

$$\left. \alpha_2 y_2 \left(1 - e^{-y_1 \lambda_2}\right) \right] + \lambda_3 \left[r_3 y_3 (1 - \beta_3 y_3) - \gamma_5 y_3 y_4 - \alpha_3 y_3 \left(1 - e^{-y_1 \lambda_3}\right) \right]$$

The necessary conditions for the minimum are as follows:

$$
\begin{bmatrix} \dot{y}_1 \\[2mm] \dot{y}_2 \\[2mm] \dot{y}_3 \\[2mm] \dot{y}_4 \end{bmatrix}
=
\underbrace{\begin{bmatrix}
u - \gamma_6 y_1 \\[2mm]
\dot{y}_{2,\text{in}} + r_2 \dfrac{y_2 y_4}{\beta_2 + y_4} - \gamma_3 y_2 y_4 - \gamma_2 y_4 - \alpha_2 y_2 \left(1 - e^{-y_1 \lambda_2}\right) \\[3mm]
r_3 y_3 (1 - \beta_3 y_3) - \gamma_5 y_3 y_4 - \alpha_3 y_3 \left(1 - e^{-y_1 \lambda_3}\right) \\[3mm]
r_1 y_4 (1 - \beta_1 y_4) - \gamma_1 y_4 y_3 - \gamma_2 y_2 y_4 - \alpha_1 y_4 \left(1 - e^{-y_1 \lambda_1}\right)
\end{bmatrix}}_{H_\lambda}
$$

$$
\begin{bmatrix} y_1 & y_2 & y_3 & y_4 \end{bmatrix}_{t=0}
=
\begin{bmatrix} y_{1,0} & y_{2,0} & y_{3,0} & y_{4,0} \end{bmatrix}
$$

$$
\begin{bmatrix} \dot{\lambda}_1 \\[2mm] \dot{\lambda}_2 \\[2mm] \dot{\lambda}_3 \\[2mm] \dot{\lambda}_4 \end{bmatrix}
=
\underbrace{\begin{bmatrix}
(1 + \lambda_4)\alpha_1 y_4 \lambda_1 e^{-y_1 \lambda_1} + \lambda_1 \gamma_6 + \lambda_2^2 \alpha_2 y_2 e^{-y_1 \lambda_2} + \lambda_3^2 \alpha_3 y_3 e^{-y_1 \lambda_3} \\[3mm]
(1 + \lambda_4)\gamma_2 y_4 - \lambda_2 \left[\dfrac{r_2 y_4}{\beta_2 + y_4} - \gamma_3 y_4 - \gamma_4 - \alpha_2 \left(1 - e^{-y_1 \lambda_2}\right) \right] \\[3mm]
(1 + \lambda_4)\gamma_1 y_4 - \lambda_3 \left[r_3 (1 - 2\beta_3 y_3) - \gamma_5 y_4 - \alpha_3 \left(1 - e^{-y_1 \lambda_3}\right) \right] \\[3mm]
\left\{ -(1 + \lambda_4)\left[r_1(1 - 2\beta_1 y_4) - \gamma_1 y_3 - \gamma_2 y_2 - \alpha_1 \left(1 - e^{-y_1 \lambda_1}\right) \right] \right. \\[2mm]
\left. - \lambda_2 \left[\dfrac{r_2 y_2}{\beta_2 + y_4} - \dfrac{r_2 y_2 y_4}{(\beta_2 + y_4)^2} - \gamma_3 y_2 \right] + \lambda_3 \gamma_5 y_3 \right\}
\end{bmatrix}}_{-H_y}
$$

$$
\begin{bmatrix} \lambda_1 & \lambda_2 & \lambda_3 & \lambda_4 \end{bmatrix}^{\mathsf{T}}_{t_f}
=
\begin{bmatrix} 0 & 0 & 0 & 0 \end{bmatrix}
$$

$$
H_u = 1 + \lambda_1
$$

$$
H(t_f) = 0
$$

$$
H(t-) = H(t+)
$$

$$\begin{bmatrix} q_1 \\ q_2 \end{bmatrix} = \begin{bmatrix} y_2 - ay_4 \\ y_3 - by_4 \end{bmatrix}_{t_1} = \begin{bmatrix} 0 \\ 0 \end{bmatrix}$$

$$\begin{bmatrix} \lambda_1 \\ \lambda_2 \\ \lambda_3 \\ \lambda_4 \end{bmatrix}_{t_{1-}} = \begin{bmatrix} \lambda_1 \\ \lambda_2 \\ \lambda_3 \\ \lambda_4 \end{bmatrix}_{t_{1+}} + \underbrace{\begin{bmatrix} \dfrac{\partial q_1}{\partial y_1} & \dfrac{\partial q_1}{\partial y_2} & \dfrac{\partial q_1}{\partial y_3} & \dfrac{\partial q_1}{\partial y_4} \\ \dfrac{\partial q_2}{\partial y_1} & \dfrac{\partial q_2}{\partial y_2} & \dfrac{\partial q_2}{\partial y_3} & \dfrac{\partial q_2}{\partial y_4} \end{bmatrix}}_{\mathbf{q_x}(t_1)}^{\mathsf{T}} \begin{bmatrix} \mu_1 \\ \mu_2 \end{bmatrix}$$

$$= \begin{bmatrix} \lambda_1 \\ \lambda_2 \\ \lambda_3 \\ \lambda_4 \end{bmatrix}_{t_{1+}} + \begin{bmatrix} 0 & 1 & 0 & -a \\ 0 & 0 & 1 & -b \end{bmatrix}^{\mathsf{T}} \begin{bmatrix} \mu_1 \\ \mu_2 \end{bmatrix}$$

□

6.6   Discontinuous Controls

Consider the problem described in Section 6.5 (p. 172), but without the interior point constraints. In this modified problem, if the controls happen to be discontinuous (i.e., have finite jump discontinuities) at time t_1, then the states defined by

$$\dot{\mathbf{y}} = \mathbf{g}(\mathbf{y}, \mathbf{u})$$

would have different left and right-hand side derivatives with respect to time at t_1. In other words, there would be corners at t_1 in the state versus time trajectories. See, for example, Figure 3.4 (p. 77).

Following the approach of Section 6.5, the necessary conditions for the minimum in the present problem can be easily shown to be

$$\dot{\boldsymbol{\lambda}} = -H_{\mathbf{y}} \qquad\qquad \dot{\mathbf{y}} = H_{\boldsymbol{\lambda}} \qquad\qquad H_{\mathbf{u}} = \mathbf{0}$$

$$\mathbf{q} = \mathbf{0} \qquad\qquad \boldsymbol{\lambda}(t_{1-}) = \boldsymbol{\lambda}(t_{1+}) \qquad\qquad H(t_{1-}) = H(t_{1+})$$

$$\boldsymbol{\lambda}(t_{\mathrm{f}}) = \mathbf{0} \qquad\qquad H(t_{\mathrm{f}}) = 0 \qquad\qquad \mathbf{y}(0) = \mathbf{y}_0$$

The necessary conditions at the time of discontinuity t_1 are

$$\boldsymbol{\lambda}(t_{1-}) = \boldsymbol{\lambda}(t_{1+}) \qquad \text{and} \qquad H(t_{1-}) = H(t_{1+})$$

which are also known as **Weierstrass–Erdmann corner conditions**. They imply the continuity of H and λ on the corners at the minimum.

6.7   Multiple Integral Problems

Optimal control problems involving multiple integrals are constrained by partial differential equations. A general theory similar to the Pontryagin's minimum principle is not available to handle these problems. To find the necessary conditions for the minimum in these problems, we assume that the variations of the involved integrals are weakly continuous* and find the equations that eliminate the variation of the augmented objective functional.

Example

To illustrate the above approach, let us consider the optimal control problem of determining the concentration-dependent diffusivity in a non-volatile liquid, as described in Section 1.3.4 (p. 9).

   The objective is to find the diffusivity function (gas diffusivity versus its concentration in liquid) that minimizes the error between the experimental and the calculated mass of gas absorbed in liquid, i. e.,

$$I = \int_0^{t_{\mathrm{f}}} (m_{\mathrm{c}} - m_{\mathrm{e}})^2 \, \mathrm{d}t = \int_0^{t_{\mathrm{f}}} \Bigg[\underbrace{\int_0^{L} cA \, \mathrm{d}z - m_{\mathrm{e}}}_{(m_{\mathrm{c}}-m_{\mathrm{e}})} \Bigg]^2 \, \mathrm{d}t$$

The calculated mass of gas in liquid is governed by the state equation, i. e., Equation (1.16) on p. 9, which is expressed as

$$G \equiv \underbrace{-\dot{c} + D\left(1 + \frac{c}{\rho}\right)c_{zz} + \left[\left(1 + \frac{c}{\rho}\right)\frac{\mathrm{d}D}{\mathrm{d}c} + \frac{D}{\rho}\right]c_z^2}_{g} = 0 \qquad (6.21)$$

along with the initial conditions

$$c(0,0) = c_{\mathrm{sat}}(t = 0) \qquad \text{(at the gas–liquid interface)}$$
$$c(z,0) = 0 \qquad\qquad \text{for } 0 < z \le L$$

---

* Recall from Section 4.3 (p. 88) that it means having continuous partial derivatives of the integrands — a precondition for the Lagrange Multiplier Rule.

and the boundary conditions

$$c(0,t) = c_{\text{sat}}(t) \left.\vphantom{\begin{array}{c}a\\b\end{array}}\right\} \quad \text{for } 0 < t \le t_f$$
$$c_z(L) = 0$$

The state equation, $G = 0$, constitutes a partial differential equation constraint. Applying the Lagrange Multiplier Rule, the equivalent problem is to find the control function $D(c)$ that minimizes the augmented objective functional

$$J = I + \int_0^{t_f} \int_0^L \lambda G \, dz \, dt$$

subject to the initial and boundary conditions. Since the partial differential equation is a series of constraints at each point in time t and depth z, the Lagrange multiplier λ is a function of t as well as z.

Applying the necessary condition for the minimum, $\delta J = 0$, we obtain

$$\underbrace{\delta I + \int_0^{t_f} \int_0^L (\delta\lambda G + \lambda \delta G) \, dz \, dt = 0}_{\delta J} \tag{6.22}$$

The above equation will result in a number of equations — all being necessary conditions for the minimum. To satisfy $\delta J = 0$, we will expand the variational terms and apply the same reasoning as was done in Section 3.2 (p. 58). Essentially, we will find the equations which when satisfied eliminate each additive term on the left-hand side of Equation (6.22).

The first term of the integrand in Equation (6.22) is eliminated with the satisfaction of $G = 0$, i.e., Equation (6.21).

Next, we have

$$\delta I = \int_0^{t_f} \left[2(m_c - m_e) \int_0^L \delta c A \, dz \right] dt = \int_0^{t_f} \int_0^L 2A(m_c - m_e)\delta c \, dz \, dt$$

and

$$\delta G = -\delta\dot{c} + g_c\delta c + g_{c_z}\delta c_z + g_{c_{zz}}\delta c_{zz} + g_D\delta D$$

where g is the right-hand side of the state equation, Equation (1.16) on p. 9, or the term as indicated in Equation (6.21). Note that g depends on c, c_z, c_{zz}, and D.

Substituting the above expressions for δI and δG in Equation (6.22), we obtain

$$\delta J = \int_0^{t_f} \int_0^L \left\{ \left[2A(m_c - m_e) + \lambda g_c \right] \delta c + \lambda \left[-\delta \dot{c} + g_{c_z} \delta c_z + g_{c_{zz}} \delta c_{zz} \right. \right.$$

$$\left. \left. + g_D \delta D \right] \right\} dz\, dt = 0 \qquad (6.23)$$

Next, we simplify the terms containing $\delta \dot{c}$, δc_z, and δc_{zz}. Integrating these terms by parts, we get

$$-\int_0^{t_f} \int_0^L \lambda \delta \dot{c}\, dz\, dt = -\int_0^L \int_0^{t_f} \lambda \delta \dot{c}\, dt\, dz = -\int_0^L \left\{ \left[\lambda \delta c \right]_0^{t_f} - \int_0^{t_f} \dot{\lambda} \delta c\, dt \right\} dz$$

$$\int_0^{t_f} \int_0^L \lambda g_{c_z} \delta c_z\, dz\, dt = \int_0^{t_f} \left\{ \left[\lambda g_{c_z} \delta c \right]_0^L - \int_0^L \frac{\partial}{\partial z}(\lambda g_{c_z}) \delta c\, dz \right\} dt$$

$$\int_0^{t_f} \int_0^L \lambda g_{c_{zz}} \delta c_{zz}\, dz\, dt = \int_0^{t_f} \left\{ \left[\lambda g_{c_{zz}} \delta c_z - \frac{\partial}{\partial z}(\lambda g_{c_{zz}} \delta c) \right]_0^L \right.$$

$$\left. + \int_0^L \frac{\partial^2}{\partial z^2}(\lambda g_{c_{zz}}) \delta c\, dz \right\} dt$$

Note that we have applied integration by parts twice to obtain the last equation. The last three equations when substituted into Equation (6.23) yield

$$\delta J = \underbrace{\int_0^{t_f} \int_0^L \left[\dot{\lambda} + 2A(m_c - m_e) + \lambda g_c - \frac{\partial}{\partial z}(\lambda g_{c_z}) + \frac{\partial^2}{\partial z^2}(\lambda g_{c_{zz}}) \right] \delta c\, dz\, dt}_{\delta J_1} -$$

$$\underbrace{\int_0^L \left[\lambda \delta c \right]_0^{t_f} dz}_{\delta J_2} - \underbrace{\int_0^{t_f} \left[\lambda g_{c_z} - \frac{\partial}{\partial z}(\lambda g_{c_{zz}}) \right]_{z=0} \delta c(0, t)\, dt}_{\delta J_3} +$$

$$\int_0^{t_f} \left[\lambda g_{c_z} - \frac{\partial}{\partial z}(\lambda g_{c_{zz}}) \right]_{z=L} \delta c(L,t)\, dt + \underbrace{\int_0^{t_f} \left[\lambda g_{c_{zz}} \delta c_z \right]_0^L dt}_{\delta J_5} +$$

$$\underbrace{\int_0^{t_f}\int_0^L \lambda g_D \delta D\, dz\, dt}_{\delta J_6} = 0$$

We will now find the equations that eliminate the terms δJ_1 to J_6, as indicated in the above equation, so that the necessary condition $\delta J = 0$ is satisfied for the minimum.

Elimination of δJ_1

This term is eliminated by defining

$$\dot{\lambda} = -2A(m_c - m_e) - \lambda g_c + \frac{\partial}{\partial z}(\lambda g_{c_z}) - \frac{\partial^2}{\partial z^2}(\lambda g_{c_{zz}})$$

After substituting the expressions for the derivatives g_c, g_{c_z}, and $g_{c_{zz}}$ in the above equation, we obtain the costate equation

$$\dot{\lambda} = -2A(m_c - m_e) - \frac{\lambda}{\rho}D_c c_z^2 - \left(1 + \frac{c}{\rho}\right)\left[\lambda D_c c_{zz} + \lambda D_{cc}c_z^2 - D\lambda_{zz}\right] \quad (6.24)$$

Elimination of δJ_2

Because the initial concentration of the gas in the liquid is known at the interface and is zero elsewhere, δc is zero for all z at $t = 0$. However, since the final gas concentration is not specified, we enforce

$$\lambda(z, t_f) = 0, \quad 0 \le z \le L \quad (6.25)$$

to eliminate δJ_2. The above equation is the final condition for Equation (6.24).

Elimination of δJ_3

Now $c(0,t)$ is specified to be the equilibrium concentration $c_{sat}(t)$ at the interface throughout the interval $[0, t_f]$. Thus, $\delta c(0, t)$ is always zero so that δJ_3 is zero.

Elimination of δJ_4

Since $c(L, t)$ is not specified, $\delta c(L, t)$ is not necessarily zero. Therefore, we eliminate δJ_4 by specifying

$$\lambda(L, t) = 0, \quad 0 \le t \le t_{\mathrm{f}} \tag{6.26}$$

which is the first boundary condition for Equation (6.24).

Elimination of δJ_5

The above equation along with the enforcement

$$\lambda(0, t) = 0, \quad 0 \le t \le t_{\mathrm{f}} \tag{6.27}$$

eliminates δJ_5. This specification is the second boundary condition for Equation (6.24).

Elimination of δJ_6

Elimination of this term requires that the coefficient of δD in the integrand be zero, i. e.,

$$\lambda g_D = \lambda \left[\left(1 + \frac{c}{\rho} \right) c_{zz} + \frac{1}{\rho} c_z^2 \right] = 0, \quad 0 \le z \le L, \quad 0 \le t \le t_{\mathrm{f}} \tag{6.28}$$

The coefficient λg_D is the variational derivative of J with respect to D.

To summarize, the necessary conditions for the minimum are

1. The state equation, Equation (6.21), and its initial and boundary conditions

2. The costate equation, Equation (6.24), and its final and boundary conditions

3. Equation (6.28), which is the stationarity condition.

Bibliography

K.J. Arrow, L. Hurwicz, and H. Uzawa. Constraint qualifications in nonlinear programming. *Nav. ResLogist.*, 8(2), 1961.

A.E. Bryson Jr. and Y.-C. Ho. *Applied Optimal Control*, Chapters 2 and 3. Taylor & Francis, Pennsylvania, 1975.

D.E. Kirk. *Optimal Control Theory — An Introduction*, Chapter 5. Dover Publications Inc., New York, 2004.

N.G. Long and D. Léonard. *Optimal Control Theory and Static Optimization in Economics*, Chapters 6 and 7. Cambridge University Press, Cambridge, UK, 1992.

A. Takayama. *Mathematical Economics*, Chapter 1. Cambridge University Press, Cambridge, UK, 1985.

Exercises

6.1 Utilize the results of Section 6.1.1 to find the necessary conditions for the minimum of

$$I = F_0[\mathbf{y}(t_f)] + \int_0^{t_f} F_1[\mathbf{y}(t), \mathbf{u}(t)]\, dt$$

subject to

$$\dot{\mathbf{y}} = \mathbf{g}[\mathbf{y}(t), \mathbf{u}(t)], \quad \mathbf{y}(0) = \mathbf{y}_0$$

6.2 Show that for L defined in Equation (6.16) on p. 163,

$$\frac{dL}{dt} = \frac{\partial L}{\partial t}$$

at the minimum.

6.3 For the batch reactor of Example 6.3 (p. 159), find the necessary conditions to maximize the final mole fraction of the intermediate product, $y_2(t_f)$, with the specification $y_3(t_f) = y_{3,f}$.

6.4 Find the necessary condition for the minimum in Example 6.7 (p. 164) with additional constraints

$$u_1 \geq u_{1,\min} \quad \text{and} \quad u_2 \leq u_{2,\max}$$

throughout the time interval $[0, t_f]$.

6.5 Repeat Exercise 6.3 in presence of the following selectivity constraints:

$$y_2 = b_1 y_1 \quad \text{and} \quad y_2 = b_2 y_3$$

instead of $y_3(t_f) = y_{3,f}$.

6.6 Find the necessary conditions for the minimum in the heat exchanger problem described in Section 1.3.3 (p. 8).

6.7 Find the necessary conditions for maximum oil production in the Vapex problem described in Section 1.3.7 (p. 13).

Chapter 7

Numerical Solution of Optimal Control Problems

The solution of an optimal control problem requires the satisfaction of differential equations subject to initial as well as final conditions. Except when the equations are linear and the objective functional is simple enough, an analytical solution is impossible. This is the reality of most of the problems for which optimal controls can only be determined using numerical methods.

In this chapter, we introduce the gradient and penalty function methods, which are quite effective in solving a wide range of optimal control problems. Given initial guesses, these methods help in determining the local optimum (i.e., minimum or maximum) of the objective functional of a problem. We cannot discount the possibility of having a number of local optima in an optimal control problem. To strengthen the globality of the optimum, we need to apply these methods with several initial guesses and compare the resulting optima. Note that finding the maximum in a problem is equivalent to finding the minimum of the negative of the objective functional.

7.1 Gradient Method

We introduce this method to determine the minimum in the optimal control problem of Section 6.1.1 (p. 153) in which both final time and final state are free.

7.1.1 Free Final Time and Free Final State

The objective of the problem is to find the vector of control functions $\mathbf{u}(t)$ and the final time t_f that minimize

$$I = \int\limits_{0}^{t_\mathrm{f}} F[\mathbf{y}(t), \mathbf{u}(t)]\,\mathrm{d}t$$

subject to

$$\dot{y} = g[y(t), u(t)], \quad y(0) = y_0$$

The above problem is equivalent to minimizing the augmented functional

$$J = \int_0^{t_f} \left[F + \lambda^{\top}(-\dot{y} + g) \right] dt \tag{6.4}$$

subject to $y(0) = y_0$. From Section 6.1.1, the necessary conditions for the minimum of I are

$$\underbrace{\dot{y} = g, \quad y(0) = y_0,}_{\substack{\text{state equations and}\\\text{initial conditions}}} \quad \underbrace{\dot{\lambda} = -H_y, \quad \lambda(t_f) = 0,}_{\substack{\text{costate equations and}\\\text{final conditions}}} \quad H_u = 0, \quad \text{and} \quad H(t_f) = 0$$

In the gradient method, the state and costate equations are solved using initial guesses for the control u and the final time t_f. The guessed u and t_f are then improved using, respectively,

1. H_u, the variational derivative of J with respect to u at each time in the interval $[0, t_f]$, and

2. $H(t_f)$, the partial derivative of J with respect to t_f.

The above partial derivatives are the components of the gradient of J. The state and costate equations are solved again using the improved u and t_f, which are further improved using the resulting gradient of J. This procedure, when repeated, leads to the satisfaction of the remaining necessary conditions, $H_u = 0$ and $H(t_f) = 0$, and thus to the minimum. The step-wise iterative procedure of the gradient method is as follows.

7.1.2   Iterative Procedure

Guess control functions and the final time in the beginning, and carry out the following steps:

1. Integrate the state equations forward to the final time using the initial conditions at $t = 0$ and the control functions.

2. Calculate the objective functional using the control functions, the final time, and the corresponding state as determined by the state equations.

3. Integrate the costate equations backward to $t = 0$ using the final conditions at the final time, the control functions, and the state determined in the previous step.

4. Improve the control functions and the final time using the gradient information and repeat Step 1 onward until there is no further reduction in the objective functional.

The last step employs the improvement strategy that is described next.

7.1.3 Improvement Strategy

With the state and costate equations satisfied at the end of Step 3 above, the variation of the augmented objective functional becomes [see Equation (6.9), p. 156]

$$\delta J = \int\limits_{0}^{t_{\mathrm{f}}} H_{\mathbf{u}}^{\top}\,\delta\mathbf{u}\,\mathrm{d}t + H(t_{\mathrm{f}})\delta t_{\mathrm{f}} \tag{7.1}$$

In the above equation, $H_{\mathbf{u}}$ or the coefficient of $\delta\mathbf{u}$ is the variational derivative of J with respect to $\mathbf{u}$ at a given time in the interval $[0, t_{\mathrm{f}}]$. Similarly, $H(t_{\mathrm{f}})$ or the coefficient of δt_{f} is the partial derivative of J with respect to the final time t_{f}. The improvement strategy is to achieve maximum reduction in J by changing $\mathbf{u}(t)$ and t_{f}, which are not necessarily optimal. The straightforward way is to change $\mathbf{u}(t)$ and t_{f}, respectively, by

$$\delta\mathbf{u}(t) = -\epsilon H_{\mathbf{u}}(t) \quad \text{and} \quad \delta t_{\mathrm{f}} = -\epsilon H(t_{\mathrm{f}}) \tag{7.2}$$

where ϵ is some positive real number. With these changes, the variation in J becomes

$$\delta J = -\epsilon\left[\int\limits_{0}^{t_{\mathrm{f}}} H_{\mathbf{u}}^{\top} H_{\mathbf{u}}(t)\,\mathrm{d}t + H^{2}(t_{\mathrm{f}})\right]$$

For sufficiently small ϵ, the changes $\delta\mathbf{u}$, and δt_{f} are small enough so that the change in J is given by

$$\underbrace{J(\mathbf{u} + \delta\mathbf{u}, t_{\mathrm{f}} + \delta t_{\mathrm{f}})}_{J_{\text{next}}} - J(\mathbf{u}, t_{\mathrm{f}}) \;=\; \delta J$$

The above change in J is the most negative for some optimal ϵ. Hence, based on this ϵ, the most improved (i. e., reduced) functional value J_{next} results from

$$\mathbf{u}_{\text{next}}(t) = \mathbf{u}(t) + \delta\mathbf{u}(t) \quad \text{and} \quad t_{\mathrm{f,next}} = t_{\mathrm{f}} + \delta t_{\mathrm{f}} \tag{7.3}$$

and the corresponding state $\mathbf{y}_{\text{next}}$. Thus, Equations (7.2) and (7.3) form the strategy to improve $\mathbf{u}$ and t_{f}. Since $\mathbf{u}$ and t_{f} (as well as $\mathbf{u}_{\text{next}}$ and $t_{\mathrm{f,next}}$) along with the corresponding states satisfy the state equations,

$$J = I \quad \text{and} \quad J_{\text{next}} = I_{\text{next}}$$

from the definition of J in Equation (6.4) on p. 154.

7.1.3.1 Numerical Implementation

To implement the improvement strategy numerically, we split the time interval into N subintervals of equal length and use numerical integration [e. g., composite Simpson's 1/3 Rule given in Section 9.12 (p. 275) for even number of subintervals] to calculate δJ as well as J. Thus, J or I is rendered into a function dependent on the vector of optimization parameters

$$\mathbf{p} \equiv \begin{bmatrix} \mathbf{u}(t_0) & \mathbf{u}(t_1) & \cdots & \mathbf{u}(t_N) & t_f \end{bmatrix}^\top \tag{7.4}$$

where $t_0, t_1, \ldots, t_N$ form the time-grid of $(N+1)$ equispaced grid points in the time interval $[0, t_f]$.

However, a complication arises with this approach. An improvement in t_f changes the time grid, thereby requiring the estimation of controls and states on the new time grid for the next round of improvements. We avoid this situation by linearly transforming the independent variable t in the variable interval $[0, t_f]$ to a new independent variable σ in the *fixed* interval $[0, 1]$.

Transformation of the Independent Variable

Let the new independent variable σ be given by the linear relation

$$\sigma = at + b$$

where a and b are some unknown constants. Substituting, respectively, the initial and final values of σ and t in the above equation, we obtain

$$0 = b \quad \text{and} \quad 1 = at_f + b \quad \implies \quad a = \frac{1}{t_f}$$

Hence

$$\sigma = \frac{t}{t_f}, \qquad \frac{\mathrm{d}\sigma}{\mathrm{d}t} = \frac{1}{t_f}, \qquad \text{and} \qquad \frac{\mathrm{d}y_i}{\mathrm{d}t} = \frac{\mathrm{d}y_i}{\mathrm{d}\sigma}\frac{\mathrm{d}\sigma}{\mathrm{d}t} = \frac{\mathrm{d}y_i}{\mathrm{d}\sigma}\frac{1}{t_f}$$

Based on the above relations, the objective of the optimal control problem is to find the control $\mathbf{u}$ and the final time t_f that minimize

$$I = \int_0^{t_f} F[\mathbf{y}(t), \mathbf{u}(t)]\frac{\mathrm{d}\sigma}{\mathrm{d}\sigma}\,\mathrm{d}t = \int_0^1 t_f F[\mathbf{y}(\sigma), \mathbf{u}(\sigma)]\,\mathrm{d}\sigma$$

subject to

$$\frac{\mathrm{d}\mathbf{y}}{\mathrm{d}\sigma} = t_f \mathbf{g}[\mathbf{y}(\sigma), \mathbf{u}(\sigma)], \quad \mathbf{y}(0) = \mathbf{y}_0$$

The Hamiltonian is then given by

$$H = t_f(F + \boldsymbol{\lambda}^\top \mathbf{g})$$

Necessary Conditions for the Minimum

The above problem is equivalent to minimizing the augmented functional

$$J = \int_0^1 \left[t_f F + \boldsymbol{\lambda}^\top \left(-\frac{\mathrm{d}\mathbf{y}}{\mathrm{d}\sigma} + t_f \mathbf{g} \right) \right] \mathrm{d}\sigma = \int_0^1 \left(H - \boldsymbol{\lambda}^\top \frac{\mathrm{d}\mathbf{y}}{\mathrm{d}\sigma} \right) \mathrm{d}\sigma$$

subject to $\mathbf{y}(0) = \mathbf{y}_0$. The variation δJ is given by [compare with Equation (6.14) on p. 161]

$$\delta J = \int_0^1 \left[(H_\mathbf{y} + \frac{d\boldsymbol{\lambda}}{d\sigma})^\top \delta\mathbf{y} + (H_\boldsymbol{\lambda} - \frac{d\mathbf{y}}{d\sigma})^\top \delta\boldsymbol{\lambda} + H_\mathbf{u}^\top \delta\mathbf{u} \right] d\sigma - \boldsymbol{\lambda}^\top(1)\,\delta\mathbf{y}(1)$$

$$+ \underbrace{\int_0^1 \frac{H}{t_f}\,d\sigma\,\delta t_f}_{J_{t_f}} \tag{7.5}$$

where J_{t_f}, as indicated above, is the partial derivative of J with respect to the final time t_f. Since $\delta J = 0$ at the minimum, the necessary conditions for the minimum are

$$\underbrace{\frac{d\mathbf{y}}{d\sigma} = t_f\mathbf{g}, \quad \mathbf{y}(0) = \mathbf{y}_0,}_{\substack{\text{state equations and} \\ \text{initial conditions}}} \qquad \underbrace{\frac{d\boldsymbol{\lambda}}{d\sigma} = -H_\mathbf{y}, \quad \boldsymbol{\lambda}(1) = \mathbf{0}}_{\substack{\text{costate equations and} \\ \text{final conditions}}}$$

$$H_\mathbf{u} = \mathbf{0}, \qquad J_{t_f} = \int_0^1 \frac{H}{t_f}\,d\sigma = 0$$

Improvements in u and t_f to Reduce J or I

These improvements are done in Step 4 of Section 7.1.2 (p. 186) where the state and costate equations are satisfied so that

$$J = I = \int_0^1 t_f F\,d\sigma \qquad \text{and} \qquad \delta J = \int_0^1 H_\mathbf{u}^\top \delta\mathbf{u}\,d\sigma + \underbrace{\int_0^1 \frac{H}{t_f}\,d\sigma\,\delta t_f}_{J_{t_f}}$$

First, we compute the integrals in the above equation numerically over the fixed σ-interval $[0, 1]$. The interval is split into N subintervals of equal length using $(N + 1)$ equi-spaced grid points

$$\sigma_0 = 0, \qquad \sigma_1, \qquad \sigma_2, \qquad \ldots, \qquad \sigma_N = 1$$

For example, using composite Simpson's 1/3 Rule (Section 9.12, p. 275) for an even N,

$$\delta J = \frac{1}{3N}\left[H_{\mathbf{u}}^{\top}(\sigma_0)\delta\mathbf{u}(\sigma_0) + 4\sum_{i=1,3,5,\ldots}^{N} H_{\mathbf{u}}^{\top}(\sigma_i)\delta\mathbf{u}(\sigma_i) + 2\sum_{i=2,4,6,\ldots}^{N} H_{\mathbf{u}}^{\top}(\sigma_i)\delta\mathbf{u}(\sigma_i)+ \right.$$

$$\left. H_{\mathbf{u}}^{\top}(\sigma_N)\delta\mathbf{u}(\sigma_N) \right] +$$

$$\underbrace{\frac{1}{3Nt_{\mathrm{f}}}\left[H(\sigma_0) + 4\sum_{i=1,3,5,\ldots}^{N} H(\sigma_i) + 2\sum_{i=2,4,6,\ldots}^{N} H(\sigma_i) + H(\sigma_N) \right]\delta t_{\mathrm{f}}}_{J_{t_{\mathrm{f}}}}$$

where the vector

$$\nabla J \equiv \begin{bmatrix} H_{\mathbf{u}}(\sigma_0) & H_{\mathbf{u}}(\sigma_1) & \cdots & H_{\mathbf{u}}(\sigma_N) & J_{t_{\mathrm{f}}} \end{bmatrix}^{\top}$$

is the gradient of J.

Next, we consider the vector of optimization parameters, $\mathbf{p}$, defined in Equation (7.4). Let a change in $\mathbf{p}$ to reduce J be given by

$$\delta\mathbf{p} \equiv \begin{bmatrix} \delta\mathbf{u}(\sigma_0) & \delta\mathbf{u}(\sigma_1) & \cdots & \delta\mathbf{u}(\sigma_N) & \delta t_{\mathrm{f}} \end{bmatrix}^{\top}$$

According to Appendix 7.A (p. 229), the reduction in J per unit size of $\delta\mathbf{p}$ is maximum along the direction of $-\nabla J$, i.e., opposite to that of the gradient ∇J. This direction is known as the direction of the **steepest descent**. It can be easily seen that Equations (7.2) and (7.3) already utilize the steepest descent direction to improve $\mathbf{u}$ as a continuous function of t, and t_{f}.

Now a change of magnitude ϵ_0 in $\mathbf{p}$ along the steepest descent direction is given by

$$\delta\mathbf{p} = \epsilon_0\underbrace{\left(-\nabla J/\|\nabla J\| \right)}_{\substack{\text{unit vector} \\ \text{along } -\nabla J}} \equiv -\epsilon\nabla J$$

where $\epsilon \equiv \epsilon_0/\|\nabla J\|$. For some optimal ϵ, the improved functional value J_{next} obtained from

$$\mathbf{p}_{\mathrm{next}} = \mathbf{p} + \delta\mathbf{p}$$

and the corresponding state $\mathbf{y}_{\mathrm{next}}$ is expected to be less than J as much as possible. Note that

$$J = I \quad \text{and} \quad J_{\mathrm{next}} = I_{\mathrm{next}}$$

because the controls and the corresponding states satisfy the state equations. The improvement in $\mathbf{p}_{\mathrm{next}}$ translates to the improvements

$$\mathbf{u}_{\mathrm{next}}(\sigma_i) = \mathbf{u}(\sigma_i) - \epsilon H_{\mathbf{u}}(\sigma_i); \quad i = 0, 1, \ldots, N; \quad \text{and}$$

$$t_{\mathrm{f,next}} = t_{\mathrm{f}} - \epsilon J_{t_{\mathrm{f}}}$$

which reduce J.

Subsequent reduction of J is achieved as follows. Based on the improved controls and final time (namely, $\mathbf{u}_{next}$ and $t_{f,next}$), Steps 1–3 of Section 7.1.2 (p. 186) are repeated, and ∇J is recalculated and utilized to repeat the improvements and reduce J. This iterative procedure continued until the reduction in J becomes insignificant or the norm of ∇J becomes negligible. This minimization procedure is known as the **gradient algorithm**. It affords a simple and effective way to solve a wide range of optimal control problems.

Remark on Gradient Improvement

A number of methods have been developed to improve the steepest descent so that the minimum in a problem is attained in a least possible number of iterations. Since the gradient of a function is zero at the minimum, one approach is to use the quadratically convergent Newton–Raphson method (see Section 9.11, p. 273) to zero out the gradient. However, the method needs the Jacobian of ∇J, which is also known as the Hessian of the J. The **Broyden–Fletcher–Goldfarb–Shanno (BFGS) method** (see Press et al., 2007) ingeniously utilizes the gradient information (i. e., $\delta \mathbf{p}$) to construct the Hessian and provides the optimal improvement $\mathbf{p}_{next}$ to bring the maximum reduction in J. The BFGS method is known to be one of the most efficient and stable optimization methods. The numerical examples in this book are solved using this method.

Number of Grid Points

Note that the above numerical implementation provides a solution of an optimal control problem as discrete values of optimal controls and states on the grid of the independent variable. The solution becomes more accurate as we increase the number of grid points, i. e., N. In the limit of N tending to infinity, the solution has optimal controls and states as continuous functions of the independent variable.

In practice, we solve an optimal control problem with increasing values of N up to a limit beyond which either there is no significant difference in the solution or the computations become very intensive. Note that larger the N the larger is number of discrete control values to be optimized and the harder it is to solve the problem.

7.1.4   Algorithm for the Gradient Method

Based on the above development, the algorithm is as follows:

1. Transform the optimal control problem from the variable time interval $[0, t_f]$ to the fixed σ-interval, $[0, 1]$

2. Assume a value for t_f and N. Discretize the σ-interval $[0, 1]$ using N

equi-spaced grid points. Assume a value of the control function **u** at each grid point.

3. Integrate state equations forward using the initial conditions and the control function values. Save the values of state variables at each grid point.

4. Evaluate the objective functional using the values of control functions and state variables. Save the objective functional value.

5. Integrate costate equations backward using the final conditions, the control function values, and the saved values of the state variables. Save the values of costate variables at each grid point.

6. Improve **u** and t_f using the gradient of the objective functional as follows. Change **u** by $-\epsilon H_\mathbf{u}$ at each discrete point and t_f by $-\epsilon J_{t_f}$ where ϵ is a positive real number, which causes maximum reduction in the objective functional.

7. Repeat computations Step 3 onward until there is no further reduction in the objective functional or the norm of the gradient becomes negligible. When either event happens, the **u**, t_f, and corresponding state variables are optimal.

The above algorithm is explained in more detail with the help of the following example:

Example 7.1
Consider the reaction $A + B \longrightarrow C$ carried out in a fed-batch reactor. The volume y_1 of the reaction mixture, and the concentrations y_2, y_3, and y_4, respectively, of A, B, and C are governed by

$$\frac{dy_1}{dt} = u_1, \qquad\qquad\qquad y_1(0) = y_{1,0}$$

$$\frac{dy_2}{dt} = \frac{u_1(y_0 - y_2)}{y_1} - k_0 e^{a/u_2} y_2 y_3, \quad y_2(0) = y_{2,0}$$

$$\frac{dy_3}{dt} = -\frac{u_1 y_3}{y_1} - k_0 e^{a/u_2} y_2 y_3, \qquad y_3(0) = y_{3,0}$$

$$\frac{dy_4}{dt} = -\frac{u_1 y_4}{y_1} + k_0 e^{a/u_2} y_2 y_3, \qquad y_4(0) = y_{4,0}$$

where the control

- u_1 is the time dependent volumetric flow rate (cm^3/min) of the reactor feed with y_0 (g/cm^3) concentration of A, and

- u_2 (K) is the reactor temperature as a function of time.

It is desired to find the optimal controls $u_1(t)$ and $u_2(t)$ that maximize $y_4(t_f)$, i. e., the final concentration of C. Thus, the objective is to minimize

$$I = -y_4(t_f) = -\int_0^{y_4(t_f)} dy_4 = -\int_0^{t_f} \frac{dy_4}{dt} dt$$

$$= \int_0^{t_f} \left(\frac{u_1 y_4}{y_1} - k_0 e^{a/u_2} y_2 y_3 \right) dt$$

subject to the state equations.

Transformation to Fixed σ-Interval

The equivalent problem in the fixed σ-interval $[0, 1]$ is to minimize

$$I = \int_0^1 t_f \underbrace{\left(\frac{u_1 y_4}{y_1} - \kappa y_2 y_3 \right)}_{F} d\sigma$$

where we have introduced

$$\kappa \equiv k_0 e^{a/u_2}$$

The state equations are

$$\frac{dy_1}{d\sigma} = t_f u_1, \qquad\qquad\qquad y_1(0) = y_{1,0}$$

$$\frac{dy_2}{d\sigma} = t_f \left[\frac{u_1(y_0 - y_2)}{y_1} - \kappa y_2 y_3 \right], \quad y_2(0) = y_{2,0}$$

$$\frac{dy_3}{d\sigma} = t_f \left(-\frac{u_1 y_3}{y_1} - \kappa y_2 y_3 \right), \qquad y_3(0) = y_{3,0}$$

$$\frac{dy_4}{d\sigma} = t_f \left(-\frac{u_1 y_4}{y_1} + \kappa y_2 y_3 \right), \qquad y_4(0) = y_{4,0}$$

The Hamiltonian is then given by

$$H = t_f \left\{ \frac{u_1 y_4}{y_1} - \kappa y_2 y_3 + \lambda_1 u_1 + \lambda_2 \left[\frac{u_1(y_0 - y_2)}{y_1} - \kappa y_2 y_3 \right] \right.$$

$$\left. + \lambda_3 \left[-\frac{u_1 y_3}{y_1} - \kappa y_2 y_3 \right] + \lambda_4 \left[-\frac{u_1 y_4}{y_1} + \kappa y_2 y_3 \right] \right\}$$

Necessary Conditions for the Minimum

The necessary conditions for the minimum of I are

$$
\begin{bmatrix} \dfrac{dy_1}{d\sigma} \\[2ex] \dfrac{dy_2}{d\sigma} \\[2ex] \dfrac{dy_3}{d\sigma} \\[2ex] \dfrac{dy_4}{d\sigma} \end{bmatrix} = t_{\mathrm{f}} \underbrace{\begin{bmatrix} u_1 \\[2ex] \dfrac{u_1(y_0 - y_2)}{y_1} - \kappa y_2 y_3 \\[2ex] -\dfrac{u_1 y_3}{y_1} - \kappa y_2 y_3 \\[2ex] -\dfrac{u_1 y_4}{y_1} + \kappa y_2 y_3 \end{bmatrix}}_{H_\lambda}, \qquad \begin{bmatrix} y_1 \\[2ex] y_2 \\[2ex] y_3 \\[2ex] y_4 \end{bmatrix}_{\sigma=0} = \begin{bmatrix} y_{1,0} \\[2ex] y_{2,0} \\[2ex] y_{3,0} \\[2ex] y_{4,0} \end{bmatrix}
$$

$$
\begin{bmatrix} \dfrac{d\lambda_1}{d\sigma} \\[2ex] \dfrac{d\lambda_2}{d\sigma} \\[2ex] \dfrac{d\lambda_3}{d\sigma} \\[2ex] \dfrac{d\lambda_4}{d\sigma} \end{bmatrix} = t_{\mathrm{f}} \underbrace{\begin{bmatrix} \dfrac{(1 - \lambda_4)u_1 y_4}{y_1^2} + \dfrac{\lambda_2 u_1(y_0 - y_2)}{y_1^2} - \dfrac{\lambda_3 u_1 y_3}{y_1^2} \\[2ex] (1 - \lambda_4)\kappa y_3 + \lambda_2\left(\dfrac{u_1}{y_1} + \kappa y_3\right) + \lambda_3 \kappa y_3 \\[2ex] (1 - \lambda_4)\kappa y_2 + \lambda_2 \kappa y_2 + \lambda_3\left(\dfrac{u_1}{y_1} + \kappa y_2\right) \\[2ex] -(1 - \lambda_4)\dfrac{u_1}{y_1} \end{bmatrix}}_{-H_y},
$$

$$
\begin{bmatrix} \lambda_1 & \lambda_2 & \lambda_3 & \lambda_4 \end{bmatrix}_{\sigma=1}^{\mathsf{T}} = \begin{bmatrix} 0 & 0 & 0 & 0 \end{bmatrix}^{\mathsf{T}}
$$

$$
\begin{bmatrix} H_{u_1} \\[2ex] H_{u_2} \end{bmatrix} = t_{\mathrm{f}} \begin{bmatrix} \dfrac{(1 - \lambda_4)y_4}{y_1} + \lambda_1 + \dfrac{\lambda_2(y_0 - y_2)}{y_1} - \dfrac{\lambda_3 y_3}{y_1} \\[2ex] (1 + \lambda_2 + \lambda_3 - \lambda_4)\dfrac{a\kappa y_2 y_3}{u_2^2} \end{bmatrix} = \begin{bmatrix} 0 \\[2ex] 0 \end{bmatrix}
$$

$$
J_{t_{\mathrm{f}}} = \int_0^1 \frac{H}{t_{\mathrm{f}}} \, d\sigma = 0
$$

The gradient algorithm to solve this problem is as follows.

Gradient Algorithm

1. Set the iteration counter $k = 0$. Assume t_f^k and an even N. Obtain the fixed σ-grid of $(N+1)$ equi-spaced grid points

$$\sigma_0(=0), \quad \sigma_1, \quad \sigma_2, \quad \ldots, \quad \sigma_{N-1}, \quad \sigma_N(=1)$$

We use k in a superscript to denote the iteration number.

At each grid point, assume control function values as follows:

$$\mathbf{u}_i^k \equiv \mathbf{u}^k(\sigma_i) = \begin{bmatrix} u_1^k(\sigma_i) \\ u_2^k(\sigma_i) \end{bmatrix}, \quad i = 0, 1, \ldots, N$$

2. Integrate state equations forward from $\sigma = 0$ to 1 using the initial conditions and the control function values $\mathbf{u}_i^k$, $i = 0, 1, \ldots, N$. Save the values of state variables at the grid points as

$$\mathbf{y}_i^k \equiv \mathbf{y}^k(\sigma_i) = \begin{bmatrix} y_1^k(\sigma_i) \\ y_2^k(\sigma_i) \\ y_3^k(\sigma_i) \\ y_4^k(\sigma_i) \end{bmatrix}, \quad i = 0, 1, \ldots, N$$

3. Evaluate the objective functional using the controls $\mathbf{u}_i^k$s and the state variables $\mathbf{y}_i^k$s. Using composite Simpson's 1/3 Rule, the objective functional value is given by

$$J^k = I^k = \frac{1}{3N}\left(A_0 + 4\sum_{\substack{N \\ 1,3,5,\ldots}} A_i + 2\sum_{\substack{N \\ i=2,4,6,\ldots}} A_i + A_N\right)$$

where

$$A_i \equiv t_f F(\mathbf{y}_i^k, \mathbf{u}_i^k) = t_f^k\left[\frac{u_1^k y_4^k}{y_1^k} - k_0 \exp\left(\frac{a}{u_2^k}\right) y_2^k y_3^k\right], \quad i = 0, 1, \ldots, N$$

4. Check the improvement in I for $k > 0$. Given a tolerable error $\varepsilon_1 > 0$, if

$$\left| I^k - I^{k-1} \right| < \varepsilon_1$$

then go to Step 10.

5. Integrate costate equations backward from $\sigma = 1$ to 0 using the final conditions, the controls $\mathbf{u}_i^k$s, and the state variables $\mathbf{y}_i^k$s. Save the values

of costate variables at the grid points as

$$\boldsymbol{\lambda}_i^k \equiv \boldsymbol{\lambda}^k(\sigma_i) = \begin{bmatrix} \lambda_1^k(\sigma_i) \\ \lambda_2^k(\sigma_i) \\ \lambda_3^k(\sigma_i) \\ \lambda_4^k(\sigma_i) \end{bmatrix}, \quad i = 0, 1, \ldots, N$$

6. Evaluate the gradient of J^k by calculating the partial derivatives

$$H_{\mathbf{u},i}^k \equiv H_{\mathbf{u}}^k(\sigma_i) = \begin{bmatrix} H_{u_1}(\mathbf{y}_i^k, \mathbf{u}_i^k, \boldsymbol{\lambda}_i^k) \\ H_{u_2}(\mathbf{y}_i^k, \mathbf{u}_i^k, \boldsymbol{\lambda}_i^k) \end{bmatrix}; \quad i = 0, 1, \ldots, N$$

and $\quad J_{t_f}^k = \dfrac{1}{3Nt_f}\left(H_0^k + 4\displaystyle\sum_{1,3,5,\ldots}^{N} H_i^k + 2\displaystyle\sum_{i=2,4,6,\ldots}^{N} H_i^k + H_N^k \right)$

where $\quad H_i^k \equiv t_f\left[F(\mathbf{y}_i^k, \mathbf{u}_i^k, \boldsymbol{\lambda}_i^k) + \boldsymbol{\lambda}_i^{\top} \mathbf{g}(\mathbf{y}_i^k, \mathbf{u}_i^k, \boldsymbol{\lambda}_i^k) \right]; \quad i = 0, 1, \ldots, N$

Check the magnitude of the gradient. Given a small positive real number ε_2, if the norm of the gradient

$$\sqrt{\sum_{i=0}^{N}\sum_{j=1}^{2}\left[H_{u_j}(\mathbf{y}_i^k, \mathbf{u}_i^k, \boldsymbol{\lambda}_i^k) \right]^2 + \left[J_{t_f}^k \right]^2} \; < \; \varepsilon_2$$

then go to Step 10.

7. Improve control functions by calculating

$$\mathbf{u}_i^{k+1} = \mathbf{u}_i^k - \epsilon H_{\mathbf{u},i}^k, \quad i = 0, 1, \ldots, N$$

where ϵ is a positive real number causing maximum reduction in I^k. Assign

$$\mathbf{u}_i^{k+1} \rightarrow \mathbf{u}_i^k; \quad i = 0, 1, \ldots, N$$

8. Improve the final time using

$$t_f^{k+1} = t_f^k - \epsilon J_{t_f}^k$$

and assign

$$t_f^{k+1} \rightarrow t_f^k$$

9. Increment k by one and repeat calculations Step 2 onward.

10. Terminate the algorithm with the following result:

- The optimal objective functional value is I^k.
- The optimal control $\hat{\mathbf{u}}(t)$ is represented by $\mathbf{u}_i^k$, $i = 0, 1, \ldots, N$.
- The optimal final time is t_{f}^k.
- The optimal state $\hat{\mathbf{y}}(t)$ is represented by $\mathbf{y}_i^k$, $i = 0, 1, \ldots, N$.

Results

Table 7.1 lists the parameters used to solve the optimal control problem with the above algorithm. In this problem, the final time and final states are free. It is desired to find two controls, $u_1(t)$ and $u_2(t)$, and the final time t_{f} that maximize the final product concentration $y_4(t_{\mathrm{f}})$.

Table 7.1 Parameters for the problem of Example 7.1

$y_{1,0}$	10^3 cm^3	$y_{4,0}$	0 g/cm^3	k_0	6×10^7 cm^3/(g · min)
$y_{2,0}$	0 g/cm^3	y_0	10 g/cm^3	N	100
$y_{3,0}$	40 g/cm^3	a	-8420 K		

The initial and optimal states are plotted in Figures 7.1 and 7.2, respectively. The corresponding controls are shown in Figure 7.3.

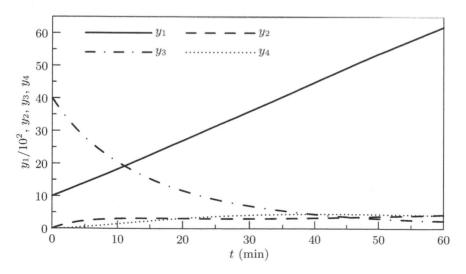

Figure 7.1 The initial states versus time for Example 7.1

With the initially guessed final time of 60 min, the initial controls provided $I = -4.12$, which corresponds to the final product concentration of 4.12 g/cm³.

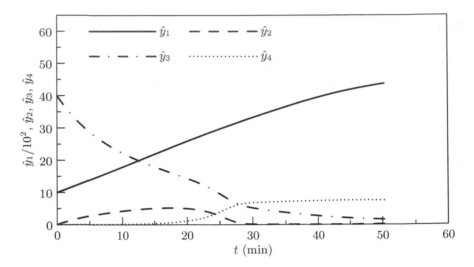

Figure 7.2 The optimal states versus time for Example 7.1. The final time has reduced

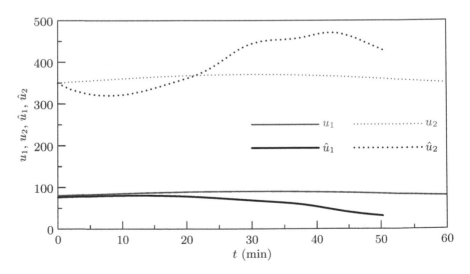

Figure 7.3 The initially guessed and optimal controls for Example 7.1

The application of the gradient algorithm minimized I, i. e., maximized the product concentration to 7.56 g/cm³. The final time was reduced from 60 to 50.3 min. Figure 7.4 plots I versus the iteration of the gradient algorithm.

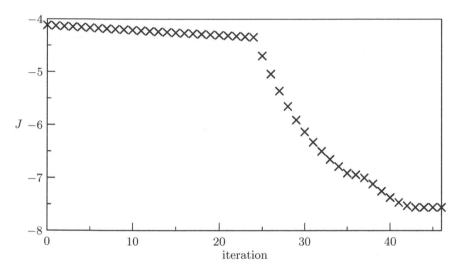

Figure 7.4 The objective functional versus iteration for Example 7.1

7.1.5   Fixed Final Time and Free Final State

The objective in this problem (see Section 6.2.1, p. 161) is to find the **u** that minimizes

$$I = \int_0^{t_f} F[\mathbf{y}(t), \mathbf{u}(t)] \, dt$$

subject to

$$\dot{\mathbf{y}} = \mathbf{g}[\mathbf{y}(t), \mathbf{u}(t)], \quad \mathbf{y}(0) = \mathbf{y}_0$$

The final time t_f is specified, i. e., fixed.

Example 7.2

Let us consider the problem of Example 7.1 (p. 192) with the final time fixed at 60 min. In this case we apply the gradient algorithm on p. 195 with t_f^k fixed at 60 min and skip Step 8 of the algorithm.

Results

For the same set of parameters and initial controls used in Example 7.1, Figure 7.5 shows the optimal states with the fixed final time. The corresponding controls are shown in Figure 7.6.

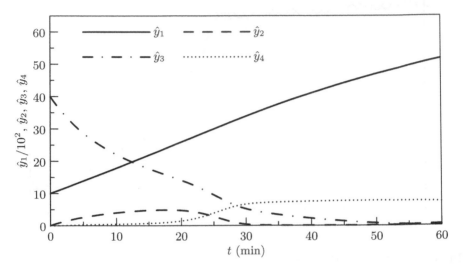

Figure 7.5 The optimal states versus time for Example 7.2. The final time is fixed

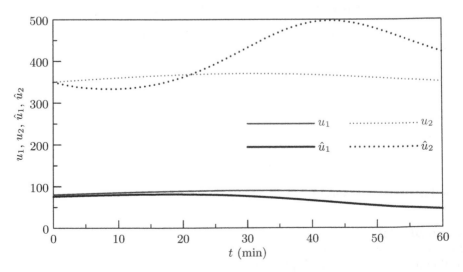

Figure 7.6 The initially guessed and optimal controls corresponding to Figure 7.5

The optimal I is -7.51, corresponding to the final product concentration of 7.51 g/cm^3. When the final time was free and available for optimization in the last example, we got better results — more production in less time. The reason is that free final time affords more freedom (than when it is fixed) in the optimal control of a process.

⧠

7.2 Penalty Function Method

This is a simple method for solving an optimal control problem with inequality constraints. As the name suggests, the method penalizes the objective functional in proportion to the violation of the constraints, which are not enforced directly. A constrained problem is solved using successive applications of an optimization method with increasing penalties for constraint violations. This strategy gradually leads to the solution, which satisfies the inequality constraints.

To illustrate the penalty function method, we consider the following optimal control problem.

7.2.1 Free Final Time and Final State on Hypersurfaces

The objective in this problem is to find the control $\mathbf{u}$ and the final time t_f that minimize

$$I = \int_0^{t_f} F[\mathbf{y}(t), \mathbf{u}(t)] \, \mathrm{d}t$$

subject to

$$\dot{\mathbf{y}} = \mathbf{g}[\mathbf{y}(t), \mathbf{u}(t)], \quad \mathbf{y}(0) = \mathbf{y}_0$$

and the l-dimensional hypersurface

$$\mathbf{q}[\mathbf{y}(t_f)] = \mathbf{0}$$

This problem is equivalent to finding the minimum of the functional

$$M = \underbrace{\int_0^{t_f} \left[F + \boldsymbol{\lambda}^\top (-\dot{\mathbf{y}} + \mathbf{g}) \right] \mathrm{d}t}_{J} + \boldsymbol{\mu}^\top \mathbf{q}$$

where $\boldsymbol{\mu}$ is the vector of Lagrange multipliers corresponding to the hypersurface. From Section 6.1.3 (p. 158), the necessary conditions for the minimum of M are

$$\dot{\mathbf{y}} = \mathbf{g} \qquad \mathbf{y}(0) = \mathbf{y}_0 \qquad \dot{\boldsymbol{\lambda}} = -H_{\mathbf{y}} \qquad \boldsymbol{\lambda}^\top(t_f) = \boldsymbol{\mu}^\top \mathbf{q}_{\mathbf{y}(t_f)}$$

$$H_{\mathbf{u}} = \mathbf{0} \qquad H(t_f) = 0 \qquad \mathbf{q} = \mathbf{0}$$

The gradient algorithm as such cannot work with the above equations. With an assumed control if we integrate the state equations forward from the initial conditions, there is no guarantee that the state will satisfy $\mathbf{q} = \mathbf{0}$ at the final time.

The penalty function method addresses this difficulty by prescribing

$$\boldsymbol{\mu} = \mathbf{W}\mathbf{q}$$

where $\mathbf{W}$ is an $l \times l$ diagonal weighting matrix with all positive diagonal elements. With this arrangement,

$$\boldsymbol{\lambda}^\top(t_f) = \underbrace{\mathbf{q}^\top \mathbf{W}}_{\boldsymbol{\mu}^\top} \mathbf{q}_{\mathbf{y}(t_f)}$$

and the augmented objective functional becomes

$$M = J + \underbrace{(\mathbf{W}\mathbf{q})^\top}_{\boldsymbol{\mu}} \mathbf{q} = J + \underbrace{\mathbf{q}^\top \mathbf{W}\mathbf{q}}_{\substack{\text{penalty} \\ \text{function}}}$$

where the last term, which is known as the penalty function, does not contribute to the augmented functional M as long as all constraints $\mathbf{q} = \mathbf{0}$ are satisfied.

In contrast, if a final state condition, say, $q_i = 0$, is violated, M is enlarged or penalized by a positive penalty term $d_{ii} q_i^2$. For sufficiently large d_{ii}, the penalty term would increase M to such an extent that its minimization would necessitate q_i to be suitably close to zero. Because of this fact, sequential minimization of M each time with increased magnitude of $\mathbf{W}$ would lead to the minimum that satisfies $\mathbf{q} = \mathbf{0}$. This minimization procedure is called the penalty function method, which fits easily with any optimization algorithm to enforce constraints. The computational algorithm combining this method with the gradient algorithm is as follows.

Computational Algorithm

To determine $\mathbf{u}$ and t_f at the minimum,

1. Initialize the outer counter $r = 0$. Choose a real number $\alpha > 1$.

2. Set the weighting matrix $\mathbf{W} = \alpha^r \mathbf{1}$ where $\mathbf{1}$ is the $l \times l$ identity matrix given by

$$\mathbf{1} = \begin{bmatrix} 1 & 0 & \cdots & 0 \\ 0 & 1 & \cdots & 0 \\ 0 & 0 & \ddots & 0 \\ 0 & 0 & \cdots & 1 \end{bmatrix}$$

3. On the problem transformed to the fixed σ-interval $[0,1]$, apply the gradient algorithm of Section 7.1.4 (p. 191) with

$$\boldsymbol{\lambda}^{\mathsf{T}}(1) = \underbrace{\mathbf{q}^{\mathsf{T}} \mathbf{W}}_{\boldsymbol{\mu}^{\mathsf{T}}} \mathbf{q_{y(1)}}$$

For $r > 0$, the initial guesses for the controls and the final time are, respectively, given by

$$\mathbf{u}_i^r; \quad i = 0, 1, \ldots, N \quad \text{and} \quad t_f^r$$

When the gradient algorithm converges in, say, k iterations, save the state, the control, and the final time, i.e.,

$$\mathbf{y}_i^k, \quad \mathbf{u}_i^k; \quad i = 0, 1, \ldots, N; \quad \text{and} \quad t_f^k$$

4. Increment the counter r by one and assign

$$\mathbf{y}_i^k \to \mathbf{y}_i^r, \quad \mathbf{u}_i^k \to \mathbf{u}_i^r, \quad i = 0, 1, \ldots, N; \quad \text{and} \quad t_f^k \to t_f^r$$

5. Check whether the constraint $\mathbf{q} = \mathbf{0}$ is satisfied. Given a positive real number ε_3 close to zero, if $\|\mathbf{q}\|$, i.e.,

$$\sqrt{\sum_{i=1}^l \left[q_i(\mathbf{y}_N^k)\right]^2} > \varepsilon_3$$

then go to Step 3. Otherwise, the constraint is satisfied, and the values

$$\mathbf{y}_i^r, \quad \mathbf{u}_i^r, \quad i = 0, 1, \ldots, N; \quad \text{and} \quad t_f^r$$

correspond to the minimum.

Example 7.3

Consider the batch reactor problem in Example 7.1 (p. 192) with selectivity constraints at the final time as the vector of two hypersurfaces

$$\mathbf{q}[\mathbf{y}(t_f)] = \begin{bmatrix} y_4 - b_1 y_2 \\ y_4 - b_2 y_3 \end{bmatrix} = \begin{bmatrix} 0 \\ 0 \end{bmatrix}$$

Following is the computational algorithm to solve this problem using the penalty function method:

Computational Algorithm

1. Set the counter $r = 0$. Initialize the weighting matrix as

$$\mathbf{W} = \begin{bmatrix} \alpha^r & 0 \\ 0 & \alpha^r \end{bmatrix}$$

2. Apply the gradient algorithm of Section 7.1.4 (p. 191) with

$$\boldsymbol{\lambda}^\top(1) = \begin{bmatrix} \lambda_1 \\ \lambda_2 \\ \lambda_3 \end{bmatrix}_{\sigma=1}^\top = \underbrace{\begin{bmatrix} y_4 - b_1 y_2 \\ y_4 - b_2 y_3 \end{bmatrix}_{\sigma=1}^\top}_{\mathbf{q}^\top} \underbrace{\begin{bmatrix} \alpha^r & 0 \\ 0 & \alpha^r \end{bmatrix}}_{\mathbf{W}} \underbrace{\begin{bmatrix} -b_1 & 0 & 1 \\ 0 & -b_2 & 1 \end{bmatrix}}_{\mathbf{q_{y(1)}}}$$

$$= \alpha^r \begin{bmatrix} -b_1(y_4 - b_1 y_2) \\ -b_2(y_4 - b_2 y_3) \\ 2y_4 - b_1 y_2 - b_2 y_3 \end{bmatrix}^\top$$

Recall that σ is the transformed time in the fixed interval $[0, 1]$, and $\sigma = 1$ corresponds to the final time t_f. When the gradient algorithm converges in, say, k iterations, save the state, the control, and the final time, i. e.,

$$\mathbf{y}_i^k, \quad T^k, \quad i = 0, 1, \ldots, N; \quad \text{and} \quad t_f^k$$

3. Increment the counter r by one and assign

$$\mathbf{y}_i^k \rightarrow \mathbf{y}_i^r, \quad T_i^k \rightarrow T_i^r; \quad i = 0, 1, \ldots, N; \quad \text{and} \quad t_f^k \rightarrow t_f^r$$

4. Check whether the constraint $\mathbf{q} = \mathbf{0}$ is satisfied. For a positive real number ε_3 close to zero, if

$$\sqrt{\sum_{i=1}^2 [q_i(\mathbf{y}_N^k)]^2} > \varepsilon_3$$

then go to Step 2. Otherwise, the constraint is satisfied, and the values

$$\mathbf{y}_i^r, \quad T_i^r; \quad i = 0, 1, \ldots, N; \quad \text{and} \quad t_f^r$$

correspond to the minimum.

Results

We present the results for $b_1 = 5$ and $b_2 = 2$ and the same set of parameters and initial controls as used in Example 7.1.

Figure 7.7 shows the convergence of the penalty function method. At each outer iteration r, the plotted objective functional value is the optimal with corresponding constraint violation quantified as $\|\mathbf{q}\|$. It was 17.2 at the very beginning (corresponding to the initial controls) and dropped finally to 5.9×10^{-8} at the convergence, which was attained in 11 outer iterations.

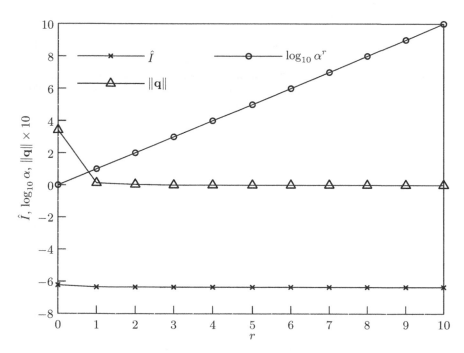

Figure 7.7 The intermediate, optimal objective functional $\hat{I}$, penalty factor α, and the constraint norm $\|\mathbf{q}\|$ versus outer iterations

Figure 7.8 shows the optimal states obtained at the convergence of the penalty function method. The states correspond to the optimal controls, which are shown in Figure 7.9. The final time was reduced from 60 to 42.8 min.

The selectivity constraints are satisfied at the optimal final time of 42.8 min. The optimal objective functional is -6.35, which corresponds to the final product concentration of 6.35 g/cm³.

Compared to the problem in the absence of the selectivity constraints (Example 7.1), the final product concentration has reduced a bit due to the imposed selectivity constraints.

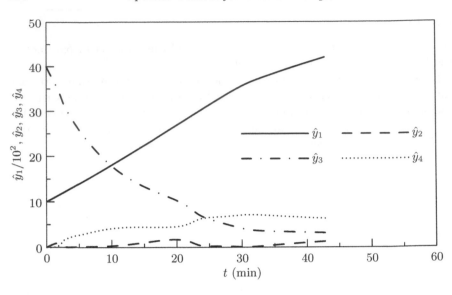

Figure 7.8 The optimal states versus time for Example 7.3

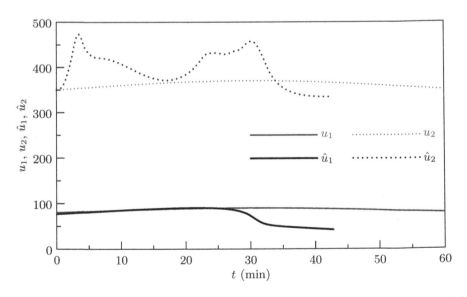

Figure 7.9 The initially guessed and optimal controls versus time for Example 7.3

7.2.2 Free Final Time but Fixed Final State

The objective in this problem is to find the control $\mathbf{u}$ that minimizes

$$I = \int_0^{t_f} F[\mathbf{y}(t), \mathbf{u}(t)] \, dt$$

subject to

$$\dot{\mathbf{y}} = \mathbf{g}[\mathbf{y}(t), \mathbf{u}(t)], \quad \mathbf{y}(0) = \mathbf{y}_0, \quad \text{and} \quad \mathbf{y}(t_f) = \mathbf{y}_f$$

The final state fixed as $\mathbf{y}(t_f) = \mathbf{y}_f$ is the vector of hypersurfaces

$$\mathbf{q}[\mathbf{y}(t_f)] \equiv \mathbf{y}(t_f) - \mathbf{y}_f = 0$$

at the final time t_f. From Section 6.1.3 (p. 158), the necessary conditions for the minimum of M are

$$\dot{\mathbf{y}} = \mathbf{g} \qquad \mathbf{y}(0) = \mathbf{y}_0 \qquad \dot{\boldsymbol{\lambda}} = -H_{\mathbf{y}} \qquad \boldsymbol{\lambda}^\top(t_f) = \boldsymbol{\mu}^\top \mathbf{q}_{\mathbf{y}(t_f)} = \boldsymbol{\mu}^\top$$
$$H_{\mathbf{u}} = \mathbf{0} \qquad H(t_f) = 0 \qquad \mathbf{q} = \mathbf{0}$$

considering the fact that $\mathbf{q}_{\mathbf{y}(t_f)}$ is an identity matrix.

Applying the penalty function method developed in Section 7.2.1 (p. 201),

$$\boldsymbol{\mu} = \mathbf{W}\mathbf{q}, \qquad \boldsymbol{\lambda}^\top(t_f) = \boldsymbol{\mu}^\top = \mathbf{q}^\top \mathbf{W}$$

and the objective functional M becomes

$$M = J + (\mathbf{W}\mathbf{q})^\top \mathbf{q} = J + \underbrace{\mathbf{q}^\top \mathbf{W}\mathbf{q}}_{\substack{\text{penalty} \\ \text{function}}}$$

The computational algorithm to determine the optimal control is the same as on p. 202.

Example 7.4

Consider the batch reactor problem in Example 7.3 with the final state fixed as

$$y_i(t_f) = y_{i,f}; \quad i = 1, 2, 3$$

instead of the final selectivity constraints.

The computational algorithm to solve this problem is as follows.

Computational Algorithm

This modified problem is solvable by the algorithm of the last example using the vector of hypersurfaces

$$\mathbf{q}[\mathbf{y}(t_f)] = \begin{bmatrix} y_1 - y_{1,f} \\ y_2 - y_{2,f} \\ y_3 - y_{3,f} \end{bmatrix} = \begin{bmatrix} 0 \\ 0 \\ 0 \end{bmatrix}$$

and the final costate vector

$$
\boldsymbol{\lambda}^\top(1) = \begin{bmatrix} \lambda_1 \\ \lambda_2 \\ \lambda_3 \end{bmatrix}_{\sigma=1}^\top = \underbrace{\begin{bmatrix} y_1 - y_{1,\mathrm{f}} \\ y_2 - y_{2,\mathrm{f}} \\ y_3 - y_{3,\mathrm{f}} \end{bmatrix}_{\sigma=1}^\top}_{\mathbf{q}^\top} \underbrace{\begin{bmatrix} \alpha^r & 0 & 0 \\ 0 & \alpha^r & 0 \\ 0 & 0 & \alpha^r \end{bmatrix}}_{\mathbf{W}} = \alpha^r \begin{bmatrix} y_1 - y_{1,\mathrm{f}} \\ y_2 - y_{2,\mathrm{f}} \\ y_3 - y_{3,\mathrm{f}} \end{bmatrix}_{\sigma=1}^\top
$$

Results

This problem was solved with

$$
y_{1,\mathrm{f}} = 5 \times 10^3 \qquad y_{2,\mathrm{f}} = 0.5 \qquad y_{3,\mathrm{f}} = 0.5 \quad (\mathrm{g/cm}^3)
$$

and the same set of parameters and initial controls as used in Example 7.3.

The constraint violation in terms of $\|\mathbf{q}\|$ was 1182 at the very beginning with initial controls and converged finally to 1.1×10^{-3} in five outer iterations. Upon convergence, the final values of y_1, y_2, and y_3 were

$$
y_{1,\mathrm{f}} = 5 \times 10^3 \qquad y_{2,\mathrm{f}} = 0.4993 \qquad y_{3,\mathrm{f}} = 0.4993 \quad (\mathrm{g/cm}^3)
$$

and the optimal objective functional was -7.5, which corresponds to the final product concentration of 7.5 g/cm^3. The optimal final time reduced from 60 to 48.1 min. Figure 7.10 shows the optimal states obtained at the convergence. The corresponding optimal controls are shown in Figure 7.11.

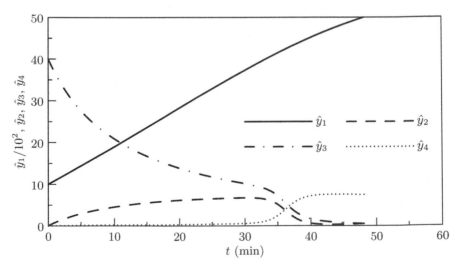

Figure 7.10 The optimal states versus time for Example 7.4 upon convergence of the penalty function method

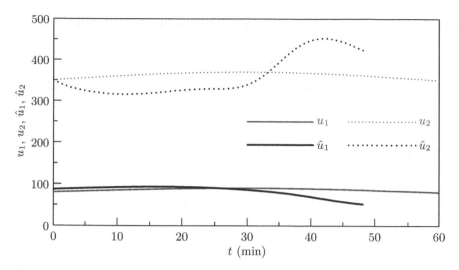

Figure 7.11 The initially guessed and optimal controls corresponding to Figure 7.10

7.2.3 Algebraic Equality Constraints

The objective in this problem is to find the control $\mathbf{u}$ that minimizes

$$I = \int_0^{t_f} F[\mathbf{y}(t), \mathbf{u}(t)]\, dt$$

subject to

$$\dot{\mathbf{y}} = \mathbf{g}[\mathbf{y}(t), \mathbf{u}(t)], \quad \mathbf{y}(0) = \mathbf{y}_0$$

and l algebraic equality constraints

$$\mathbf{f}(\mathbf{y}, \mathbf{u}) = \mathbf{0}$$

It is assumed that the partial derivatives $\mathbf{f_y}$ and $\mathbf{f_u}$ are continuous, and the dimension of $\mathbf{f}$ is less than that of $\mathbf{u}$. As shown in Section 6.3.1 (p. 163), the equivalent problem is to find the minimum of the augmented functional

$$M = \int_0^{t_f} \left[F + \boldsymbol{\lambda}^\top(-\dot{\mathbf{y}} + \mathbf{g}) + \boldsymbol{\mu}^\top \mathbf{f} \right] dt = \int_0^{t_f} \left(L - \boldsymbol{\lambda}^\top \dot{\mathbf{y}} \right) dt$$

where $\boldsymbol{\mu}$ are the time dependent multipliers associated with the algebraic equalities and

$$L = F + \boldsymbol{\lambda}^\top \mathbf{g} + \boldsymbol{\mu}^\top \mathbf{f} = H + \boldsymbol{\mu}^\top \mathbf{f}$$

is the Lagrangian. The necessary conditions for the minimum of M, and equivalently of I, are

$$\dot{\mathbf{y}} = \mathbf{g} \qquad \mathbf{y}(0) = \mathbf{y}_0 \qquad \dot{\boldsymbol{\lambda}} = -L_\mathbf{y} \qquad \boldsymbol{\lambda}(t_\mathrm{f}) = \mathbf{0}$$
$$\mathbf{f} = \mathbf{0} \qquad L_\mathbf{u} = \mathbf{0} \qquad L(t_\mathrm{f}) = 0$$

Since the algebraic equations $\mathbf{f} = \mathbf{0}$ are not expected to be satisfied with an initially assumed control, we penalize the Lagrangian. The approach is similar to that in Section 7.2.1 (p. 201). We set $\boldsymbol{\mu} = \mathbf{W}\mathbf{f}$ at each time in the interval $[0, t_\mathrm{f}]$. Thus, the objective functional

$$M = \int_0^{t_\mathrm{f}} \left[F + \boldsymbol{\lambda}^\top(-\dot{\mathbf{y}} + \mathbf{g}) + \underbrace{\mathbf{f}^\top \mathbf{W} \mathbf{f}}_{\substack{\text{penalty} \\ \text{function}}} \right] \mathrm{d}t$$

is penalized through the penalty function whenever any algebraic constraint is not satisfied. The computational algorithm to determine the minimum is as follows.

Computational Algorithm

To determine $\mathbf{u}$ and t_f at the minimum,

1. Initialize the outer counter $r = 0$. Choose a real number $\alpha > 1$.

2. Set the $l \times l$ diagonal weighting matrix

$$\mathbf{W} = \alpha^r \mathbf{1}$$

   for the $(N + 1)$ grid points of the σ-interval to be considered in the gradient algorithm. $\mathbf{1}$ is the $l \times l$ identity matrix.

3. Apply the gradient algorithm of Section 7.1.4 (p. 191) with the following initial guesses if $r > 0$:

$$\mathbf{u}_i^r; \quad i = 0, 1, \ldots, N \quad \text{and} \quad t_\mathrm{f}^r$$

   Note In this case, the augmented objective functional is M whose gradient in the transformed σ-interval $[0, 1]$ is

$$\nabla M \equiv \begin{bmatrix} L_\mathbf{u}(\sigma_0) & L_\mathbf{u}(\sigma_1) & \cdots & L_\mathbf{u}(\sigma_N) & M_{t_\mathrm{f}} \end{bmatrix}^\top$$

where

$$L(\sigma_i) = t_f\Big[F(\mathbf{y}_i, \mathbf{u}_i) + \boldsymbol{\lambda}_i^\top \mathbf{g}(\mathbf{y}_i, \mathbf{u}_i)\Big]$$

$$+ \underbrace{[\mathbf{f}(\mathbf{y}_i, \mathbf{u}_i)]^\top \mathbf{W}}_{\boldsymbol{\mu}_i^\top} \mathbf{f}(\mathbf{y}_i, \mathbf{u}_i); \quad i = 0, 1, \ldots, N$$

Using composite Simpson's 1/3 Rule, given that the state equations are satisfied,

$$M = \frac{1}{3N}\Big(A_0 + 4\sum_{1,3,5,\ldots}^{N} A_i + 2\sum_{i=2,4,6,\ldots}^{N} A_i + A_N\Big)$$

where $A_i \equiv t_f F(\mathbf{y}_i, \mathbf{u}_i) + [\mathbf{f}(\mathbf{y}_i, \mathbf{u}_i)]^\top \mathbf{W}\, \mathbf{f}(\mathbf{y}_i, \mathbf{u}_i); \quad i = 0, 1, \ldots, N$

The improvements in $\mathbf{u}$ and t_f are, respectively, $-\epsilon L_\mathbf{u}$ at each grid point and $-\epsilon M_{t_f}$ where

$$M_{t_f} = \int_0^1 \frac{H}{t_f}\, d\sigma = \frac{1}{3Nt_f}\Big(H_0 + 4\sum_{1,3,5,\ldots}^{N} H_i + 2\sum_{i=2,4,6,\ldots}^{N} H_i + H_N\Big)$$

and $H_i \equiv t_f\Big[F(\mathbf{y}_i, \mathbf{u}_i, \boldsymbol{\lambda}_i) + \boldsymbol{\lambda}_i^\top \mathbf{g}(\mathbf{y}_i, \mathbf{u}_i, \boldsymbol{\lambda}_i)\Big]; \quad i = 0, 1, \ldots, N$

4. When the gradient algorithm converges in, say, k iterations, save the state, the control, and the final time, i.e.,

$$\mathbf{y}_i^k, \quad \mathbf{u}_i^k; \quad i = 0, 1, \ldots, N; \quad \text{and} \quad t_f^k$$

5. Increment the counter r by one and assign

$$\mathbf{y}_i^k \to \mathbf{y}_i^r, \quad \mathbf{u}_i^k \to \mathbf{u}_i^r; \quad i = 0, 1, \ldots, N; \quad \text{and} \quad t_f^k \to t_f^r$$

6. Check whether the constraints $\mathbf{f} = \mathbf{0}$ are satisfied throughout the time grid. Given a positive real number ε_3 close to zero if

$$\sqrt{\sum_{i=0}^{N}\sum_{j=1}^{l}[f_j(\mathbf{y}_i^k, \mathbf{u}_i^k)]^2} > \varepsilon_3$$

then go to Step 3. Otherwise, the constraints are satisfied, and the values

$$\mathbf{y}_i^r, \quad \mathbf{u}_i^r; \quad i = 0, 1, \ldots, N; \quad \text{and} \quad t_f^r$$

correspond to the minimum.

Example 7.5

Consider the batch reactor problem in Example 7.1 (p. 192) subject to the following algebraic equality constraint throughout the time interval $[0, t_f]$:

$$u_1 = b_1 y_3^2 + b_2 y_3$$

In this case,

$$f \equiv u_1 - b_1 y_3^2 - b_2 y_3 = 0$$

and the weighting matrix is just an element α^r at the r-th outer iteration. The corresponding Lagrange multiplier at a grid point σ_i in the σ-interval is

$$\mu_i = f_i \alpha^r = (u_{1,i} - b_1 y_{3,i}^2 - b_2 y_{3,i}) \alpha^r$$

Results

Using the above algorithm, this problem was solved with $b_1 = -0.1$, $b_2 = 5$, and the same set of parameters and initial controls as used in Example 7.1.

With initial controls, the constraint violation in terms of $\|\mathbf{q}\|$ was 55.7, which reduced and converged to 3.3×10^{-2} in 10 outer iterations. At convergence, the optimal objective functional was -4.45, which corresponds to the final product concentration of 4.45 g/cm^3. The optimal final time reduced from 60 to 33.3 min.

Figure 7.12 plots the initial and final relations between u_1 and y_3. The final relation is obtained at the convergence and matches with the imposed constraint, $f = 0$.

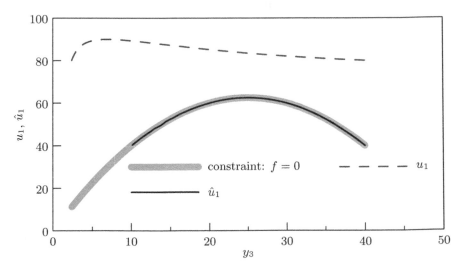

Figure 7.12 The relation between u_1 and y_3. The thin solid line at convergence is seen to overlap with the imposed constraint

The optimal states obtained at the convergence are shown in Figure 7.13. Figure 7.14 shows the corresponding optimal controls.

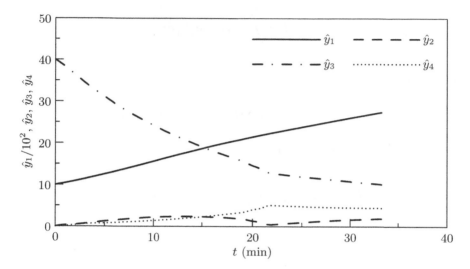

Figure 7.13 The optimal states when the algebraic constraints are satisfied upon convergence in Example 7.5

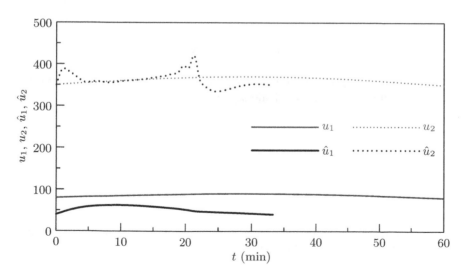

Figure 7.14 The initially guessed and optimal controls corresponding to Figure 7.13

7.2.4 Integral Equality Constraints

The objective in this problem is to find the control $\mathbf{u}$ that minimizes

$$I = \int_0^{t_f} F[\mathbf{y}(t), \mathbf{u}(t)] \, dt$$

subject to

$$\dot{\mathbf{y}} = \mathbf{g}[\mathbf{y}(t), \mathbf{u}(t)], \quad \mathbf{y}(0) = \mathbf{y}_0$$

and l integral equality constraints

$$\int_0^{t_f} F_j(\mathbf{y}, \mathbf{u}, t) \, dt = k_j, \quad j = 1, 2, \ldots, l$$

As shown in Section 6.4.1 (p. 168), the equivalent problem is to find the minimum of the augmented functional

$$M = \int_0^{t_f} (L - \boldsymbol{\lambda}^\top \dot{\mathbf{y}}) \, dt - \boldsymbol{\mu}^\top \mathbf{k}$$

where $\boldsymbol{\mu}$ is the vector of time invariant Lagrange multipliers, μ_is, corresponding to the integral equality constraints, L is the Lagrangian given by

$$L = H + \boldsymbol{\mu}^\top \mathbf{F} = F + \boldsymbol{\lambda}^\top \mathbf{g} + \boldsymbol{\mu}^\top \mathbf{F}$$

and $\mathbf{F}$ and $\mathbf{k}$ are respective vectors of F_js and k_js. The necessary conditions for the minimum of M, and equivalently of I are

$$\dot{\boldsymbol{\lambda}} = -L_\mathbf{y} \qquad \dot{\mathbf{y}} = L_\boldsymbol{\lambda} = \mathbf{g} \qquad L_\mathbf{u} = \mathbf{0}$$

$$\boldsymbol{\lambda}(t_f) = \mathbf{0} \qquad L(t_f) = 0 \qquad \int_0^{t_f} \mathbf{F} \, dt = \mathbf{k}$$

along with $\mathbf{y}(0) = \mathbf{y}_0$.

When solving this problem using the penalty function method, the integral constraints are enforced by specifying

$$\mu_j = \left(\int_0^{t_f} F_j(\mathbf{y}, \mathbf{u}, t) \, dt - k_j \right) \alpha, \quad j = 1, 2, \ldots, l$$

where α is a real number greater than unity. With this specification, the augmented functional becomes

$$M = \int_0^{t_f} \left(H - \boldsymbol{\lambda}^\top \dot{\mathbf{y}}\right) dt + \underbrace{\sum_{j=1}^{l} \left(\int_0^{t_f} F_j \, dt - k_j\right)^2 \alpha}_{\text{penalty function}}$$

The summation term above is the penalty function, which is positive whenever any integral equality constraint is violated.

Example 7.6

Consider the batch reactor problem in Example 7.1 (p. 192) subject to the following integral equality constraints:

$$\frac{1}{t_f} \int_0^{t_f} u_1 \, dt = \bar{u}_1 \quad \text{and} \quad \frac{1}{t_f} \int_0^{t_f} u_2 \, dt = \bar{u}_2$$

which specify the average values of the controls over the operation time of the reactor. In the transformed σ-interval $[0, 1]$, the constraints become

$$\int_0^1 u_1 \, d\sigma = \bar{u}_1 \quad \text{and} \quad \int_0^1 u_2 \, d\sigma = \bar{u}_2$$

The augmented objective functional is then given by

$$M = \int_0^1 \left(H - \boldsymbol{\lambda}^\top \frac{d\mathbf{y}}{d\sigma}\right) d\sigma + \underbrace{\sum_{j=1}^{l} \left(\int_0^1 u_j \, d\sigma - \bar{u}_j\right)^2 \alpha}_{\text{penalty function}}$$

where H is provided in Example 7.1.

Results

Using the algorithm on p. 210, this problem was solved with $\bar{u}_1 = 60 \text{ cm}^3/\text{min}$, $\bar{u}_2 = 350$ K, and the same set of parameters and initial controls as used in Example 7.1.

With initial controls, the constraint violation in terms of $\|\mathbf{q}\|$ was 29.3, which reduced and converged to 1.4×10^{-5} in nine outer iterations. At convergence, the optimal objective functional was -7.3, which corresponds to the final product concentration of 7.3 g/cm^3. The optimal final time in this case increased from 60 to 64 min.

The optimal states obtained at convergence are shown in Figure 7.15. Figure 7.16 shows the corresponding optimal controls.

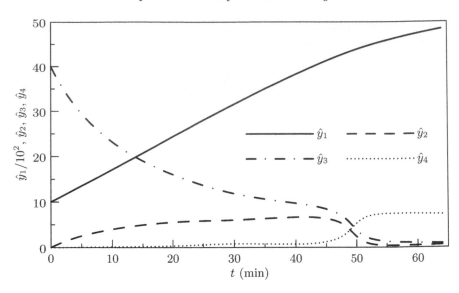

Figure 7.15 The optimal states when the integral equality constraints are satisfied upon convergence in Example 7.6

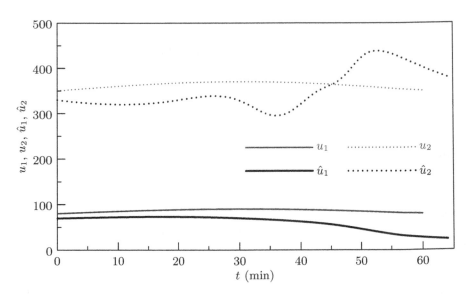

Figure 7.16 The initially guessed and optimal controls corresponding to Figure 7.15

⬚

7.2.5 Algebraic Inequality Constraints

This problem is similar to that in Section 7.2.3 (p. 209) except that the algebraic equality constraints are replaced with the inequalities

$$\mathbf{f}(\mathbf{y}, \mathbf{u}) \leq \mathbf{0}$$

Recall from Section 6.3.2 (p. 166) that the necessary conditions for the minimum are

$$\dot{\mathbf{y}} = \mathbf{g} \qquad \mathbf{y}(0) = \mathbf{y}_0 \qquad \dot{\boldsymbol{\lambda}} = -L_{\mathbf{y}} \qquad \boldsymbol{\lambda}(t_{\mathrm{f}}) = \mathbf{0}$$
$$\mathbf{f} \leq \mathbf{0} \qquad L_{\mathbf{u}} = \mathbf{0} \qquad L(t_{\mathrm{f}}) = 0 \qquad \boldsymbol{\mu} \geq \mathbf{0}, \quad \boldsymbol{\mu}^{\mathsf{T}} L_{\boldsymbol{\mu}} = \mathbf{0}$$

where $L = F + \boldsymbol{\lambda}^{\mathsf{T}}\mathbf{g} + \boldsymbol{\mu}^{\mathsf{T}}\mathbf{f}$.

To handle the inequalities, the penalty function method is slightly modified as follows. At any time, the multipliers corresponding to the inequalities are prescribed as

$$\boldsymbol{\mu} = \mathbf{W}\mathbf{e}$$

where $\mathbf{e}$ is vector of l elements

$$e_j = \begin{cases} 0 & \text{if} \quad f_j(t) \leq 0 \\ 1 & \text{if} \quad f_j(t) > 0 \end{cases} \qquad j = 1, 2, \ldots l$$

Thus, the augmented objective functional

$$M = \int_0^{t_{\mathrm{f}}} \left[F + \boldsymbol{\lambda}^{\mathsf{T}}(-\dot{\mathbf{y}} + \mathbf{g}) + \underbrace{\mathbf{e}^{\mathsf{T}}\mathbf{W}}_{\boldsymbol{\mu}^{\mathsf{T}}}\mathbf{f} \right] \mathrm{d}t$$

has the penalty function $\mathbf{e}^{\mathsf{T}}\mathbf{W}\mathbf{f}$, which is positive and enlarges M whenever any inequality is violated. The computational algorithm to find the minimum is as follows.

Computational Algorithm

To determine $\mathbf{u}$ and t_{f} at the minimum,

1. Initialize the counter $r = 0$. Choose a real number $\alpha > 1$.

2. Set the $l \times l$ diagonal weighting matrix

$$\mathbf{W} = \alpha^r \mathbf{1}$$

for the $(N + 1)$ grid points of the σ-interval to be considered in the gradient algorithm. $\mathbf{1}$ is the $l \times l$ identity matrix.

3. Apply the gradient algorithm of Section 7.1.4 (p. 191). Use the following initial guesses if $r > 0$:

$$\mathbf{u}_i^r; \quad i = 0, 1, \ldots, N \quad \text{and} \quad t_f^r$$

Note In this case, the augmented objective functional is M whose gradient in the transformed σ-interval $[0, 1]$ is

$$\nabla M \equiv \begin{bmatrix} L_{\mathbf{u}}(\sigma_0) & L_{\mathbf{u}}(\sigma_1) & \cdots & L_{\mathbf{u}}(\sigma_N) & M_{t_f} \end{bmatrix}^\top$$

where

$$L(\sigma_i) = t_f \Big[F(\mathbf{y}_i, \mathbf{u}_i) + \boldsymbol{\lambda}_i^\top \mathbf{g}(\mathbf{y}_i, \mathbf{u}_i) \Big]$$

$$+ \underbrace{[\mathbf{f}(\mathbf{y}_i, \mathbf{u}_i)]^\top \mathbf{W}}_{\boldsymbol{\mu}_i^\top} \mathbf{f}(\mathbf{y}_i, \mathbf{u}_i); \quad i = 0, 1, \ldots, N$$

Using composite Simpson's 1/3 Rule, given that the state equations are satisfied,

$$M = \frac{1}{3N} \Big(A_0 + 4 \sum_{\substack{N \\ 1,3,5,\ldots}} A_i + 2 \sum_{\substack{N \\ i=2,4,6,\ldots}} A_i + A_N \Big)$$

where $A_i \equiv t_f F(\mathbf{y}_i, \mathbf{u}_i) + [\mathbf{f}(\mathbf{y}_i, \mathbf{u}_i)]^\top \mathbf{W} \mathbf{f}(\mathbf{y}_i, \mathbf{u}_i); \quad i = 0, 1, \ldots, N$

The improvements in $\mathbf{u}$ and t_f are, respectively, $-\epsilon L_{\mathbf{u}}$ at each grid point and $-\epsilon M_{t_f}$ where

$$M_{t_f} = \int_0^1 \frac{H}{t_f} \, d\sigma = \frac{1}{3N t_f} \Big(H_0 + 4 \sum_{\substack{N \\ 1,3,5,\ldots}} H_i + 2 \sum_{\substack{N \\ i=2,4,6,\ldots}} H_i + H_N \Big)$$

and $H_i \equiv t_f \Big[F(\mathbf{y}_i, \mathbf{u}_i, \boldsymbol{\lambda}_i) + \boldsymbol{\lambda}_i^\top \mathbf{g}(\mathbf{y}_i, \mathbf{u}_i, \boldsymbol{\lambda}_i) \Big]; \quad i = 0, 1, \ldots, N$

The elements of $\mathbf{e}_i$ $(i = 0, 1, \ldots, N)$ are determined as follows:

$$e_{j,i} = \begin{cases} 0 & \text{if } f_j(t) \leq 0 \\ 1 & \text{if } f_j(t) > 0 \end{cases}; \quad j = 1, 2, \ldots l; \quad i = 0, 1, \ldots, N$$

4. When the gradient algorithm converges in, say, k iterations, save the state, the control, and the final time, i.e.,

$$\mathbf{y}_i^k, \quad \mathbf{u}_i^k; \quad i = 0, 1, \ldots, N; \quad \text{and} \quad t_f^k$$

5. Increment the counter r by one and assign

$$\mathbf{y}_i^k \to \mathbf{y}_i^r, \quad \mathbf{u}_i^k \to \mathbf{u}_i^r, \quad \mathbf{e}_i^k \to \mathbf{e}_i^r; \quad i = 0, 1, \ldots, N; \quad \text{and} \quad t_f^k \to t_f^r$$

6. Given a positive real number ε_3 close to zero, if the error

$$E = \sum_{i=0}^{N} [\mathbf{e}_i^r]^{\mathsf{T}} \mathbf{f}(\mathbf{y}_i^r, \mathbf{u}_i^r) > \varepsilon_3 \tag{7.6}$$

then some constraints are violated. Therefore, go to Step 3.

Otherwise, the constraints $\mathbf{f} \leq \mathbf{0}$ are satisfied, and the values

$$\mathbf{y}_i^r, \quad \mathbf{u}_i^r, \quad i = 0, 1, \ldots, N; \quad \text{and} \quad t_{\mathrm{f}}^r$$

correspond to the minimum.

Example 7.7

Consider the batch reactor problem in Example 7.1 (p. 192) subject to the following inequality constraints throughout the time interval $[0, t_{\mathrm{f}}]$:

$$u_1 \geq u_{1,\min} \quad \text{and} \quad u_{2,\min} \leq u_2 \leq u_{2,\max}$$

These constraints can be expressed as

$$\mathbf{f} = \begin{bmatrix} f_1 \\ f_2 \\ f_3 \end{bmatrix} \equiv \begin{bmatrix} -u_1 + u_{1,\min} \\ -u_2 + u_{2,\min} \\ u_2 - u_{2,\max} \end{bmatrix} \leq \begin{bmatrix} 0 \\ 0 \\ 0 \end{bmatrix}$$

Results

Using the above algorithm, this problem was solved with

$$u_{1,\min} = 50 \ \mathrm{cm^3/min}, \quad u_{2,\min} = 300 \ \mathrm{K}, \quad u_{2,\max} = 400 \ \mathrm{K}$$

and the same set of parameters and initial controls used in Example 7.1.

As observed from Figure 7.3 (p. 198), the initial controls do not violate the inequality constraints of the present problem. Hence, the initial constraint violation in terms of E in Equation (7.6) was zero. Finally, when the penalty function algorithm converged in six outer iterations, the value of E was 4.5×10^{-5}. The effect of the penalty function was to suppress any violation of the constraints during the iterations of the computational algorithm. At the convergence, the optimal objective functional was -5.9, which corresponds to the final product concentration of 5.9 g/cm^3. The optimal final time decreased slightly from 60 to 58.3 min.

The optimal states obtained at convergence are shown in Figure 7.17. Figure 7.18 shows the corresponding optimal controls, which satisfy the inequality constraints. In the absence of the inequality constraints, the optimal controls would be those in Figure 7.3 (p. 198), which violate two of the constraints.

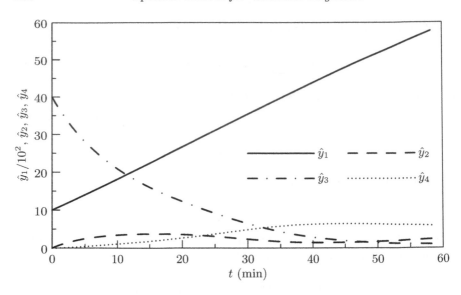

Figure 7.17 The optimal states when the inequality constraints are satisfied upon convergence in Example 7.7

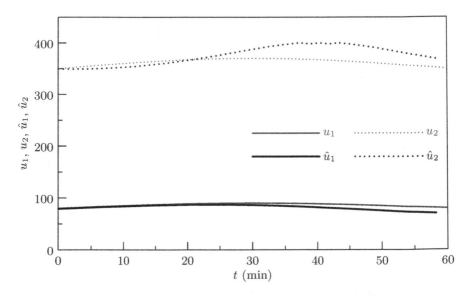

Figure 7.18 The initially guessed and optimal controls corresponding to Figure 7.17

⬚

7.2.6 Integral Inequality Constraints

This problem is similar to that in Section 7.2.4 (p. 214) except that the integral equality constraints are replaced with the inequalities

$$I_j \equiv \int_0^{t_f} F_j(\mathbf{y}, \mathbf{u}, t)\, dt \le k_j, \quad j = 1, 2, \dots, l$$

From Section 6.4.2 (p. 171), the necessary conditions for the minimum are

$$
\begin{array}{cccc}
\dot{\mathbf{y}} = \mathbf{g} & \mathbf{y}(0) = \mathbf{y}_0 & \dot{\boldsymbol{\lambda}} = -L_\mathbf{y} & \boldsymbol{\lambda}(t_f) = \mathbf{0} \\
L_\mathbf{u} = \mathbf{0} & L(t_f) = 0 & \mathbf{I} \le \mathbf{k} & \boldsymbol{\mu} \ge \mathbf{0}
\end{array}
$$

and $\boldsymbol{\mu}^\top(\mathbf{I} - \mathbf{k}) = 0$ where $\mathbf{I}$ is the vector of the integrals I_js.

Similar to Section 7.2.5 (p. 217), we prescribe $\boldsymbol{\mu} = \mathbf{We}$ where $\mathbf{e}$ is a vector of l elements

$$e_j = \begin{cases} 0 & \text{if } I_j - k_j \le 0 \\ 1 & \text{if } I_j - k_j > 0 \end{cases} \quad j = 1, 2, \dots l$$

Thus, M has the penalty function $\mathbf{e}^\top \mathbf{W}(\mathbf{I} - \mathbf{k})$, which is positive and enlarges M whenever any integral inequality is violated.

Computational Algorithm

The algorithm is the same as that on p. 217 except for the following changes:

- $\boldsymbol{\mu}$, $\mathbf{W}$, and $\mathbf{e}$ are uniform across the σ-interval.

- $\mathbf{f}$ appearing in the algebraic inequalities is replaced with $(\mathbf{I} - \mathbf{k})$.

Example 7.8

Consider the batch reactor problem in Example 7.1 (p. 192) subject to the following integral inequality constraints:

$$\frac{1}{t_f} \int_0^{t_f} u_1\, dt \le \bar{u}_1 \quad \text{and} \quad \frac{1}{t_f} \int_0^{t_f} u_2\, dt \le \bar{u}_2$$

which specify the upper limits to the average control values over the operation time of the reactor.

Results

Using the above algorithm, this problem was solved with $\bar{u}_1 = 60$ cm^3/min, $\bar{u}_2 = 350$ K, and the same set of parameters and initial controls as used in Example 7.1. The optimal states obtained at convergence are shown in Figure 7.19. Figure 7.20 shows the corresponding optimal controls.

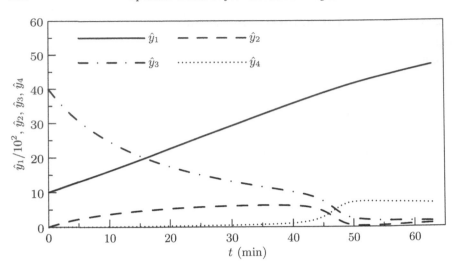

Figure 7.19 The optimal states when the integral inequality constraints are satisfied upon convergence in Example 7.8

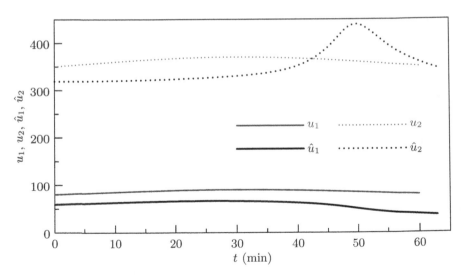

Figure 7.20 The initially guessed and optimal controls corresponding to Figure 7.19

The initial constraint violation in terms of E in Equation (7.6) was 39.1. Upon convergence in just one iteration, E reduced to zero. At the convergence, the optimal objective functional was -6.9, which corresponds to the final product concentration of 6.9 g/cm^3. The optimal final time in this case increased from 60 to 62.9 min.

□

Problems with Fixed Final Time

The final time is not an optimization parameter in these problems. Thus, these problems are straightaway solvable by the algorithms in the previous sections using the specified final time and discarding the improvements to it.

7.3   Shooting Newton–Raphson Method

Some times the stationarity condition $H_\mathbf{u} = \mathbf{0}$ of an optimal control problem can be solved to obtain an explicit expression for $\mathbf{u}$ in terms of the state $\mathbf{y}$ and the costate $\boldsymbol{\lambda}$. That expression, when substituted into state and costate equations, couples them. Thus, state equations become dependent on $\boldsymbol{\lambda}$ and must be integrated simultaneously with the costate equations. The simultaneous integration constitutes a two point boundary value problem in which

1. the state is specified at the first boundary, i. e., at the initial time;

2. the costate is specified at the second boundary, i. e., at the final time; and

3. the state and costate equations depend on state as well as costate variables.

The shooting Newton–Raphson method enables the solution of this problem. With a guessed initial costate, both state and costate equations are integrated forward or "shot" to the final time. The discrepancy between the final costate obtained in this way and that specified is improved using the Newton–Raphson method.

We explain the shooting Newton–Raphson method with the help of an optimal control problem having one state, one control, and fixed final time. The objective of the problem is to find the control function $u(t)$ that minimizes the functional

$$I = \int_0^{t_f} F[y(t), u(t)]\, \mathrm{d}t$$

subject to

$$y = g[y(t), u(t)], \quad y(0) = y_0$$

The necessary conditions for the minimum are

$$\underbrace{\frac{\mathrm{d}y}{\mathrm{d}t} = g, \quad y(0) = y_0,}_{\substack{\text{state equation and} \\ \text{initial condition}}} \quad \underbrace{\frac{\mathrm{d}\lambda}{\mathrm{d}t} = -H_y, \quad \lambda(t_f) = 0,}_{\substack{\text{costate equation and} \\ \text{final condition}}} \quad \text{and} \quad \underbrace{H_u = 0}_{\substack{\text{stationarity} \\ \text{condition}}}$$

Two Point Boundary Value Problem

Suppose that it is possible to solve $H_u = 0$ and obtain an explicit expression, $u = u(y, \lambda)$. Utilizing this expression, the state and costate equations for the minimum turn into the two point boundary value problem

$$\frac{\mathrm{d}y}{\mathrm{d}t} = g[y, u(y, \lambda)] = g(y, \lambda), \qquad y(0) = y_0,$$

$$\frac{\mathrm{d}\lambda}{\mathrm{d}t} = \underbrace{-H_y[y, u(y, \lambda), \lambda]}_{h} = h(y, \lambda), \qquad \lambda(t_f) = \lambda_f$$

Application of the Newton–Raphson Method

Consider the final costate obtained from the simultaneous forward integration of the above equations with $\lambda(0)$ guessed as λ_0. The final costate is a function of λ_0, i.e., $\lambda(\lambda_0, t_f)$. Its difference from the specified costate λ_f is the discrepancy function

$$f(\lambda_0) \equiv \lambda(\lambda_0, t_f) - \lambda_f \tag{7.7}$$

which can be zeroed out by iteratively improving λ_0 using the Newton–Raphson method (Section 9.11, p. 273). Thus, the improved value $\lambda_{0,\text{next}}$ is given by

$$\lambda_{0,\text{next}} = \lambda_0 - \frac{f(\lambda_0)}{\left[\dfrac{\partial f}{\partial \lambda_0} \right]_{t_f}}$$

where, from Equation (7.7), the derivative

$$\left[\frac{\partial f}{\partial \lambda_0} \right]_{t_f} = \left[\frac{\partial \lambda}{\partial \lambda_0} \right]_{t_f}$$

Derivative Evaluation

Observe that the above derivative is the final-time value of the time dependent variable

$$\frac{\partial \lambda}{\partial \lambda_0}(t)$$

We will now find the equations governing this variable.

Equations Governing $\partial\lambda/\partial\lambda_0$

Differentiating the state and costate equations with respect to λ_0, we get

$$\frac{\partial}{\partial\lambda_0}\left(\frac{dy}{dt}\right) = \frac{\partial g}{\partial\lambda_0} \qquad\qquad \frac{\partial}{\partial\lambda_0}\left(\frac{d\lambda}{dt}\right) = \frac{\partial h}{\partial\lambda_0}$$

or $\quad\dfrac{d}{dt}\left(\dfrac{\partial y}{\partial\lambda_0}\right) = \dfrac{\partial g}{\partial y}\dfrac{\partial y}{\partial\lambda_0} + \dfrac{\partial g}{\partial\lambda}\dfrac{\partial\lambda}{\partial\lambda_0} \qquad \dfrac{d}{dt}\left(\dfrac{\partial\lambda}{\partial\lambda_0}\right) = \dfrac{\partial h}{\partial y}\dfrac{\partial y}{\partial\lambda_0} + \dfrac{\partial h}{\partial\lambda}\dfrac{\partial\lambda}{\partial\lambda_0}$

where $\quad\dfrac{\partial y}{\partial\lambda_0}(0) = 0 \qquad\qquad\qquad \dfrac{\partial\lambda}{\partial\lambda_0}(0) = 1$

The last two equations arise from differentiating with respect to λ_0 the initial conditions for the state and costate equations, which are, respectively,

$$y(0) = y_0 \quad\text{and}\quad \lambda(0) = \lambda_0$$

Based on the above development, the computational algorithm for the shooting Newton–Raphson method is as follows.

Computational Algorithm

To determine **u** at the minimum,

1. Guess the initial costate, λ_0.

2. Integrate simultaneously to the final time the differential equations for the state, costate, $\partial y/\partial\lambda_0$, and $\partial\lambda/\partial\lambda_0$ using the respective initial conditions.

3. Use $\partial\lambda/\partial\lambda_0$ so obtained at the final time to improve λ_0 using the Newton–Raphson method.

4. Go to Step 2 until the improvement is negligible. When that happens, the state and control are optimal.

Example 7.9

Consider an isothermal liquid-phase reaction A $\longrightarrow$ B in a CSTR in the presence of a solid catalyst. The process model is given by

$$\frac{dy}{dt} = u(y_f - y) - ky^2, \quad y(0) = y_0$$

where y is the concentration of species A, u is the time dependent volumetric throughput per unit reactor volume, y_f is y in the feed, and k is the reaction rate coefficient. In a given time t_f, it is desired to find the $u(t)$ that minimizes the deviation in y and u from the reference condition (y_s, u_s), i.e.,

$$I = \int_0^{t_f} \left[(y - y_s)^2 + (u - u_s)^2\right] dt$$

The Hamiltonian for this problem is given by

$$H = (y - y_\mathrm{s})^2 + (u - u_\mathrm{s})^2 + \lambda\left[u(y_\mathrm{f} - y) - ky^2\right]$$

The necessary conditions to be satisfied at the minimum are the process model or the state equation with the initial condition, the costate equation with the final condition, i. e.,

$$\frac{\mathrm{d}\lambda}{\mathrm{d}t} = -2(y - y_\mathrm{s}) + \lambda(u + 2ky), \quad \lambda(t_\mathrm{f}) = 0$$

and the stationarity condition

$$H_u = 2(u - u_\mathrm{s}) + \lambda(y_\mathrm{f} - y) = 0$$

Two Point Boundary Value Problem

The last equation yields the expression for the optimal control

$$u = \frac{\lambda(y - y_\mathrm{f})}{2} + u_\mathrm{s}$$

which, when substituted into the state and costate equations, results in the two point boundary value problem

$$\frac{\mathrm{d}y}{\mathrm{d}t} = \left[\frac{\lambda(y - y_\mathrm{f})}{2} + u_\mathrm{s}\right](y_\mathrm{f} - y) - ky^2, \qquad y(0) = y_0$$

$$\frac{\mathrm{d}\lambda}{\mathrm{d}t} = -2(y - y_\mathrm{s}) + \lambda\left[\frac{\lambda(y - y_\mathrm{f})}{2} + u_\mathrm{s} + 2ky\right], \qquad \lambda(t_\mathrm{f}) = 0$$

The optimal control solution is essentially the solution of the above differential equations with boundary conditions at opposite end points of the time interval. In order to solve the equations using the shooting Newton–Raphson method, we need the derivative state equations to provide λ_{λ_0}.

Equations for λ_{λ_0}

The shooting Newton–Raphson method will use the derivative λ_{λ_0} at t_f to improve the guess $\lambda(0) = \lambda_0$, thereby zeroing out $\lambda(t_\mathrm{f})$. We differentiate with respect to λ_0 the latest state and costate equations as well as the initial boundary conditions

$$y(0) = y_0 \quad \text{and} \quad \lambda(0) = \underbrace{\lambda_0}_{\text{assumed}}$$

to obtain the following derivative equations:

$$\frac{\mathrm{d}y_{\lambda_0}}{\mathrm{d}t} = -\left[\frac{\lambda(y_{\mathrm{f}} - y)}{2} + u_{\mathrm{s}}\right]y_{\lambda_0} + \left[(y - y_{\mathrm{f}})\lambda_{\lambda_0} + \lambda y_{\lambda_0}\right]\frac{(y_{\mathrm{f}} - y)}{2} - 2ky_{\lambda_0}$$

$$\frac{\mathrm{d}\lambda_{\lambda_0}}{\mathrm{d}t} = -2y_{\lambda_0} + \left[\frac{\lambda(y_{\mathrm{f}} - y)}{2} + u_{\mathrm{s}} + 2ky\right]\lambda_{\lambda_0} + \left[(y - y_{\mathrm{f}})\lambda_{\lambda_0} + (\lambda + 4k)y_{\lambda_0}\right]\frac{\lambda}{2}$$

along with the respective initial conditions

$$y_{\lambda_0}(0) = 0 \quad \text{and} \quad \lambda_{\lambda_0}(0) = 1$$

Computational Algorithm

1. Set the iteration counter, $k = 0$. Guess the initial costate, λ_0^k.

2. Integrate simultaneously to the final time the last four differential equations for the state, costate, y_{λ_0}, and λ_{λ_0} using the respective initial conditions.

3. Improve λ_0^k by calculating

$$\lambda_0^{k+1} = \lambda_0^k - \frac{f(\lambda_0^k)}{\lambda_{\lambda_0}(\lambda_0^k, t_{\mathrm{f}})}$$

4. Check the improvement in λ_0. Given a small real number $\epsilon > 0$, if

$$\left|\lambda_0^{k+1} - \lambda_0^k\right| > \epsilon$$

then go to Step 2. Otherwise the state and control provide the minimum.

Results

Using the above algorithm, this problem was solved with the parameters listed in Table 7.2. The initial guess of $\lambda(0) = 0$ provided $\lambda(t_{\mathrm{f}}) = 15.7$. Figure 7.21 shows the optimal state, costate, and control obtained at the convergence, which was obtained in nine iterations.

The minimum objective functional value was 2.4 with final $\lambda(t_{\mathrm{f}})$ of $O(10^{-11})$. The initial and optimal costates are shown in Figure 7.22.

Table 7.2 Parameters for the problem of Example 7.9

y_0	5 g/cm^3	y_{s}	8 g/cm^3	y_{f}	10 g/cm^3
u_{s}	5 min^{-1}	k	10^{-3} cm^3/(g · min)	t_{f}	1 min

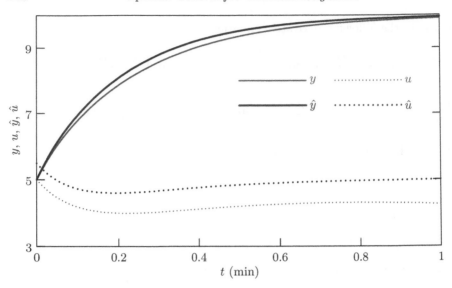

Figure 7.21 The initial and optimal states and controls in Example 7.9

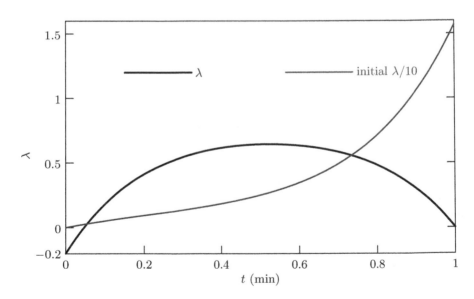

Figure 7.22 The initial and optimal costates in Example 7.9. The initial costate is in gray

7.A   Derivation of Steepest Descent Direction

Consider a continuously differentiable function $f = f(\mathbf{x})$ where $\mathbf{x}$ is the variable vector

$$\begin{bmatrix} x_1 & x_2 & \dots & x_n \end{bmatrix}^{\mathsf{T}}$$

For a small change

$$\mathrm{d}\mathbf{x} = \begin{bmatrix} \mathrm{d}x_1 & \mathrm{d}x_2 & \dots & \mathrm{d}x_n \end{bmatrix}^{\mathsf{T}}$$

the change in f is given by

$$\mathrm{d}f = \sum_{i=1}^{n} \frac{\partial f}{\partial x_i} \mathrm{d}x_i$$

Let the magnitude of $\mathrm{d}\mathbf{x}$ be denoted by $\mathrm{d}s$. Thus,

$$\mathrm{d}s = \sqrt{\sum_{i=1}^{n} \mathrm{d}x_i^2}$$

Then the change in f per unit change in $\mathrm{d}s$ is

$$\frac{\mathrm{d}f}{\mathrm{d}s} = \sum_{i=1}^{n} \underbrace{\frac{\partial f}{\partial x_i}}_{v_i} \underbrace{\frac{\mathrm{d}x_i}{\mathrm{d}s}}_{w_i} = \sum_{i=1}^{n} v_i w_i = \mathbf{v}^{\mathsf{T}} \mathbf{w}$$

where $\mathbf{v}$ is the gradient of f, and $\mathbf{w}$ is the unit vector along $\mathrm{d}\mathbf{x}$.

7.A.1   Objective

Next, consider the problem of minimization of f by changing $\mathbf{x}$ by $\mathrm{d}\mathbf{x}$ of length $\mathrm{d}s$ along a direction. For a decrease in f due to the change, the rate $\mathrm{d}f/\mathrm{d}s$ has to be negative. The objective is to find the direction or a unit vector $\mathbf{w}$ that results in the most negative rate $\mathrm{d}f/\mathrm{d}s$ or, equivalently, minimizes the rate, which from the last equation is

$$\sum_{i=1}^{n} v_i w_i$$

This problem is subject to the constraint that $\mathbf{w}$, being a unit vector, must have unit magnitude, i. e.,

$$K \equiv \sum_{i=1}^{n} w_i^2 - 1 = 0$$

Applying the Lagrange Multiplier Rule, the equivalent problem is to find the **w** that minimizes

$$J = \sum_{i=1}^{n} v_i w_i + \lambda \left(\sum_{i=1}^{n} w_i^2 - 1 \right)$$

where λ is a Lagrange multiplier. Now, at the minimum of J

$$\mathrm{d}J = \sum_{i=1}^{n} (v_i + 2\lambda w_i) \mathrm{d}w_i + \mathrm{d}\lambda \left(\sum_{i=1}^{n} w_i^2 - 1 \right) = 0$$

Hence, the necessary conditions for the minimum are

$$v_i + 2\lambda w_i = 0; \quad i = 1, 2, \ldots, n \quad \text{and} \quad \sum_{i=1}^{n} w_i^2 - 1 = 0$$

which yield, respectively,

$$w_i = -\frac{v_i}{2\lambda}; \quad i = 1, 2, \ldots, n \quad \text{and} \quad \lambda = \frac{1}{2} \sqrt{\sum_{i=1}^{n} v_i^2} = \frac{\|\mathbf{v}\|}{2}$$

Substituting the expression for λ in that for w_i above, we obtain

$$w_i = -\frac{v_i}{\|\mathbf{v}\|}; \quad i = 1, 2, \ldots, n$$

at the minimum. From the above result, the optimal **w** or $\hat{\mathbf{w}}$ is the unit vector opposite to the direction of **v**, i.e., the gradient of the function f.

7.A.2 Sufficiency Check

To ensure that the above result guarantees the minimum, we consider the second order Taylor expansion of f with respect to **w** at $\hat{\mathbf{w}}$, keeping the corresponding λ fixed at $\|\mathbf{v}\|/2$. Thus, for sufficiently small d**w**,

$$J(\hat{\mathbf{w}} + \mathrm{d}\mathbf{w}, \lambda) - J(\hat{\mathbf{w}}, \lambda) = \mathrm{d}J(\hat{\mathbf{w}}, \lambda) + \frac{1}{2} \mathrm{d}^2 J(\hat{\mathbf{w}}, \lambda)$$

Since $\mathrm{d}J(\hat{\mathbf{w}}, \lambda)$ is zero being the necessary condition, and the constraint $K = 0$ implies $\mathrm{d}K = 0$, we obtain

$$J(\hat{\mathbf{w}} + \mathrm{d}\hat{\mathbf{w}}, \lambda) - J(\hat{\mathbf{w}}, \lambda) = \frac{1}{2} \mathrm{d}^2 J(\hat{\mathbf{w}}, \lambda) = \lambda \sum_{i=1}^{n} \mathrm{d}w_i^2 + \underbrace{\mathrm{d}\lambda \sum_{i=1}^{n} \hat{w}_i \mathrm{d}w_i}_{(\mathrm{d}K)/2 = 0}$$

$$= \frac{\|\mathbf{v}\|}{2} \sum_{i=1}^{n} \mathrm{d}w_i^2 > 0$$

Thus, $J(\hat{\mathbf{w}}, \lambda)$, i.e., df/ds corresponding to $\hat{\mathbf{w}}$, is indeed the minimum. The above result shows that df/ds is minimized by

$$
\hat{\mathbf{w}} = -\left[\frac{v_1}{\|\mathbf{v}\|} \quad \frac{v_2}{\|\mathbf{v}\|} \quad \cdots \quad \frac{v_n}{\|\mathbf{v}\|} \right]^{\top}
$$

which is a unit vector in the direction opposite to the gradient of the function $f(\mathbf{x})$. Simply put, the rate of function decrease is maximum along the **steepest descent** direction, i.e., opposite to the gradient of the function.

<hr />

Bibliography

M.I. Kamien and N.L. Schwartz. *Dynamic Optimization*, Chapter 6, page 149. Elsevier, Amsterdam, 1991.

H.J. Kelley. Method of gradients. In G. Leitman, editor, *Optimization Techniques*, Chapter 6. Academic Press, New York, 1962.

D.E. Kirk. *Optimal Control Theory — An Introduction*, Chapter 6, pages 329–343. Dover Publications Inc., New York, 2004.

R. Luus and L. Lapidus. The control of nonlinear systems Part II: Convergence by combined first and second variations. *AIChE J.*, 13:108–113, 1967.

W.H. Press, S.A. Teukolsky, W.T. Vetterlin, and B.P. Flannery. *Numerical Recipes — The Art of Scientific Computing*, Chapter 10, page 521. Cambridge University Press, New York, 3rd edition, 2007.

S.S. Rao. *Engineering Optimization Theory and Practice*, Chapter 7, pages 428–455. John Wiley & Sons, Inc., New Jersey, 4th edition, 2009.

A.P. Sage and C.C. III White. *Optimum Systems Control*, Chapter 10, pages 300–325. Prentice Hall Inc., New Jersey, 2nd edition, 1977.

<hr />

Exercises

7.1 Simplify the calculation of I^k and $J_{t_f}^k$ in the gradient algorithm for Example 7.1 (p. 192).

7.2 Following Kelley (1962), modify the algorithm for the gradient method on p. 191 to determine the final time t_f using the following conditions for the minimum of I with respect to t_f:

$$\frac{dI}{dt} = F(\mathbf{y}, \mathbf{u}) = 0$$

$$\frac{d^2 I}{dt^2} = \sum_{i=1}^{n} \frac{\partial F}{\partial y_i} \underbrace{\frac{dy_i}{dt}}_{g_i} + \sum_{i=1}^{m} \frac{\partial F}{\partial u_i} \frac{du_i}{dt} > 0$$

7.3 A CSTR carrying out a first order exothermic reaction is described by (Luus and Lapidus, 1967)

$$\frac{dy_1}{dt} = -2y_1 - a_1 + (y_1 + a_1)\exp\left(\frac{a_2 y_1}{y_1 + 4a_1}\right) - (y_1 + 0.5a_1)u, \quad y_1(0) = y_{1,0}$$

$$\frac{dy_2}{dt} = a_1 - y_2 - (y_2 + a_1)\exp\left(\frac{a_2 y_1}{y_1 + 4a_1}\right), \quad y_2(0) = y_{2,0}$$

where y_1 and y_2 are, respectively, the deviations of temperature and concentration from the steady state values. The control u is related to the coolant flow rate. Find the u that minimizes state deviations and control action given by

$$I = \int_0^{t_f} (y_1^2 + y_2^2 + a_3 u^2)\, dx$$

for the following set of parameters:

$y_{1,0} = 0.05$	$y_{2,0} = 0$	$t_f = 0.78$
$a_1 = 0.5$	$a_2 = 25$	$a_3 = 0.1$

7.4 Repeat Exercise 7.3 with the following additional constraints:

$$0.4 \le u \le 1.4$$

7.5 Consider the process of production planning described by

$$\frac{dy}{dt} = u, \quad y(0) = 0$$

where y is the product inventory, and u is the production rate. It is desired to find the u that minimizes the cost functional

$$I = \int_0^{t_f} (a_1 u^2 + a_2 y)\, dt$$

subject to the following constraints:

$$y(t_f) = y_f \quad \text{and} \quad u \ge 0$$

Using a suitable computational algorithm, solve the above problem for

$$a_1 = a_2 = y_f = t_f = 1$$

Compare the optimal u with the analytical solution (Kamien and Schwartz, 1991) given by

$$u = \frac{a_2(2t - t_f)}{4a_1} + \frac{y_f}{t_f}, \quad u \geq 0$$

7.6 The integral equality constraints of the optimal control problem in Section 7.2.4 (p. 214) may be transformed into differential equations. Using this approach,

    a. derive the necessary conditions for the minimum,

    b. formulate an algorithm using the penalty function method to find the minimum, and

    c. solve Example 7.6 (p. 215).

7.7 Repeat Exercise 7.6 to solve Example 7.8 (p. 221).

7.8 Solve Example 7.8 (p. 221) in the presence of the following additional constraint:

$$u_2 \leq 400 \text{ K}$$

throughout the time interval $[0, t_f]$.

Chapter 8

Optimal Periodic Control

Optimal periodic control involves a periodic process, which is characterized by a repetition of its state over a fixed time period. Examples from nature include the circadian rhythm of the core body temperature of mammals and the cycle of seasons. Man-made processes are run periodically by enforcing periodic control inputs such as periodic feed rate to a chemical reactor or cyclical injection of steam to heavy oil reservoirs inside the earth's crust. The motivation is to obtain performance that would be better than that under optimal steady state conditions.

In this chapter, we first describe how to solve an optimal periodic control problem. Next, we derive the pi criterion to determine whether better periodic operation is possible in the vicinity of an optimal steady state operation.

8.1 Optimality of Periodic Controls

Consider the objective to minimize the functional

$$I = \int_0^{t_f} F[\mathbf{y}(t), \mathbf{u}(t)] \, dt$$

where $t_f > 0$ is the time period of the control $\mathbf{u}$, and the state is governed by

$$\dot{\mathbf{y}} = \mathbf{g}[\mathbf{y}(t), \mathbf{u}(t)]$$

with the periodicity condition

$$\mathbf{y}(0) = \mathbf{y}(t_f)$$

The above problem is equivalent to minimizing the augmented functional

$$J = \int_0^{t_f} \left[F + \boldsymbol{\lambda}^\top (-\dot{\mathbf{y}} + \mathbf{g}) \right] dt$$

subject to $\mathbf{y}(0) = \mathbf{y}(t_f)$.

8.1.1 Necessary Conditions

Similar to Section 6.1.1 (p. 153), the variation of J is

$$\delta J = \int_0^{t_f} \left[(H_\mathbf{y} + \dot{\boldsymbol{\lambda}})^\top \delta\mathbf{y} + (H_\boldsymbol{\lambda} - \dot{\mathbf{y}})^\top \delta\boldsymbol{\lambda} + H_\mathbf{u}^\top \delta\mathbf{u} \right] dt + \boldsymbol{\lambda}^\top(0)\,\delta\mathbf{y}(0)$$
$$- \boldsymbol{\lambda}^\top(t_f)\,\delta\mathbf{y}_f + H(t_f)\delta t_f$$

Note that the above equation has one additional term, $\boldsymbol{\lambda}^\top(0)\,\delta\mathbf{y}(0)$, when compared to Equation (6.9) on p. 156. The reason is that $\delta\mathbf{y}(0)$ in the present problem is not zero since $\mathbf{y}(0)$ is not fixed but equal to $\mathbf{y}(t_f)$. This periodicity condition for the general case of unspecified t_f means that $\mathbf{y}(0)$ is equal to the final state $\mathbf{y}_f$. As a result

$$\delta\mathbf{y}(0) = \delta\mathbf{y}_f$$

Considering this fact, the necessary conditions for the minimum of J, and equivalently of I, are

$$\dot{\boldsymbol{\lambda}} = -H_\mathbf{y} \qquad\qquad \dot{\mathbf{y}} = H_\boldsymbol{\lambda} = \mathbf{g} \qquad\qquad H_\mathbf{u} = \mathbf{0}$$

$$\boldsymbol{\lambda}(0) = \boldsymbol{\lambda}(t_f) \qquad\qquad \mathbf{y}(0) = \mathbf{y}(t_f) \qquad\qquad H(t_f) = 0$$

From now on, we will use τ instead of t_f in the periodic problems.

Example 8.1

Consider the periodic operation of the batch reactor in Example 2.10 (p. 45) with the objective to minimize

$$I = -\int_0^\tau ckx^a \, dt, \quad k = k_0 \exp\left(-\frac{E}{RT}\right)$$

subject to

$$\dot{x} = -akx^a, \quad x(0) = x(\tau)$$

using the temperature $T(t)$ as the control over a fixed time period τ.

The Hamiltonian for this problem is

$$H = -ckx^a - \lambda akx^a$$

Thus, for I to be minimum, the necessary conditions are as follows:

$$\dot{x} = \underbrace{-akx^a}_{H_\lambda}, \quad x(0) = x(\tau)$$

$$\dot{\lambda} = \underbrace{akx^{a-1}(c + \lambda)}_{H_x}, \quad \lambda(0) = \lambda(\tau)$$

$$H_T = -\frac{kE}{RT^2}x^a(c + a\lambda) = 0$$

Note that if τ is not fixed, then we have the additional necessary condition

$$H(\tau) = -\Big[ckx^a + \lambda akx^a \Big]_\tau = 0$$

▯

Example 8.2

Consider the CSTR described in Example 6.7 (p. 164) but under periodic operation in which the objective is to minimize

$$I = \int_0^\tau \sum_{i=1}^2 \big[(y_i - y_i^s)^2 + (u_i - u_i^s)^2 \big]\, dt$$

where τ is a fixed time period. The state is governed by

$$\dot{y}_1 = u_1(y_f - y_1) - ky_1y_2, \quad y_1(0) = y_1(\tau)$$

$$\dot{y}_2 = u_2 - u_1y_2, \qquad\qquad y_2(0) = y_2(\tau)$$

as well as the equality constraint

$$\frac{1}{\tau} \int_0^\tau u_1\, dt = a \qquad \text{or} \qquad \int_0^\tau u_1\, dt = a\tau$$

which requires the average CSTR throughput to be some desired value a. The Lagrangian for this problem [see Equation (6.19), p. 169] is given by

$$L = \sum_{i=1}^2 \big[(y_i - y_i^s)^2 + (u_i - u_i^s)^2 \big] + \lambda_1[u_1(y_f - y_1) - ky_1y_2]$$

$$+ \lambda_2[u_2 - u_1y_2] + \mu u_1$$

Note that the Lagrange multipliers λ_1 and λ_2 are time dependent, while the multiplier μ is time invariant. The necessary conditions for the minimum of

I are as follows:

$$\begin{bmatrix} \dot{y}_1 \\ \dot{y}_2 \end{bmatrix} = \underbrace{\begin{bmatrix} u_1(y_f - y_1) - k y_1 y_2 \\ u_2 - u_1 y_2 \end{bmatrix}}_{L_\lambda}, \qquad \begin{bmatrix} y_1 \\ y_2 \end{bmatrix}_{t=0} = \begin{bmatrix} y_1 \\ y_2 \end{bmatrix}_\tau$$

$$\begin{bmatrix} \dot{\lambda}_1 \\ \dot{\lambda}_2 \end{bmatrix} = \underbrace{\begin{bmatrix} -2(y_1 - y_1^s) + \lambda_1(u_1 + k y_2) \\ -2(y_2 - y_2^s) + \lambda_1 k y_1 + \lambda_2 u_1 \end{bmatrix}}_{-L_y}, \qquad \begin{bmatrix} \lambda_1 \\ \lambda_2 \end{bmatrix}_\tau^\mathsf{T} = \begin{bmatrix} \lambda_1 \\ \lambda_2 \end{bmatrix}_\tau^\mathsf{T}$$

$$\begin{bmatrix} L_{u_1} \\ L_{u_2} \end{bmatrix} = \begin{bmatrix} 2(u_1 - u_1^s) + \lambda_1(y_f - y_1) - \lambda_2 y_2 + \mu \\ 2(u_2 - u_2^s) + \lambda_2 \end{bmatrix} = \begin{bmatrix} 0 \\ 0 \end{bmatrix}$$

$$\int_0^\tau L_\mu \, dt = \int_0^\tau u_1 \, dt = a\tau$$

Inequality Constraint

If we do not want the CSTR throughput to surpass some maximum value a, then the equality constraint in the above problem has to be replaced with the inequality

$$\frac{1}{\tau} \int_0^\tau u_1 \, dt \le a \quad \text{or} \quad \int_0^\tau u_1 \, dt \le a\tau$$

In this case, the necessary conditions for the minimum are given by (see Section 6.4.2, p. 171)

- the ones above except $\int_0^\tau L_\mu \, dt = a\tau$, which is replaced with

$$\int_0^\tau L_\mu \, dt = \int_0^\tau u_1 \, dt \le a\tau$$

- $\mu \geq 0$, and the complimentary slackness condition

$$\mu\left(\int_0^\tau u_1 \, dt - a\tau \right) = 0$$

☐

8.2 Solution Methods

The solution of a optimal periodic control problem requires the integration of state and costate equations, both subject to periodicity conditions. Other than this integration aspect, the solution methods for optimal periodic control problems are similar to those for non-periodic problems. Therefore, we will focus on the methods to integrate state and costate equations under periodicity conditions.

A periodicity condition implies that the initial and final values of a state (or costate) variable are equal to a single value. Thus, in a optimal periodic control problem, the set of state as well as costate equations poses a two point boundary value problem. Either successive substitution or the shooting Newton–Raphson method may be used to integrate the periodic state and costate equations.

8.2.1 Successive Substitution Method

This is a simple but slow method in which a set of state (or costate) equations is integrated assuming the initial conditions. The final conditions obtained from integration are then substituted for the initial conditions in the next round of integration. This procedure is repeated until the initial and final conditions match.

8.2.2 Shooting Newton–Raphson Method

This method was introduced in Section 7.3 (p. 223) to solve a two point boundary value problem. We illustrate this method to solve a set of state equations with periodicity conditions. They require zeroing out the discrepancy function

$$\mathbf{f}(\mathbf{y}_0) \;=\; \underbrace{\mathbf{y}(\mathbf{y}_0)\Big|_\tau}_{\substack{\text{final state} \\ \text{vector}}} \;-\; \underbrace{\mathbf{y}_0}_{\substack{\text{initial state} \\ \text{vector}}}$$

where we use $\mathbf{y}_0$ to denote the initial state vector $\mathbf{y}(0)$. Starting with some guess, $\mathbf{y}(0)$ is improved iteratively using the Newton–Raphson method, i. e.,

$$\mathbf{y}_{0,\text{next}} = \mathbf{y}_0 - [\mathbf{f}_{\mathbf{y}_0}]_\tau^{-1}\mathbf{f}(\mathbf{y}_0) \tag{8.1}$$

where $[\mathbf{f}_{\mathbf{y}_0}]_\tau$ is the Jacobian

$$\begin{bmatrix} \dfrac{\partial y_1}{\partial y_{1,0}} - 1 & \dfrac{\partial y_1}{\partial y_{2,0}} & \cdots & \dfrac{\partial y_1}{\partial y_{n,0}} \\[3mm] \dfrac{\partial y_2}{\partial y_{1,0}} & \dfrac{\partial y_2}{\partial y_{2,0}} - 1 & \cdots & \dfrac{\partial y_2}{\partial y_{n,0}} \\[3mm] \vdots & \vdots & \vdots & \vdots \\[3mm] \dfrac{\partial y_n}{\partial y_{1,0}} & \dfrac{\partial y_n}{\partial y_{2,0}} & \cdots & \dfrac{\partial y_n}{\partial y_{n,0}} - 1 \end{bmatrix}_\tau \tag{8.2}$$

which is obtained by differentiating $\mathbf{f}(\mathbf{y}_0)$ with respect to $\mathbf{y}_0$, i. e., $\mathbf{y}(0)$.

Evaluation of the Jacobian

The above Jacobian is the matrix $\mathbf{f}_{\mathbf{y}_0}$ evaluated at the final time τ. The time dependent elements of $\mathbf{f}_{\mathbf{y}_0}$ are governed by differential equations that arise from the state equations. For example, differentiating the i-th state equation and the initial condition,

$$y_i = g_i \quad \text{and} \quad y_i(0) = y_{i,0}$$

with respect to $y_{j,0}$ yields the following differential equation:

$$\frac{d}{dt}\left(\frac{\partial y_i}{\partial y_{j,0}}\right) = \sum_{k=1}^{n} \frac{\partial g_i}{\partial y_k}\frac{\partial y_k}{\partial y_{j,0}}, \qquad \left[\frac{\partial y_i}{\partial y_{j,0}}\right]_{t=0} = \begin{cases} 1 & \text{if} \quad i = j \\ 0 & \text{if} \quad i \neq j \end{cases}$$

whose integration provides the desired value of $\partial y_i/\partial y_{j,0}$ at the final time.

The computational algorithm to solve state equations with periodicity conditions is as follows.

Computational Algorithm

1. Assume the initial conditions for the state equations.

2. Using the initial conditions, integrate simultaneously the differential equations for state variables and $\partial y_i/\partial y_{j,0}$s, which constitute $\mathbf{f}_{\mathbf{y}_0}$.

3. Improve the initial state using

$$\mathbf{y}_{0,\text{next}} = \mathbf{y}_0 - [\mathbf{f}_{\mathbf{y}_0}]_\tau^{-1}\mathbf{f}(\mathbf{y}_0)$$

Go to Step 2 until the improvement is negligible.

4. Repeat the above steps for costate equations.

Example 8.3

Consider a CSTR under periodic operation governed by the state equations

$$\frac{dy_1}{dt} = u_1(y_f - y_1) - \mathcal{R}, \qquad y_1(0) = y_1(\tau)$$

$$\frac{dy_2}{dt} = u_2 - u_1 y_2, \qquad y_2(0) = y_2(\tau)$$

where y_1 and y_2 are concentrations of the reactant and catalyst, $\mathcal{R} = k_0 y_1^2 y_2$ is the rate of production, y_f is the reactant concentration in the feed, and u_1 and u_2 are, respectively, the volumetric throughput and catalyst mass flow rate per unit reactor volume. The controls u_1 and u_2 have the time period τ. The objective is to find the controls and the time period that maximize the average $\mathcal{R}$, i.e., minimize

$$I = -\frac{1}{\tau} \int_0^\tau k_0 y_1^2 y_2 \, dt$$

Transformation to Fixed σ-interval

We begin by transforming the problem to the fixed σ-interval $[0, 1]$, as was done in Section 7.1.3.1 (p. 187). The equivalent transformed problem on this interval is to find the controls and the time period that minimize

$$I = \int_0^1 \underbrace{-k_0 y_1^2 y_2}_{F} \, d\sigma$$

subject to the state equations

$$\frac{dy_1}{d\sigma} = \tau\left[u_1(y_f - y_1) - k_0 y_1^2 y_2\right], \quad y_1(0) = y_1(1)$$

$$\frac{dy_2}{d\sigma} = \tau\left[u_2 - u_1 y_2\right], \qquad y_2(0) = y_2(1)$$

where the controls $u_1(\sigma)$ and $u_2(\sigma)$ have the time period of unity. The Hamiltonian for this problem is

$$H = -k_0 y_1^2 y_2 + \tau\left\{\lambda_1\left[u_1(y_f - y_1) - k_0 y_1^2 y_2\right] + \lambda_2(u_2 - u_1 y_2)\right\}$$

In terms of H, the augmented objective functional is given by

$$J = \int_0^1 \left(H - \lambda_1 \frac{dy_1}{d\sigma} - \lambda_2 \frac{dy_2}{d\sigma}\right) d\sigma$$

The necessary conditions for the minimum are

1. the state equations in the σ-interval

2. the costate equations

$$\frac{d\lambda_1}{d\sigma} = 2k_0 y_1 y_2 + \tau \lambda_1 (u_1 + 2k_0 y_1 y_2), \quad \lambda_1(0) = \lambda_1(1)$$

$$\frac{d\lambda_2}{d\sigma} = k_0 y_1^2 + \tau (\lambda_1 k_0 y_1^2 + \lambda_2 u_1), \qquad \lambda_2(0) = \lambda_2(1)$$

3. the stationarity conditions with respect to the controls u_1 and u_2, i.e.,

$$\begin{bmatrix} H_{u_1} \\ H_{u_2} \end{bmatrix} = \tau \begin{bmatrix} \lambda_1 (y_f - y_1) \\ \lambda_2 \end{bmatrix} = \begin{bmatrix} 0 \\ 0 \end{bmatrix}$$

4. the stationarity with respect to the time period τ, i.e.,

$$J_\tau = \int_0^1 \left(\frac{H - F}{\tau} \right) d\sigma = 0 \tag{8.3}$$

Satisfaction of Periodicity Conditions

In order to integrate the state and costate equations satisfying the periodicity conditions, we need the respective derivative differential equations for the shooting Newton–Raphson method.

Derivative State Equations

For each $i = \{1, 2\}$, taking the partial derivatives of

1. the state equation for y_i and

2. the initial condition $y_i(0) = y_{i,0}$

with respect to all guessed initial states — $y_{1,0}$ and $y_{2,0}$ — yields the following derivative state equations:

$$\frac{d}{d\sigma}\left(\frac{\partial y_1}{\partial y_{1,0}} \right) = \tau \left[-(u_1 + 2k_0 y_1 y_2)\frac{\partial y_1}{\partial y_{1,0}} - k_0 y_1^2 \frac{\partial y_2}{\partial y_{1,0}} \right], \quad \frac{\partial y_1}{\partial y_{1,0}}(0) = 1$$

$$\frac{d}{d\sigma}\left(\frac{\partial y_1}{\partial y_{2,0}} \right) = \tau \left[-(u_1 + 2k_0 y_1 y_2)\frac{\partial y_1}{\partial y_{2,0}} - k_0 y_1^2 \frac{\partial y_2}{\partial y_{2,0}} \right], \quad \frac{\partial y_1}{\partial y_{2,0}}(0) = 0$$

$$\frac{d}{d\sigma}\left(\frac{\partial y_2}{\partial y_{1,0}} \right) = -\tau u_1 \frac{\partial y_2}{\partial y_{1,0}}, \qquad\qquad\qquad\quad \frac{\partial y_2}{\partial y_{1,0}}(0) = 0$$

$$\frac{d}{d\sigma}\left(\frac{\partial y_2}{\partial y_{2,0}} \right) = -\tau u_1 \frac{\partial y_2}{\partial y_{2,0}}, \qquad\qquad\qquad\quad \frac{\partial y_2}{\partial y_{2,0}}(0) = 1$$

Derivative Costate Equations

Similar differentiation of the costate equations yields the following derivative costate equations:

$$\frac{d}{d\sigma}\left(\frac{\partial\lambda_1}{\partial\lambda_{1,0}}\right) = \tau(u_1 + 2k_0y_1y_2)\frac{\partial\lambda_1}{\partial\lambda_{1,0}}, \qquad \frac{\partial\lambda_1}{\partial\lambda_{1,0}}(0) = 1$$

$$\frac{d}{d\sigma}\left(\frac{\partial\lambda_1}{\partial\lambda_{2,0}}\right) = \tau(u_1 + 2k_0y_1y_2)\frac{\partial\lambda_1}{\partial\lambda_{2,0}}, \qquad \frac{\partial\lambda_1}{\partial\lambda_{2,0}}(0) = 0$$

$$\frac{d}{d\sigma}\left(\frac{\partial\lambda_2}{\partial\lambda_{1,0}}\right) = \tau\left[k_0y_1^2\frac{\partial\lambda_1}{\partial\lambda_{1,0}} + u_1\frac{\partial\lambda_2}{\partial\lambda_{1,0}}\right], \qquad \frac{\partial\lambda_2}{\partial\lambda_{1,0}}(0) = 0$$

$$\frac{d}{d\sigma}\left(\frac{\partial\lambda_2}{\partial\lambda_{2,0}}\right) = \tau\left[k_0y_1^2\frac{\partial\lambda_1}{\partial\lambda_{2,0}} + u_1\frac{\partial\lambda_2}{\partial\lambda_{2,0}}\right], \qquad \frac{\partial\lambda_2}{\partial\lambda_{2,0}}(0) = 1$$

The integration of the derivative state and costate equations provides the derivative values at the endpoint $\sigma = 1$. These values are needed in the Jacobian of the Newton–Raphson method [see Equations (8.1) and (8.2)].

Following is the computational algorithm of the shooting Newton–Raphson method to solve the optimal periodic control problem.

Computational Algorithm

1. Set the iteration counter $k = 0$. Assume τ^k and an even N. Obtain the fixed σ-grid of $(N + 1)$ equi-spaced grid points

$$\sigma_0(= 0), \quad \sigma_1, \quad \sigma_2, \quad \ldots, \quad \sigma_{N-1}, \quad \sigma_N(= 1)$$

At each grid point, assume a value for controls, initial state, and initial costate as follows:

$$\mathbf{u}_i^k \equiv \mathbf{u}^k(\sigma_i) = \begin{bmatrix} u_1^k(\sigma_i) \\ u_2^k(\sigma_i) \end{bmatrix}, \quad i = 0, 1, \ldots, (N - 1); \quad \text{and} \quad \mathbf{u}_N^k = \mathbf{u}_0^k$$

$$\mathbf{y}_0^k \equiv \mathbf{y}^k(\sigma_0) = \begin{bmatrix} y_1^k(\sigma_0) \\ y_2^k(\sigma_0) \end{bmatrix}, \quad \boldsymbol{\lambda}_0^k \equiv \lambda^k(\sigma_0) = \begin{bmatrix} \lambda_1^k(\sigma_0) \\ \lambda_2^k(\sigma_0) \end{bmatrix}$$

2. Integrate state equations as follows:

   2.a Set the counter $s = 0$. Assume the initial state

$$\mathbf{y}_0^s = \begin{bmatrix} y_{1,0}^k & y_{2,0}^k \end{bmatrix}^{\mathsf{T}}$$

2.b Integrate forward from $\sigma = 0$ to 1, the state equations along with derivative state equations using the control functions $\mathbf{u}_i^k$s.

2.c Improve the initial state by applying Equations (8.1) and (8.2), i.e.,

$$
\underbrace{\begin{bmatrix} y_{1,0}^{s+1} \\ y_{2,0}^{s+1} \end{bmatrix}}_{\mathbf{y}_0^{s+1}} = \underbrace{\begin{bmatrix} y_{1,0}^{s} \\ y_{2,0}^{s} \end{bmatrix}}_{\mathbf{y}_0^{s}} - \underbrace{\begin{bmatrix} \dfrac{\partial y_1}{\partial y_{1,0}} - 1 & \dfrac{\partial y_1}{\partial y_{2,0}} \\ \dfrac{\partial y_2}{\partial y_{1,0}} & \dfrac{\partial y_2}{\partial y_{2,0}} - 1 \end{bmatrix}_{\sigma=1}^{-1}}_{[\mathbf{f}_{\mathbf{y}_0}]_{\sigma=1}^{-1}} \underbrace{\begin{bmatrix} y_1(1) - y_{1,0}^{s} \\ y_2(1) - y_{2,0}^{s} \end{bmatrix}}_{\mathbf{f}(\mathbf{y}_0)}
$$

2.d Calculate the vector of change in $\mathbf{y}_0$,

$$
\mathbf{e} = \left[(y_{1,0}^{s+1} - y_{1,0}^{s}) \quad (y_{2,0}^{s+1} - y_{2,0}^{s}) \right]^{\top}
$$

Given a small real number $\epsilon > 0$, if

$$
\text{either} \quad \|\mathbf{e}\| \quad \text{or} \quad \left\| \mathbf{f}(\mathbf{y}^0) \right\| \ > \ \epsilon
$$

then assign $\mathbf{y}_0^{s+1} \to \mathbf{y}_0^{s}$, increment s by one, and go to Step 2b. Otherwise, the integration of the state equations is complete.

3. Save the values of state variables at the grid points.

$$
\mathbf{y}_i^k \equiv \mathbf{y}^k(\sigma_i) = \begin{bmatrix} y_1^s(\sigma_i) \\ y_2^s(\sigma_i) \end{bmatrix}, \quad i = 0, 1, \ldots, N
$$

4. Evaluate the objective functional. For example, using composite Simpson's 1/3 Rule,

$$
J^k = I^k = \frac{1}{3N} \left(F_0 + 4 \sum_{1,3,5,\ldots}^{N} F_i + 2 \sum_{i=2,4,6,\ldots}^{N} F_i + F_N \right)
$$

where

$$
F_i \equiv -k_0 (y_{1,i}^k)^2 y_{2,i}^k; \quad i = 0, 1, \ldots, N
$$

5. Check the improvement in I^k for $k > 0$. Given a tolerable error $\varepsilon_1 > 0$, if

$$
\left| I^k - I^{k-1} \right| < \varepsilon_1
$$

then go to Step 12.

6. Integrate costate equations as follows:

6.a Set the counter $s = 0$. Assume the initial costate

$$\boldsymbol{\lambda}_0^s = \begin{bmatrix} \lambda_{1,0}^k & \lambda_{2,0}^k \end{bmatrix}^\top$$

6.b Integrate forward from $\sigma = 0$ to 1, the costate equations along with the derivative costate equations using the control functions $\mathbf{u}_i^k$s and the state variables $\mathbf{y}_i^k$s.

6.c Improve the initial costate by applying Equations (8.1) and (8.2), i. e.,

$$\underbrace{\begin{bmatrix} \lambda_{1,0}^{s+1} \\ \lambda_{2,0}^{s+1} \end{bmatrix}}_{\boldsymbol{\lambda}_0^{s+1}} = \underbrace{\begin{bmatrix} \lambda_{1,0}^s \\ \lambda_{2,0}^s \end{bmatrix}}_{\boldsymbol{\lambda}_0^s} - \underbrace{\begin{bmatrix} \dfrac{\partial \lambda_1}{\partial \lambda_{1,0}} - 1 & \dfrac{\partial \lambda_1}{\partial \lambda_{2,0}} \\[3mm] \dfrac{\partial \lambda_2}{\partial \lambda_{1,0}} & \dfrac{\partial \lambda_2}{\partial \lambda_{2,0}} - 1 \end{bmatrix}_{\sigma=1}^{-1}}_{[\mathbf{f}_{\boldsymbol{\lambda}_0}]_{\sigma=1}^{-1}} \underbrace{\begin{bmatrix} \lambda_1(1) - \lambda_{1,0}^s \\ \lambda_2(1) - \lambda_{2,0}^s \end{bmatrix}}_{\mathbf{f}(\boldsymbol{\lambda}_0)}$$

6.d Calculate the vector of change in $\boldsymbol{\lambda}_0$,

$$\mathbf{e} = \begin{bmatrix} (\lambda_{1,0}^{s+1} - \lambda_{1,0}^s) & (\lambda_{2,0}^{s+1} - \lambda_{2,0}^s) \end{bmatrix}^\top$$

Given a small real number $\epsilon > 0$, if

$$\text{either} \quad \|\mathbf{e}\| \quad \text{or} \quad \|\mathbf{f}(\boldsymbol{\lambda}_0)\| \; > \; \epsilon$$

then assign $\boldsymbol{\lambda}_0^{s+1} \to \boldsymbol{\lambda}_0^s$, increment s by one, and go to Step 6b. Otherwise, the integration of the costate equations is complete.

7. Save the values of costate variables at the grid points.

$$\boldsymbol{\lambda}_i^k \equiv \boldsymbol{\lambda}^k(\sigma_i) = \begin{bmatrix} \lambda_1^s(\sigma_i) \\ \lambda_2^s(\sigma_i) \end{bmatrix}, \quad i = 0, 1, \ldots, N$$

8. Evaluate the gradient by calculating the partial derivatives

$$H_{\mathbf{u},i}^k = H_{\mathbf{u}}^k(\sigma_i) = \begin{bmatrix} H_{u_1}(\mathbf{y}_i^k, \mathbf{u}_i^k, \boldsymbol{\lambda}_i^k) \\ H_{u_2}(\mathbf{y}_i^k, \mathbf{u}_i^k, \boldsymbol{\lambda}_i^k) \end{bmatrix}; \quad i = 0, 1, \ldots, N$$

and $\quad J_\tau^k = \dfrac{1}{3N} \left(A_0^k + 4 \sum_{1,3,5,\ldots}^{N} A_i^k + 2 \sum_{i=2,4,6,\ldots}^{N} A_i^k + A_N^k \right)$

where $\quad A_i^k = \dfrac{H(\mathbf{y}_i^k, \mathbf{u}_i^k, \boldsymbol{\lambda}_i^k) - F(\mathbf{y}_i^k, \mathbf{u}_i^k, \boldsymbol{\lambda}_i^k)}{\tau^k}; \quad i = 0, 1, \ldots, N$

Check the magnitude of the gradient. Given a small positive real number ε_2, if the norm of the gradient

$$\sqrt{\sum_{i=0}^{N}\sum_{j=1}^{2}\left[H_{u_j}(\mathbf{y}_i^k, \mathbf{u}_i^k, \boldsymbol{\lambda}_i^k)\right]^2 + \left[J_\tau^k\right]^2} < \varepsilon_2$$

then go to Step 12.

9. Improve control functions by calculating

$$\mathbf{u}_i^{k+1} = \mathbf{u}_i^k - \epsilon H_{\mathbf{u},i}^k, \quad i = 0, 1, \ldots, N$$

where ϵ is a positive real number causing maximum reduction in I^k.

10. Improve the time period using

$$\tau^{k+1} = \tau^k - \epsilon J_\tau^k$$

11. Repeat calculations Step 2 onward after assigning

$$\tau_i^{k+1} \to \tau_i^k \quad \text{and} \quad \mathbf{u}_i^{k+1} \to \mathbf{u}_i^k, \quad i = 0, 1, \ldots, N$$

and incrementing k by one.

12. Terminate the algorithm with the following result:

- The optimal objective functional value is I^k.
- The optimal control $\hat{\mathbf{u}}(t)$ is represented by $\mathbf{u}_i^k$, $i = 0, 1, \ldots, N$.
- The optimal time period is τ^k.
- The optimal state $\hat{\mathbf{y}}(t)$ is represented by $\mathbf{y}_i^k$, $i = 0, 1, \ldots, N$.

Results

Using the above algorithm, the optimal periodic control problem was solved for the parameters listed in Table 8.1. The objective is to find two control functions and the time period that maximize the average reaction rate.

Table 8.1 Parameters for the problem of Example 8.3

y_f	5 g/cm^3
k_0	$6 \times 10^7 \ (\text{cm}^3/\text{g})^2/\text{min}$
N	40

The initial and optimal states are plotted in Figure 8.1. The corresponding controls are shown in Figure 8.2.

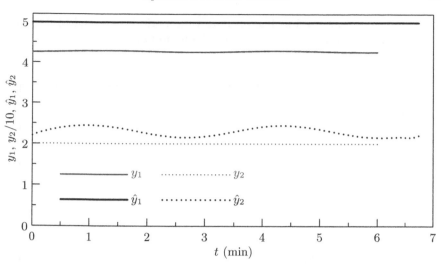

Figure 8.1 The initial and optimal states for Example 8.3

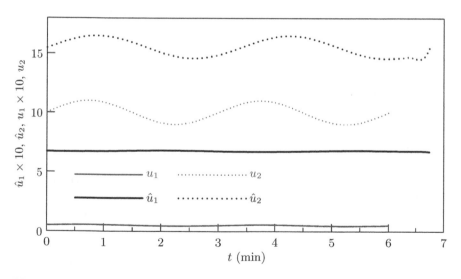

Figure 8.2 The initially guessed and optimal controls for Example 8.3

With the initially guessed time period of 6 min, the initial controls provided $I = -7.8 \times 10^{-2}$, which corresponds to the average production rate of 7.8×10^{-2} g/min.

The application of the algorithm minimized I, i.e., maximized the product concentration to 0.12 g/min. The optimal time period increased to 6.8 min. Figure 8.3 shows the convergence of I to the minimum.

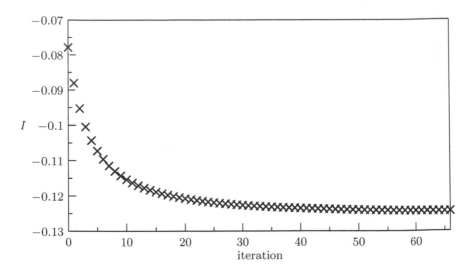

Figure 8.3 The objective functional versus iteration for Example 8.3

⬚

8.3 Pi Criterion

The pi criterion is a sufficient condition for the existence of a periodic solution that is better than the neighboring optimal steady state solution of an optimal periodic control problem. Using the criterion, we would like to know, for example, whether the time-averaged product concentration in a periodic process can be more than what the optimal steady state operation can provide. In other words, we would like to check if oscillating the optimal steady state control with some frequency and time period improves on the steady state solution.

We derive the criterion for a general periodic problem of finding the minimum of the functional

$$I = \frac{1}{\tau} \int_0^\tau F(\mathbf{y}, \mathbf{u})\, \mathrm{d}t, \quad \tau > 0 \tag{8.4}$$

subject to the following constraints:

$$\dot{\mathbf{y}} = \mathbf{g}(\mathbf{y}, \mathbf{u}), \quad \mathbf{y}(0) = \mathbf{y}(\tau) \tag{8.5}$$

$$\frac{1}{\tau} \int_0^\tau \mathbf{v}(\mathbf{y}, \mathbf{u}) \, dt = \mathbf{0} \tag{8.6}$$

$$\frac{1}{\tau} \int_0^\tau \mathbf{w}(\mathbf{y}, \mathbf{u}) \, dt \le \mathbf{0} \tag{8.7}$$

Integrating Equation (8.5) after dividing it by τ yields

$$\frac{1}{\tau} \int_0^\tau \mathbf{g}(\mathbf{y}, \mathbf{u}) \, dt = \mathbf{0}$$

Using the following notation for the integrals in the problem

$$I(\mathbf{y}, \mathbf{u}) \equiv \frac{1}{\tau} \int_0^\tau F(\mathbf{y}, \mathbf{u}) \, dt$$

$$I_i(\mathbf{y}, \mathbf{u}) \equiv \frac{1}{\tau} \int_0^\tau g_i(\mathbf{y}, \mathbf{u}) \, dt \qquad i = 1, 2, \ldots, n \tag{8.8}$$

$$I_i(\mathbf{y}, \mathbf{u}) \equiv \frac{1}{\tau} \int_0^\tau v_i(\mathbf{y}, \mathbf{u}) \, dt \qquad i = n+1, n+2, \ldots, p \tag{8.9}$$

$$I_i(\mathbf{y}, \mathbf{u}) \equiv \frac{1}{\tau} \int_0^\tau w_i(\mathbf{y}, \mathbf{u}) \, dt \qquad i = p+1, p+2, \ldots, q \tag{8.10}$$

we introduce the following functional

$$J(\mathbf{y}, \mathbf{u}, \boldsymbol{\lambda}, \boldsymbol{\mu}, \boldsymbol{\nu}) \equiv I(\mathbf{y}, \mathbf{u}) + \sum_{i=1}^n \lambda_i I_i(\mathbf{y}, \mathbf{u}) + \sum_{i=n+1}^p \mu_i I_i(\mathbf{y}, \mathbf{u}) + \sum_{i=p+1}^q \nu_i I_i(\mathbf{y}, \mathbf{u})$$

incorporating the multipliers λ_is, μ_is, and ν_is. Now in terms of the Lagrangian

$$L(\mathbf{x}, \mathbf{u}, \boldsymbol{\lambda}, \boldsymbol{\mu}, \boldsymbol{\nu}) \equiv F(\mathbf{y}, \mathbf{u}) + \sum_{i=1}^n \lambda_i g_i(\mathbf{y}, \mathbf{u}) + \sum_{i=n+1}^p \mu_i v_i(\mathbf{y}, \mathbf{u}) + \sum_{i=p+1}^q \nu_i w_i(\mathbf{y}, \mathbf{u}) \tag{8.11}$$

the functional J can be written as

$$J(\mathbf{y}, \mathbf{u}, \boldsymbol{\lambda}, \boldsymbol{\mu}, \boldsymbol{\nu}) = \frac{1}{\tau} \int_0^\tau L(\mathbf{y}, \mathbf{u}, \boldsymbol{\lambda}, \boldsymbol{\mu}, \boldsymbol{\nu}) \, dt$$

J at Optimal Steady State

Under the steady state, both $\mathbf{u}$ and $\mathbf{y}$ are time independent. Then the above problem reduces to the minimization of

$$I = F(\mathbf{y}, \mathbf{u})$$

subject to

$$\dot{\mathbf{y}} \;=\; \mathbf{g}(\mathbf{y}, \mathbf{u}) \;=\; \mathbf{0}$$

$$\frac{1}{\tau} \int_0^\tau \mathbf{v}(\mathbf{y}, \mathbf{u})\,dt \;=\; \mathbf{v}(\mathbf{y}, \mathbf{u}) \;=\; \mathbf{0}$$

$$\frac{1}{\tau} \int_0^\tau \mathbf{w}(\mathbf{y}, \mathbf{u})\,dt \;=\; \mathbf{w}(\mathbf{y}, \mathbf{u}) \;\leq\; \mathbf{0}$$

Let the pair $(\bar{\mathbf{y}}, \bar{\mathbf{u}})$ denote the optimal steady state solution with the corresponding multipliers $\bar{\boldsymbol{\lambda}}$, $\bar{\boldsymbol{\mu}}$, and $\bar{\boldsymbol{\nu}}$. The necessary conditions for this solution are provided in Appendix 8.A (p. 260). Since this solution minimizes J, satisfying the state equation and the constraints,

$$J(\bar{\mathbf{y}}, \bar{\mathbf{u}}, \bar{\boldsymbol{\lambda}}, \bar{\boldsymbol{\mu}}, \bar{\boldsymbol{\nu}}) \;=\; I(\bar{\mathbf{y}}, \bar{\mathbf{u}}) \;=\; \frac{1}{\tau} \int_0^\tau L(\bar{\mathbf{y}}, \bar{\mathbf{u}}, \bar{\boldsymbol{\lambda}}, \bar{\boldsymbol{\mu}}, \bar{\boldsymbol{\nu}})\,dt$$

In deriving the above equation, we have considered the fact that all equality constraints are satisfied at the optimal steady state, and the multipliers are zero for inactive inequality constraints (see Section 4.5, p. 109).

Further Reduction of I

We ultimately need to determine whether in the vicnity of the optimal steady state pair $(\bar{\mathbf{y}}, \bar{\mathbf{u}})$, the objective functional I could reduce further for some admissible pair of state and corresponding control, $(\mathbf{y}, \mathbf{u})$, where

$$\mathbf{y} = \bar{\mathbf{y}} + \delta\mathbf{y} \quad \text{and} \quad \mathbf{u} = \bar{\mathbf{u}} + \delta\mathbf{u}$$

The pair $(\bar{\mathbf{y}}, \bar{\mathbf{u}})$, as well as the control $\bar{\mathbf{u}}$, is called **proper**.

Assumptions

We assume that

1. the Lagrange Multiplier Rule is applicable

2. F is twice differentiable, and the derivatives of g_i, v_i, and w_i are continuous in the neighborhood of the optimal steady state solution $(\bar{\mathbf{y}}, \bar{\mathbf{u}})$

The first assumption implies that the constraint qualification or normality condition (see Section 4.3.2, p. 96) is satisfied. It ensures that an infinite number of solution pairs exist near the optimal steady state solution pair $(\bar{\mathbf{y}}, \bar{\mathbf{u}})$. This pair, as well as $\bar{\mathbf{u}}$, is called **normal** when the normality condition is satisfied.

Next, we retain the multipliers at the optimal steady state and obtain the change in J from its optimal steady state value, i.e., $J(\mathbf{y}, \mathbf{u}, \bar{\boldsymbol{\lambda}}, \bar{\boldsymbol{\mu}}, \bar{\boldsymbol{\nu}}) - J(\bar{\mathbf{y}}, \bar{\mathbf{u}}, \bar{\boldsymbol{\lambda}}, \bar{\boldsymbol{\mu}}, \bar{\boldsymbol{\nu}})$. This change is given by

$$I + \sum_{i=1}^{n} \bar{\lambda}_i I_i(\mathbf{y}, \mathbf{u}) + \sum_{i=n+1}^{p} \bar{\mu}_i I_i(\mathbf{y}, \mathbf{u}) + \sum_{i=p+1}^{q_0} \bar{\nu}_i I_i(\mathbf{y}, \mathbf{u}) - \bar{I}$$

$$= \frac{1}{\tau} \int_0^{\tau} (L - \bar{L}) \, dt = \frac{1}{\tau} \int_0^{\tau} \left(\delta\bar{L} + \frac{1}{2} \delta^2 \bar{L} \right) dt$$

where q_0 is the number of inequalities satisfied as equalities and the overbar as in $\bar{I}$ indicates the steady state value. In the last equation, we have used Equation (2.33) on p. 52 and considered a sufficiently small change along $\delta\mathbf{y}$ and $\delta\mathbf{u}$ at $\bar{\mathbf{y}}$ and $\bar{\mathbf{u}}$, respectively. Note that the multipliers are zero for inactive inequalities, which are satisfied as strict inequalities. Also, all $\bar{I}_i$s are zero for the q_0 equalities since they are satisfied by the steady state solution.

From the result proved in Appendix 8.B (p. 261) for the normal steady state pair $(\bar{\mathbf{y}}, \bar{\mathbf{u}})$, we have

$$I_j(\mathbf{y}, \mathbf{u}) = \delta\bar{I}_j \equiv \delta I_j(\bar{\mathbf{y}}, \bar{\mathbf{u}}; \delta\mathbf{y}, \delta\mathbf{u}); \quad j = 1, 2, \ldots, n + p + q_0 \tag{8.12}$$

Therefore, from the last two equations, we obtain

$$I + \sum_{i=1}^{n} \bar{\lambda}_i \delta\bar{I}_i + \sum_{i=n+1}^{p} \bar{\mu}_i \delta\bar{I}_i + \sum_{i=p+1}^{q_0} \bar{\nu}_i \delta\bar{I}_i - \bar{I} = \frac{1}{\tau} \int_0^{\tau} \left(\delta\bar{L} + \frac{1}{2} \delta^2 \bar{L} \right) dt \tag{8.13}$$

In terms of the variations of L and all I_is obtained from their definitions in Equations (8.11) and (8.8)–(8.10), we have

$$\frac{1}{\tau} \int_0^{\tau} \delta\bar{L} \, dt = \frac{1}{\tau} \int_0^{\tau} \left(\delta\bar{F} + \sum_{i=1}^{n} \bar{\lambda}_i \delta\bar{g}_i + \sum_{i=n+1}^{p} \bar{\mu}_i \delta\bar{v}_i + \sum_{i=p+1}^{q_0} \bar{\nu}_i \delta\bar{w}_i \right) dt$$

$$= \frac{1}{\tau} \int_0^{\tau} \delta\bar{F} \, dt + \sum_{i=1}^{n} \bar{\lambda}_i \delta\bar{I}_i + \sum_{i=n+1}^{p} \bar{\mu}_i \delta\bar{I}_i + \sum_{i=p+1}^{q_0} \bar{\nu}_i \delta\bar{I}_i$$

Substituting the last equation in Equation (8.13), we obtain

$$I - \bar{I} = \underbrace{\frac{1}{\tau} \int_0^{\tau} \delta\bar{F} \, dt}_{\delta\bar{I}} + \underbrace{\frac{1}{2\tau} \int_0^{\tau} \delta^2 \bar{L} \, dt}_{\delta^2 \bar{I}} \tag{8.14}$$

We shall return to the above equation after coming up with an admissible pair $(\mathbf{y}, \mathbf{u})$, or equivalently, an admissible variational pair $(\delta\mathbf{y}, \delta\mathbf{u})$.

Admissibility Criteria

The pair $(\mathbf{y}, \mathbf{u})$ is admissible if and only if it satisfies the constraints — Equations (8.5) and (8.6) and Inequality (8.7).

If $(\mathbf{y}, \mathbf{u})$ is admissible, then the related variational pair $(\delta\mathbf{y}, \delta\mathbf{u})$ is called admissible as well. This pair must satisfy some constraints, which are determined as follows. For $(\mathbf{y}, \mathbf{u})$ the constraints in terms of sufficiently small variations $\delta\mathbf{y}$ and $\delta\mathbf{u}$ near $(\bar{\mathbf{y}}, \bar{\mathbf{u}})$ become

$$\dot{\bar{\mathbf{y}}} + \delta\dot{\mathbf{y}} = \bar{\mathbf{g}} + \bar{\mathbf{g}}_\mathbf{y}\delta\mathbf{y} + \bar{\mathbf{g}}_\mathbf{u}\delta\mathbf{u}, \qquad \bar{\mathbf{y}}(0) + \delta\mathbf{y}(0) = \bar{\mathbf{y}}(\tau) + \delta\mathbf{y}(\tau)$$

$$\frac{1}{\tau}\int_0^\tau (\bar{\mathbf{v}} + \bar{\mathbf{v}}_\mathbf{y}\delta\mathbf{y} + \bar{\mathbf{v}}_\mathbf{u}\delta\mathbf{u})\,\mathrm{d}t = \mathbf{0}, \qquad \frac{1}{\tau}\int_0^\tau (\bar{\mathbf{w}} + \bar{\mathbf{w}}_\mathbf{y}\delta\mathbf{y} + \bar{\mathbf{w}}_\mathbf{u}\delta\mathbf{u})\,\mathrm{d}t \leq \mathbf{0}$$

The above constraints respectively simplify to

$$\delta\dot{\mathbf{y}} = \bar{\mathbf{g}}_\mathbf{y}\delta\mathbf{y} + \bar{\mathbf{g}}_\mathbf{u}\delta\mathbf{u}, \qquad \delta\mathbf{y}(0) = \delta\mathbf{y}(\tau) \qquad (8.15)$$

$$\frac{1}{\tau}\int_0^\tau (\bar{\mathbf{v}}_\mathbf{y}\delta\mathbf{y} + \bar{\mathbf{v}}_\mathbf{u}\delta\mathbf{u})\,\mathrm{d}t = \mathbf{0} \qquad\qquad (8.16)$$

$$\frac{1}{\tau}\int_0^\tau (\bar{\mathbf{w}}_\mathbf{y}\delta\mathbf{y} + \bar{\mathbf{w}}_\mathbf{u}\delta\mathbf{u})\,\mathrm{d}t \leq \mathbf{0} \qquad\qquad (8.17)$$

which are the equations that govern $\delta\mathbf{y}$ and $\delta\mathbf{u}$. Note that the above simplification is due to

$$\dot{\bar{\mathbf{y}}} = \bar{\mathbf{g}}, \quad \bar{\mathbf{y}}(0) = \bar{\mathbf{y}}(\tau), \quad \frac{1}{\tau}\int_0^\tau \bar{\mathbf{v}}\,\mathrm{d}t = \mathbf{0} \quad \text{and} \quad \frac{1}{\tau}\int_0^\tau \bar{\mathbf{w}}\,\mathrm{d}t \leq \mathbf{0}$$

which are obviously satisfied by the optimal steady state pair $(\bar{\mathbf{y}}, \bar{\mathbf{u}})$. As a consequence, the admissible variation pair $(\delta\mathbf{y}, \delta\mathbf{u})$ is constrained by Equations (8.15) and (8.16) and Inequality (8.17).

Admissible $(\delta\mathbf{y}, \delta\mathbf{u})$ at the Optimal Steady State

We propose the following control variation

$$\delta\mathbf{u}(t) = 2\mathbf{U}\cos(\omega t) \quad \text{where} \quad \omega \equiv 2\pi/\tau \qquad (8.18)$$

at the optimal steady state control $\bar{\mathbf{u}}$. After showing that the corresponding variational pair $(\delta\mathbf{y}, \delta\mathbf{u})$ is admissible, we will derive the sufficient condition for the resulting objective functional $I(\mathbf{y}, \mathbf{u})$ to be lower than the steady state $I(\bar{\mathbf{y}}, \bar{\mathbf{u}})$ or $\bar{I}$.

We now apply the admissibility criteria of the previous section. For the variation pair $(\delta\mathbf{y}, \delta\mathbf{u})$ to be admissible, $\delta\mathbf{u}(t)$ should generate a state variation $\delta\mathbf{y}$ governed by Equation (8.15). Moreover, the variation pair must satisfy the remaining constraints, i.e., Equation (8.16) and Inequality (8.17).

In the following treatment, we use the Fourier transform, the details of which are provided in Appendix 8.C (p. 264). Using Euler's formula

$$e^{ix} = \cos x + i \sin x$$

for $x \equiv \omega t$, the control variation can be expressed as

$$\delta\mathbf{u}(t) = \mathbf{U}\left(e^{-i\omega t} + e^{i\omega t}\right) \tag{8.19}$$

From the definition of the Fourier transform and its inverse

$$\delta\mathbf{u}(t) = \delta\mathbf{u}_{\langle-1\rangle}e^{-i\omega t} + \delta\mathbf{u}_{\langle1\rangle}e^{i\omega t}$$

where

$$\delta\mathbf{u}_{\langle-1\rangle} = \delta\mathbf{u}_{\langle1\rangle} = \mathbf{U}$$

are the Fourier coefficients of the proposed control variation $\delta\mathbf{u}$.

Satisfaction of Equation (8.15)

Taking the Fourier transform on both sides of the differential equation given by Equation (8.15), we obtain

$$\delta\dot{\mathbf{y}}_{\langle k\rangle} = (ik\omega)\delta\mathbf{y}_{\langle k\rangle} = \bar{\mathbf{g}}_{\mathbf{y}}\delta\mathbf{y}_{\langle k\rangle} + \bar{\mathbf{g}}_{\mathbf{u}}\delta\mathbf{u}_{\langle k\rangle}$$

Upon rearranging the transformed equation, we get

$$\delta\mathbf{y}_{\langle k\rangle} = \underbrace{(ik\omega\mathbf{I} - \bar{\mathbf{g}}_{\mathbf{y}})^{-1}\bar{\mathbf{g}}_{\mathbf{u}}}_{\mathbf{G}(ik\omega)} \delta\mathbf{u}_{\langle k\rangle} \tag{8.20}$$

For the proposed $\delta\mathbf{u}$, there are two Fourier coefficients $\delta\mathbf{u}_{\langle k=\pm1\rangle}$, each being equal to $\mathbf{U}$. Making these substitutions in the above equation and taking the inverse Fourier transform yields the corresponding

$$\delta\mathbf{y} = \overline{\mathbf{G}}\delta\mathbf{u}_{\langle-1\rangle}e^{-i\omega t} + \mathbf{G}\delta\mathbf{u}_{\langle1\rangle}e^{i\omega t} = \overline{\mathbf{G}}\mathbf{U}e^{-i\omega t} + \mathbf{G}\mathbf{U}e^{i\omega t} \tag{8.21}$$

where $\mathbf{G} \equiv \mathbf{G}(i\omega)$ and $\overline{\mathbf{G}} \equiv \mathbf{G}(-i\omega)$. Observe that $\delta\mathbf{y}(0) = \delta\mathbf{y}(\tau)$. Hence, the variational pair $(\delta\mathbf{y}, \delta\mathbf{u})$ satisfies Equation (8.15).

Satisfaction of Remaining Constraints

The variational pair $(\delta\mathbf{y}, \delta\mathbf{u})$ should also satisfy Equation (8.16) and Inequality (8.17) in order to be admissible. The left-hand side of an i-th constraint from Equation (8.16) or Inequality (8.17) is

$$\delta\bar{I}_i = \frac{1}{\tau}\int_0^\tau \left(\bar{a}_{\mathbf{y}}^\top\delta\mathbf{y} + \bar{a}_{\mathbf{u}}^\top\delta\mathbf{u}\right)dt; \quad i = (n+1), (n+2), \ldots, q \tag{8.22}$$

where $\bar{a}$ is, respectively, $\bar{v}_i$ or $\bar{w}_i$. Substituting for $\delta\mathbf{u}$ and $\delta\mathbf{y}$ from Equations (8.19) and (8.21), we get

$$\delta\bar{I}_i = \frac{1}{\tau}\int_0^\tau (\bar{a}_\mathbf{y}^\top[\overline{\mathbf{G}}\mathbf{U}e^{-i\omega t} + \mathbf{G}\mathbf{U}e^{i\omega t}] + \bar{a}_\mathbf{u}^\top\mathbf{U}[e^{-i\omega t} + e^{i\omega t}])\,\mathrm{d}t = 0$$

after integrating and applying the limits. Basically, we have applied the result

$$\int_0^\tau ce^{ki\omega t}\,\mathrm{d}t = \frac{c}{ki\omega}\left[e^{ki(2\pi/\tau)t}\right]_0^\tau = 0 \tag{8.23}$$

for a constant c and a non-zero integer k. Hence, the pair $(\delta\mathbf{y}, \delta\mathbf{u})$ satisfies Equation (8.16) and Inequality (8.17) as well, and thus is admissible.

Sufficient Condition or Pi Criterion

We now utilize Equation (8.14), which provides the change in I from its optimal steady state value. For the present problem, $\delta\bar{F}$ is zero since $(\bar{\mathbf{y}},\ \bar{\mathbf{u}})$ is optimal. Expanding $\delta^2\bar{L}$ in Equation (8.14), we obtain

$$I - \bar{I} = \frac{1}{2\tau}\int_0^\tau\left[(\delta\mathbf{y}^\top\bar{L}_{\mathbf{yy}} + \delta\mathbf{u}^\top\bar{L}_{\mathbf{yu}}^\top)\delta\mathbf{y} + (\delta\mathbf{y}^\top\bar{L}_{\mathbf{yu}} + \delta\mathbf{u}^\top\bar{L}_{\mathbf{uu}})\delta\mathbf{u}\right]\mathrm{d}t \tag{8.24}$$

$$= \mathbf{U}^\top\underbrace{\left(\overline{\mathbf{G}}^\top\bar{L}_{\mathbf{yy}}\mathbf{G} + \bar{L}_{\mathbf{yu}}^\top\mathbf{G} + \overline{\mathbf{G}}^\top\bar{L}_{\mathbf{yu}} + \bar{L}_{\mathbf{uu}}\right)}_{\boldsymbol{\Pi}(\omega)}\mathbf{U} \tag{8.25}$$

where $\boldsymbol{\Pi}(\omega)$ is an $(m \times m)$ matrix known as the **pi matrix**. Appendix 8.D (p. 265) provides the derivation of Equation (8.25). For I to be lower than $\bar{I}$, obviously

$$\boxed{I - \bar{I} = \mathbf{U}^\top\boldsymbol{\Pi}(\omega)\mathbf{U} < 0}$$

or, in other words, the pi matrix has to be negative-definite. This result is known as the pi criterion.

Example 8.4

Consider a second order reaction A $\longrightarrow$ B of rate coefficient k carried out in an isothermal CSTR. Assuming no volume change of mixing, the concentrations x_1 and x_2, respectively, of the reactant A and the product B are governed by the state equations

$$\dot{x}_1 = u_2(u_1 - x_1) - kx_1^2$$
$$\dot{x}_2 = kx_1^2 - u_2x_2$$

where the two controls are

1. u_1, the time dependent concentration of A fed to the reactor and

2. u_2, the volumetric flow rate in and out of the reactor

both per unit reactor volume. For the periodic operation of time period τ, the objective is to maximize the average concentration of the product B, i.e., to find the minimum of

$$I = -\frac{1}{\tau} \int_0^\tau x_2 \, dt$$

subject to the following two equality constraints:

1. The average mass of A fed to the reactor should be some value α, i.e.,

$$\frac{1}{\tau} \int_0^\tau u_1 u_2 \, dt = \alpha \qquad \text{or} \qquad \frac{1}{\tau} \int_0^\tau \underbrace{(u_1 u_2 - \alpha)}_{v_1} \, dt = 0$$

2. The average volumetric rate through the CSTR should be some value β, i.e.,

$$\frac{1}{\tau} \int_0^\tau u_2 \, dt = \beta \qquad \text{or} \qquad \frac{1}{\tau} \int_0^\tau \underbrace{(u_2 - \beta)}_{v_2} \, dt = 0$$

The Lagrangian for this problem is given by

$$L = \underbrace{-x_2}_{F} + \lambda_1 \underbrace{[u_2(u_1 - x_1) - kx_1^2]}_{g_1} + \lambda_2 \underbrace{[kx_1^2 - u_2 x_2]}_{g_2}$$
$$+ \mu_1 \underbrace{(u_1 u_2 - \alpha)}_{v_1} + \mu_2 \underbrace{(u_2 - \beta)}_{v_2}$$

Let the optimal state and control under steady state conditions be

$$\bar{\mathbf{x}} = \begin{bmatrix} \bar{x}_1 \\ \bar{x}_2 \end{bmatrix} \qquad \text{and} \qquad \bar{\mathbf{u}} = \begin{bmatrix} \bar{u}_1 \\ \bar{u}_2 \end{bmatrix}$$

Sufficient Condition for $(\bar{\mathbf{x}}, \bar{\mathbf{u}})$ to Be Proper

The optimal steady state pair $(\bar{\mathbf{x}}, \bar{\mathbf{u}})$ is proper if there exists a neighboring pair $(\mathbf{x}, \mathbf{u})$ that provides a lower objective functional value I than $\bar{I}$, which is given by $(\bar{\mathbf{x}}, \bar{\mathbf{u}})$. Assuming that the pair $(\bar{\mathbf{x}}, \bar{\mathbf{u}})$ is normal, the sufficient condition for its properness is that the right-hand side of Equation (8.25) should be less than zero, i.e., the pi criterion

$$\mathbf{U}^\top \left(\overline{\mathbf{G}}^\top \bar{L}_{\mathbf{xx}} \mathbf{G} + \bar{L}_{\mathbf{xu}}^\top \mathbf{G} + \overline{\mathbf{G}}^\top \bar{L}_{\mathbf{xu}} + \bar{L}_{\mathbf{uu}} \right) \mathbf{U} < 0$$

where $\mathbf{U} = \begin{bmatrix} U_1 & U_1 \end{bmatrix}^\top$ is the amplitude of $\delta \mathbf{u} = \begin{bmatrix} \delta u_1 & \delta u_2 \end{bmatrix}^\top$, which is the periodic control variation given by Equation (8.18) for some frequency $\omega = 2\pi/\tau$. Both $\mathbf{U}$ and ω need to be specified. The matrices in the above inequality are as follows.

The matrices $\mathbf{G}$ and $\overline{\mathbf{G}}$ are given by

$$\mathbf{G} = \mathbf{G}(i\omega) \quad \text{and} \quad \overline{\mathbf{G}} = \mathbf{G}(-i\omega)$$

where for $k = \pm 1$, we have from Equation (8.20)

$$\mathbf{G}(ik\omega) = \left\{ \underbrace{\begin{bmatrix} ik\omega & 0 \\ 0 & ik\omega \end{bmatrix}}_{\mathbf{I}(ik\omega)} - \underbrace{\begin{bmatrix} -\bar{u}_2 - 2k\bar{x}_1 & 0 \\ 2k\bar{x}_1 & -\bar{u}_2 \end{bmatrix}}_{\bar{\mathbf{g}}_\mathbf{x}} \right\}^{-1} \times$$

$$\underbrace{\begin{bmatrix} \bar{u}_2 & \bar{u}_1 - \bar{x}_1 \\ 0 & -\bar{x}_2 \end{bmatrix}}_{\bar{\mathbf{g}}_\mathbf{u}}$$

$\bar{L}_{\mathbf{xx}}$, $\bar{L}_{\mathbf{xu}}$, and $\bar{L}_{\mathbf{uu}}$ are the matrices of second order partial derivatives

$$\bar{L}_{\mathbf{xx}} = \begin{bmatrix} -2\bar{\lambda}_1 k + 2\bar{\lambda}_2 k & 0 \\ 0 & 0 \end{bmatrix} \qquad \bar{L}_{\mathbf{xu}} = \begin{bmatrix} 0 & -\bar{\lambda}_1 \\ 0 & -\bar{\lambda}_2 \end{bmatrix}$$

$$\bar{L}_{\mathbf{uu}} = \begin{bmatrix} 0 & \bar{\lambda}_1 + \bar{\mu}_1 \\ \bar{\lambda}_1 + \bar{\mu}_1 & 0 \end{bmatrix}$$

which arise from the vectors of partial derivatives

$$L_\mathbf{x} = \begin{bmatrix} \lambda_1(-u_2 - 2kx_1) + 2\lambda_2 kx_1 \\ -1 - \lambda_2 u_2 \end{bmatrix}$$

$$L_\mathbf{u} = \begin{bmatrix} \lambda_1 u_2 + \mu_1 u_2 \\ \lambda_1(u_1 - x_1) - \lambda_2 x_2 + \mu_1 u_1 + \mu_2 \end{bmatrix}$$

$\square$

8.4 Pi Criterion with Control Constraints

As long as the optimal steady state controls are normal and lie within but not at the boundaries of the set of admissible control values, the negative-definiteness of the pi matrix is the sufficient condition for the neighboring controls to be optimally better. This condition does not hold if an optimal steady state control $\bar{u}_i$ lies at its boundary $u_{i,\max}$, as shown in Figure 8.4. In this case, δu_i given by Equation (8.19) becomes invalid since $u_i = \bar{u}_i + \delta u_i$ trespasses the boundary for any non-zero U_0. As shown in the figure, this situation can be avoided by adding a suitable constant v_i to the right-hand side of Equation (8.19) such that u_i lies within the boundary.

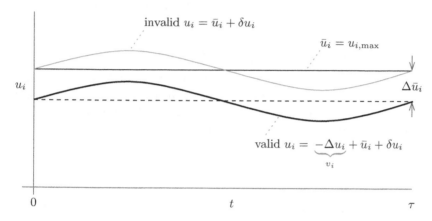

Figure 8.4 Controls near the upper limit $u_{i,\max}$

However, we need to derive the sufficient condition again. Let the control variation be given by

$$\delta \mathbf{u} = \mathbf{V} + 2\mathbf{U}\cos(\omega t) = \mathbf{V} + \mathbf{U}\big(e^{-i\omega t} + e^{i\omega t}\big), \quad \omega = \frac{2\pi}{\tau} \tag{8.26}$$

where an element V_i of $\mathbf{V}$ is

- zero if $\bar{u}_i$ is within and away from the control boundary, and

- non-zero if $\bar{u}_i$ is at the control boundary — the value of V_i being such that u_i is within the control boundary.

From the definition of the Fourier transform and its inverse (see Appendix 8.C, p. 264)

$$\delta \mathbf{u}(t) = \delta \mathbf{u}_{\langle -1\rangle} e^{-i\omega t} + \delta \mathbf{u}_{\langle 0\rangle} + \delta \mathbf{u}_{\langle 1\rangle} e^{i\omega t}$$

where

$$\delta \mathbf{u}_{\langle 0\rangle} = \mathbf{V} \quad \text{and} \quad \delta \mathbf{u}_{\langle -1\rangle} = \delta \mathbf{u}_{\langle 1\rangle} = \mathbf{U}$$

are the Fourier coefficients of the proposed control variation $\delta \mathbf{u}$.

Satisfaction of Equation (8.15)

Substituting the above Fourier coefficients of $\delta\mathbf{u}$ in Equation (8.20), we obtain

$$\delta\mathbf{y} = \overline{\mathbf{G}}\mathbf{U}e^{-i\omega t} + \widehat{\mathbf{G}}\mathbf{V} + \mathbf{G}\mathbf{U}e^{i\omega t} \tag{8.27}$$

where

$$\widehat{\mathbf{G}} \equiv \mathbf{G}(0) = -\bar{\mathbf{g}}_{\mathbf{y}}^{-1}\bar{\mathbf{g}}_{\mathbf{u}}$$

Observe that $\delta\mathbf{y}$ so obtained satisfies Equation (8.15) as well as the periodicity condition $\delta\mathbf{y}(0) = \delta\mathbf{y}(\tau)$.

Satisfaction of Remaining Constraints

To be admissible, the variational pair $(\delta\mathbf{y}, \delta\mathbf{u})$ thus obtained should satisfy the constraints given by Equation (8.16) or Inequality (8.17). Thus, the corresponding $\delta\bar{I}_i$ given by Equation (8.22) should be zero for the equality constraints and non-positive for the inequality constraints. In the foregoing treatment, we assume that these constraints are satisfied so that $(\delta\mathbf{y}, \delta\mathbf{u})$ is admissible. In other words, the corresponding pair $(\mathbf{y}, \mathbf{u})$ satisfies Equation (8.6) and Inequality (8.7) on p. 249.

Sufficient Condition or Pi Criterion with Control Constraints

Switching to Equation (8.14) on p. 251 for the change in I from its optimal steady state value $\bar{I}$, we should appreciate that $\delta\bar{F}$ is not necessarily zero in the presence of control constraints. For example, if all controls under the optimal steady state lie within the control boundaries except $\bar{u}_1$, which is at its upper limit $u_{1,\max}$, then for sufficiently small variations $(\delta\mathbf{y}, \delta\mathbf{u})$ for which $\delta^2\bar{I}$ tends to zero, Equation (8.14) yields

$$I - \bar{I} = \frac{1}{\tau}\int_0^\tau \delta\bar{F}\,dt = \frac{1}{\tau}\int_0^\tau \sum_{i=1}^n \bar{F}_{u_i}\delta u_i\,dt = \frac{1}{\tau}\int_0^\tau \bar{F}_{u_1}\delta u_1\,dt \geq 0$$

because of the following reasons:

- Inside the control boundaries, $\bar{F}_{u_i}$ for $i > 1$ must be zero as δu_i can be either positive or negative. Otherwise, for a non-zero $\bar{F}_{u_i}$, one could choose a suitable δu_i to contradict the above inequality, i.e., the minimality of $\bar{I}$.

- $I - \bar{I} \geq 0$ since $\bar{I}$ is the minimum.

The term $\bar{F}_{u_1}$ is not necessarily zero. With $\bar{u}_1$ at its upper limit, δu_1 can only be negative. Its coefficient $\bar{F}_{u_1}$ thus could be either zero or negative for $\bar{I}$ to be minimum. Consequently, $\delta\bar{F}$ is not necessarily zero.

Properness of $(\bar{\mathbf{y}}, \bar{\mathbf{u}})$

For the pair $(\bar{\mathbf{y}}, \bar{\mathbf{u}})$ to be proper, $I - \bar{I} \leq 0$. From Equation (8.14)

$$I - \bar{I} = \underbrace{\frac{1}{\tau} \int_0^\tau \delta \bar{F} \, dt}_{\delta \bar{I}} + \underbrace{\frac{1}{2\tau} \int_0^\tau \delta^2 \bar{L} \, dt}_{\delta^2 \bar{I}}$$

Then $(I - \bar{I})$ is less than zero if

- either $\delta \bar{I} = 0$ and $\delta^2 \bar{I}$ is less than zero, or

- both $\delta \bar{I}$ and $\delta^2 \bar{I}$ are less than zero.

Upon expanding $\delta \bar{F}$ and $\delta^2 \bar{L}$, we obtain

$$I - \bar{I} = \frac{1}{\tau} \int_0^\tau (\bar{F}_{\mathbf{y}} \delta \mathbf{y} + \bar{F}_{\mathbf{u}} \delta \mathbf{u}) \, dt \ +$$

$$\frac{1}{2\tau} \int_0^\tau \left[(\delta \mathbf{y}^\top \bar{L}_{\mathbf{yy}} + \delta \mathbf{u}^\top \bar{L}_{\mathbf{yu}}^\top) \delta \mathbf{y} + (\delta \mathbf{y}^\top \bar{L}_{\mathbf{yu}} + \delta \mathbf{u}^\top \bar{L}_{\mathbf{uu}}) \delta \mathbf{u} \right] \, dt$$

Integrating the right-hand side of the above equation after substituting for $\delta \mathbf{u}$ and $\delta \mathbf{y}$, respectively, from Equations (8.26) and (8.27), we finally get

$$I - \bar{I} = \underbrace{(F_{\mathbf{y}} \widehat{\mathbf{G}} + F_{\mathbf{u}}) \mathbf{V}}_{\delta \bar{I}} + \underbrace{\frac{\mathbf{V}^\top \mathbf{\Pi}(0) \mathbf{V}}{2} + \mathbf{U}^\top \mathbf{\Pi}(\omega) \mathbf{U}}_{\delta^2 \bar{I}}$$

Hence the sufficient condition for $(\bar{\mathbf{y}}, \bar{\mathbf{u}})$ to be proper is that

$$\boxed{(F_{\mathbf{y}} \widehat{\mathbf{G}} + F_{\mathbf{u}}) \mathbf{V} \leq 0 \qquad \text{and} \qquad \frac{\mathbf{V}^\top \mathbf{\Pi}(0) \mathbf{V}}{2} + \mathbf{U}^\top \mathbf{\Pi}(\omega) \mathbf{U} < 0} \qquad (8.28)$$

where the matrices $\mathbf{\Pi}(0)$ and $\mathbf{\Pi}(\omega)$ are given by, respectively,

$$\mathbf{\Pi}(0) = \widehat{\mathbf{G}}^\top \bar{L}_{\mathbf{yy}} \widehat{\mathbf{G}} + \bar{L}_{\mathbf{yu}}^\top \widehat{\mathbf{G}} + \widehat{\mathbf{G}}^\top \bar{L}_{\mathbf{yu}} + \bar{L}_{\mathbf{uu}} \quad \text{and}$$

$$\mathbf{\Pi}(\omega) = \overline{\mathbf{G}}^\top \bar{L}_{\mathbf{yy}} \mathbf{G} + \bar{L}_{\mathbf{yu}}^\top \mathbf{G} + \overline{\mathbf{G}}^\top \bar{L}_{\mathbf{yu}} + \bar{L}_{\mathbf{uu}}$$

Example 8.5

Consider Example 8.4 (p. 254) with the additional control inequality constraint $\mathbf{u} \leq \mathbf{u}_{\max}$.

The sufficient conditions for properness of the optimal steady state pair $(\bar{\mathbf{x}}, \bar{\mathbf{u}})$ stay the same as long as $\bar{\mathbf{u}}$ is less than $\mathbf{u}_{\max}$. However, if a $\bar{u}_i = u_{i,\max}$, then an appropriate control variation is given by Equation (8.26) in which the vector $\mathbf{V} = \begin{bmatrix} V_1 & V_2 \end{bmatrix}^\top$ ensures that $\mathbf{u} = \bar{\mathbf{u}} + \delta\mathbf{u}$ does not exceed $\mathbf{u}_{\max}$. Hence, V_i is

- zero if the corresponding $\bar{u}_i < u_{i,\max}$, or

- some suitable value to ensure that the neighboring perturbed control $u_i < u_{i,\max}$ if $\bar{u}_i = u_{i,\max}$.

Thus, provided that the pair $(\mathbf{y}, \mathbf{u})$ satisfies the constraints of the problem, the sufficient conditions for properness, from Inequalities (8.28), are

$$(F_{\mathbf{x}}\widehat{\mathbf{G}} + F_{\mathbf{u}})\mathbf{V} \leq 0 \qquad \text{and} \qquad \frac{\mathbf{V}^\top \mathbf{\Pi}(0)\mathbf{V}}{2} + \mathbf{U}^\top \mathbf{\Pi}(\omega)\mathbf{U} < 0$$

where

$$F_{\mathbf{x}} = \begin{bmatrix} 0 \\ -1 \end{bmatrix} \quad \text{and} \quad F_{\mathbf{u}} = \begin{bmatrix} 0 \\ 0 \end{bmatrix}$$

$$\widehat{\mathbf{G}} = \mathbf{G}(0) = -\underbrace{\begin{bmatrix} -\bar{u}_2 - 2k\bar{x}_1 & 0 \\ 2k\bar{x}_1 & -\bar{u}_2 \end{bmatrix}^{-1}}_{\bar{\mathbf{g}}_{\mathbf{x}}} \underbrace{\begin{bmatrix} \bar{u}_2 & \bar{u}_1 - \bar{x}_1 \\ 0 & -\bar{x}_2 \end{bmatrix}}_{\bar{\mathbf{g}}_{\mathbf{u}}}$$

$$\mathbf{\Pi}(0) = \widehat{\mathbf{G}}^\top \bar{L}_{\mathbf{xx}}\widehat{\mathbf{G}} + \bar{L}_{\mathbf{xu}}^\top \widehat{\mathbf{G}} + \widehat{\mathbf{G}}^\top \bar{L}_{\mathbf{xu}} + \bar{L}_{\mathbf{uu}}$$

$$\mathbf{\Pi}(\omega) = \overline{\mathbf{G}}^\top \bar{L}_{\mathbf{xx}}\mathbf{G} + \bar{L}_{\mathbf{xu}}^\top \mathbf{G} + \overline{\mathbf{G}}^\top \bar{L}_{\mathbf{xu}} + \bar{L}_{\mathbf{uu}}$$

The matrices $\mathbf{G}$, $\overline{\mathbf{G}}$, $\bar{L}_{\mathbf{xx}}$, $\bar{L}_{\mathbf{xu}}$, and $\bar{L}_{\mathbf{uu}}$ are the same as derived earlier in Example 8.4. □

8.A   Necessary Conditions for Optimal Steady State

We present the necessary conditions for the steady state minimum in the problem described by Equations (8.4)–(8.7) on p. 248.

Let $\bar{\mathbf{u}}$ be the optimal control under steady state with the corresponding state $\bar{\mathbf{y}}$, and multipliers $\bar{\boldsymbol{\lambda}}$, $\bar{\boldsymbol{\mu}}$, and $\bar{\boldsymbol{\nu}}$; all of which are time invariant. According to the John Multiplier Theorem (Section 4.5.1, p. 113), the necessary conditions for the minimum of I subject to the equality and inequality constraints are as follows:

$$\mathbf{g}(\bar{\mathbf{y}}, \bar{\mathbf{u}}) = \mathbf{0}$$
$$\mathbf{v}(\bar{\mathbf{y}}, \bar{\mathbf{u}}) = \mathbf{0}$$

$$\left.\begin{array}{ll} \bar{\nu}_i \geq 0 & \text{when} \quad w_i(\bar{\mathbf{y}}, \bar{\mathbf{u}}) = 0 \\ \text{and } \bar{\nu}_i = 0 & \text{when} \quad w_i(\bar{\mathbf{y}}, \bar{\mathbf{u}}) < 0 \end{array}\right\} \quad \text{for } i = p+1, p+2, \ldots, q$$

$$L_{\mathbf{y}}(\bar{\mathbf{y}}, \bar{\mathbf{u}}, \bar{\boldsymbol{\lambda}}, \bar{\boldsymbol{\mu}}, \bar{\boldsymbol{\nu}}) = H_{\mathbf{y}}(\bar{\mathbf{y}}, \bar{\mathbf{u}}, \bar{\boldsymbol{\lambda}}, \bar{\boldsymbol{\mu}}, \bar{\boldsymbol{\nu}}) + \bar{\boldsymbol{\mu}}^\top \mathbf{v}_{\mathbf{y}}(\bar{\mathbf{y}}, \bar{\mathbf{u}}) + \bar{\boldsymbol{\nu}}^\top \mathbf{w}_{\mathbf{y}}(\bar{\mathbf{y}}, \bar{\mathbf{u}}) = 0$$

$$L_{\mathbf{u}}(\bar{\mathbf{y}}, \bar{\mathbf{u}}, \bar{\boldsymbol{\lambda}}, \bar{\boldsymbol{\mu}}, \bar{\boldsymbol{\nu}}) = H_{\mathbf{u}}(\bar{\mathbf{y}}, \bar{\mathbf{u}}, \bar{\boldsymbol{\lambda}}, \bar{\boldsymbol{\mu}}, \bar{\boldsymbol{\nu}}) + \bar{\boldsymbol{\mu}}^\top \mathbf{v}_{\mathbf{u}}(\bar{\mathbf{y}}, \bar{\mathbf{u}}) + \bar{\boldsymbol{\nu}}^\top \mathbf{w}_{\mathbf{u}}(\bar{\mathbf{y}}, \bar{\mathbf{u}}) = 0$$

8.B Derivation of Equation (8.12)

If we have
1. an admissible and normal pair, $(\bar{\mathbf{y}}, \bar{\mathbf{u}})$,
2. an admissible variation pair $(\delta\mathbf{z}, \delta\mathbf{v})$, and
3. continuous first and second order partial derivatives of the integrands of I_j, $j = 1, 2, \ldots, q$

then there is an infinite number of admissible pairs $(\mathbf{y}, \mathbf{u})$
1. whose variations, $\delta\mathbf{y}$ and $\delta\mathbf{u}$, are, respectively, equal to $\delta\mathbf{z}$ and $\delta\mathbf{v}$; and
2. which satisfy Equation (8.12) on p. 251, i.e.,

$$I_j(\mathbf{y}, \mathbf{u}) = \delta\bar{I}_j \equiv \delta I_j(\bar{\mathbf{y}}, \bar{\mathbf{u}}; \delta\mathbf{y}, \delta\mathbf{u}), \quad j = 1, 2, \ldots, n+p+q_0$$

where $\delta\bar{I}_j$ is the variation of I_j at $(\bar{\mathbf{y}}, \bar{\mathbf{u}})$ along $(\delta\mathbf{y}, \delta\mathbf{u})$, and q_0 is the number of inequality constraints satisfied as the equalities

$$I_j(\bar{\mathbf{y}}, \bar{\mathbf{u}}) = 0, \quad j = p+1, p+2, \ldots, q_0$$

We prove this result in five steps.

Step 1 Let $k \equiv n + p + q_0$. Then since $(\bar{\mathbf{y}}, \bar{\mathbf{u}})$ is normal, there exists a set of variation pairs

$$(\delta\mathbf{y}_j, \delta\mathbf{u}_j); \quad j = 1, 2, \ldots, k$$

such that the determinant

$$D = \begin{vmatrix} \delta\bar{I}_1(\delta\mathbf{y}_1, \delta\mathbf{u}_1) & \delta\bar{I}_1(\delta\mathbf{y}_2, \delta\mathbf{u}_2) & \cdots & \delta\bar{I}_1(\delta\mathbf{y}_k, \delta\mathbf{u}_k) \\ \delta\bar{I}_2(\delta\mathbf{y}_1, \delta\mathbf{u}_1) & \delta\bar{I}_2(\delta\mathbf{y}_2, \delta\mathbf{u}_2) & \cdots & \delta\bar{I}_2(\delta\mathbf{y}_k, \delta\mathbf{u}_k) \\ \vdots & \vdots & \vdots & \vdots \\ \delta\bar{I}_k(\delta\mathbf{y}_1, \delta\mathbf{u}_1) & \delta\bar{I}_k(\delta\mathbf{y}_2, \delta\mathbf{u}_2) & \cdots & \delta\bar{I}_k(\delta\mathbf{y}_k, \delta\mathbf{u}_k) \end{vmatrix} \neq 0 \quad (8.29)$$

Let there be a control

$$\mathbf{u} \equiv \bar{\mathbf{u}} + \sum_{j=1}^{k} \beta_j \delta\mathbf{u}_j + \epsilon\delta\mathbf{v}$$

for some real numbers ϵ and β_1, β_2, ..., β_k. Dependent on these numbers there are infinite such controls. From the Embedding Theorem (Hestenes, 1966) there exists for each $\mathbf{u}$ a unique solution $\mathbf{y}$ of the differential equations

$$\dot{\mathbf{y}} = \mathbf{g}(\mathbf{y}, \mathbf{u}); \quad \mathbf{y}(0) = \bar{\mathbf{y}}(0) + \sum_{j=1}^{k} \beta_j \delta \mathbf{y}_j(0) + \epsilon \delta \mathbf{z}(0) \qquad (8.30)$$

Thus, there are infinitely many admissible pairs $(\mathbf{y}, \mathbf{u})$. Each $\mathbf{y}$ has continuous first and second partial derivatives with respect to

$$\boldsymbol{\alpha} \equiv \begin{bmatrix} \beta_1 & \beta_2 & \cdots & \beta_k & \epsilon \end{bmatrix}^{\top}$$

for the non-zero norm $\|\boldsymbol{\alpha}\| < \delta_0$ for some $\delta_0 > 0$. Moreover, these partial derivatives are piecewise continuous so that*

$$\left.\frac{\partial \mathbf{y}}{\partial \epsilon}\right|_{\boldsymbol{\alpha}=0} = \delta \mathbf{z}, \quad \text{and} \quad \left.\frac{\partial \mathbf{y}}{\partial \beta_j}\right|_{\boldsymbol{\alpha}=0} = \delta \mathbf{y}_j; \quad j = 1, 2, \ldots, k \qquad (8.31)$$

Step 2 Next, consider the equations

$$F_j(\boldsymbol{\beta}, \epsilon) \equiv I_j(\mathbf{y}, \mathbf{u}) - \epsilon \delta I_j(\bar{\mathbf{y}}, \bar{\mathbf{u}}; \delta \mathbf{z}, \delta \mathbf{v}) = 0; \quad j = 1, 2, \ldots, k \qquad (8.32)$$

They are obviously satisfied at $\boldsymbol{\alpha} = 0$ for which each $I_j = I_j(\bar{\mathbf{y}}, \bar{\mathbf{u}}) = 0$ as well as $\epsilon = 0$. Now the partial derivative of F_1 with respect to β_2 is

$$\frac{\partial F_1}{\partial \beta_2} = \frac{\partial I_1}{\partial \beta_2} = \frac{1}{\tau} \int_0^\tau \frac{\partial g_1(\mathbf{y}, \mathbf{u})}{\partial \beta_2}\, dt = \frac{1}{\tau} \int_0^\tau \sum_{i=1}^n \left(\frac{\partial g_1}{\partial y_i}\frac{\partial y_i}{\partial \beta_2} + \frac{\partial g_1}{\partial u_i}\frac{\partial u_i}{\partial \beta_2} \right) dt$$

$$= \frac{1}{\tau} \int_0^\tau \sum_{i=1}^n \left(\frac{\partial g_1}{\partial y_i}\delta y_{2i} + \frac{\partial g_1}{\partial u_i}\delta u_{2i} \right) dt = \delta I_1(\mathbf{y}, \mathbf{u}; \delta \mathbf{y}_2, \delta \mathbf{u}_2)$$

Therefore,

$$\left[\frac{\partial F_1}{\partial \beta_2}\right]_{\boldsymbol{\alpha}=0} = \delta I_1(\bar{\mathbf{y}}, \bar{\mathbf{u}}; \delta \mathbf{y}_2, \delta \mathbf{u}_2) = \delta \bar{I}_1(\delta \mathbf{y}_2, \delta \mathbf{u}_2)$$

In general,

$$\frac{\partial F_i}{\partial \beta_j} = \delta \bar{I}_i(\delta \mathbf{y}_j, \delta \mathbf{u}_j) \quad \text{at} \quad \boldsymbol{\alpha} = 0; \quad i, j = 1, 2, \ldots, k \qquad (8.33)$$

---

* For example, $\delta \mathbf{y}_1(t) = \frac{\partial}{\partial \beta_1}\mathbf{y}(t)$ at $\boldsymbol{\alpha} = 0$ is defined by the equations

$$\frac{\mathrm{d}}{\mathrm{d}t}\left(\frac{\partial \mathbf{y}}{\partial \beta_1}\right) = \sum_{i=1}^n \frac{\partial \mathbf{g}}{\partial y_i}\frac{\partial y_i}{\partial \beta_1} + \sum_{i=1}^m \frac{\partial \mathbf{g}}{\partial u_i}\frac{\partial u_i}{\partial \beta_1}; \quad \left.\frac{\partial \mathbf{y}}{\partial \beta_1}\right|_{t=0} = \delta \mathbf{y}_1(0)$$

which are obtained from Equations (8.30) at $\boldsymbol{\alpha} = 0$.

As a result, the Jacobian of $\mathbf{F} \equiv \begin{bmatrix} F_1 & F_2 & \ldots & F_k \end{bmatrix}^{\mathsf{T}}$ with respect to $\boldsymbol{\beta} \equiv \begin{bmatrix} \beta_1 & \beta_2 & \ldots & \beta_k \end{bmatrix}^{\mathsf{T}}$ at $\boldsymbol{\alpha} = \mathbf{0}$ is the non-zero determinant D in Equation (8.29). From the Implicit Function Theorem (Section 9.16, p. 277), Equations (8.32) are valid in an open region around $\boldsymbol{\beta} = \mathbf{0}$ and have solutions

$$\beta_j = b_j(\epsilon); \quad |\epsilon| < \delta_1 > 0; \quad j = 1, 2, \ldots, k \tag{8.34}$$

where the partial derivatives $\partial b_j / \partial \epsilon$ are continuous, and

$$b_j(0) = 0; \quad j = 1, 2, \ldots, k$$

Step 3 Differentiating Equations (8.32) with respect to ϵ at $\epsilon = 0$, and utilizing Equation (8.33) as well as the similarly derived equations

$$\frac{\mathrm{d}I_j}{\mathrm{d}\epsilon} = \delta I_j(\bar{\mathbf{y}}, \bar{\mathbf{u}}; \delta \mathbf{z}, \delta \mathbf{v}) \quad \text{at} \quad \epsilon = 0; \quad j = 1, 2, \ldots, k$$

we obtain

$$\begin{bmatrix} \delta \bar{I}_1(\delta \mathbf{y}_1, \delta \mathbf{u}_1) & \delta \bar{I}_1(\delta \mathbf{y}_2, \delta \mathbf{u}_2) & \ldots & \delta \bar{I}_1(\delta \mathbf{y}_k, \delta \mathbf{u}_k) \\ \delta \bar{I}_2(\delta \mathbf{y}_1, \delta \mathbf{u}_1) & \delta \bar{I}_2(\delta \mathbf{y}_2, \delta \mathbf{u}_2) & \ldots & \delta \bar{I}_2(\delta \mathbf{y}_k, \delta \mathbf{u}_k) \\ \vdots & \vdots & \vdots & \vdots \\ \delta \bar{I}_k(\delta \mathbf{y}_1, \delta \mathbf{u}_1) & \delta \bar{I}_k(\delta \mathbf{y}_2, \delta \mathbf{u}_2) & \ldots & \delta \bar{I}_k(\delta \mathbf{y}_k, \delta \mathbf{u}_k) \end{bmatrix} \begin{bmatrix} \mathrm{d}b_1/\mathrm{d}\epsilon \\ \mathrm{d}b_2/\mathrm{d}\epsilon \\ \vdots \\ \mathrm{d}b_k/\mathrm{d}\epsilon \end{bmatrix}_{\epsilon=0} = \begin{bmatrix} 0 \\ 0 \\ \vdots \\ 0 \end{bmatrix}$$

Since the determinant of the matrix in the above equation is non-zero from Equation (8.29), we have, in terms of $\mathbf{b} = \begin{bmatrix} b_1 & b_2 & \ldots & b_k \end{bmatrix}^{\mathsf{T}}$,

$$\frac{\partial \mathbf{b}}{\partial \epsilon} = \mathbf{0} \quad \text{at} \quad \epsilon = 0$$

Step 4 Using the above result, the variations of $\mathbf{y}$ and $\mathbf{u}$, are, respectively,

$$\delta \mathbf{y} = \frac{\mathrm{d}}{\mathrm{d}\epsilon} \mathbf{y}[\mathbf{b}(\epsilon), \epsilon] \bigg|_{\epsilon=0} = \sum_{i=1}^{k} \frac{\partial \mathbf{y}}{\partial b_i} \underbrace{\frac{\mathrm{d}b_i}{\mathrm{d}\epsilon} \bigg|_{\epsilon=0}}_{= 0} + \delta \mathbf{z} = \delta \mathbf{z} \quad \text{and}$$

$$\delta \mathbf{u} = \delta \mathbf{v} \quad \text{similarly}$$

As a consequence, $(\delta \mathbf{z}, \delta \mathbf{v})$ is the variation pair of the family of infinite pairs represented by $(\mathbf{y}, \mathbf{u})$. Therefore, from Equations (8.32), which have solutions given by Equations (8.34), the infinitely many pairs $(\mathbf{y}, \mathbf{u})$ satisfy

$$I_j(\mathbf{y}, \mathbf{u}) = \epsilon \delta I_j(\bar{\mathbf{y}}, \bar{\mathbf{u}}; \delta \mathbf{y}, \delta \mathbf{u}); \quad j = 1, 2, \ldots, k$$

Since the variation of a functional is homogeneous,

$$\epsilon \delta I_j(\bar{\mathbf{y}}, \bar{\mathbf{u}}; \delta \mathbf{y}, \delta \mathbf{u}) = \delta I_j(\bar{\mathbf{y}}, \bar{\mathbf{u}}; \underbrace{\epsilon \delta \mathbf{y}}_{\delta \mathbf{y}_0}, \underbrace{\epsilon \delta \mathbf{u}}_{\delta \mathbf{u}_0}); \quad j = 1, 2, \ldots, k$$

In terms of the variations $\delta \mathbf{y}_0$ and $\delta \mathbf{u}_0$ as indicated, we obtain

$$I_j(\mathbf{y}, \mathbf{u}) = \delta I_j(\bar{\mathbf{y}}, \bar{\mathbf{u}}; \delta \mathbf{y}_0, \delta \mathbf{u}_0); \quad j = 1, 2, \ldots, k$$

Since $\delta \mathbf{y}_0$ and $\delta \mathbf{u}_0$ are, after all, respective variations in $\mathbf{y}$ and $\mathbf{u}$, we can rewrite the above equations as

$$I_j(\mathbf{y}, \mathbf{u}) = \delta I_j(\bar{\mathbf{y}}, \bar{\mathbf{u}}; \delta \mathbf{y}, \delta \mathbf{u}); \quad j = 1, 2, \ldots, k \qquad (8.12)$$

Step 5   Finally, we need to show that the family of infinite pairs $(\mathbf{y}, \mathbf{u})$ satisfies the inactive inequalities, i. e., the remaining strict inequalities

$$I_j(\mathbf{y}, \mathbf{u}) < 0; \quad j = k+1, k+2, \ldots, q \qquad (8.35)$$

that are satisfied by the pair $(\bar{\mathbf{y}}, \bar{\mathbf{u}})$. Now for sufficiently small ϵ

$$I_j(\mathbf{y}, \mathbf{u}) = I_j(\bar{\mathbf{y}}, \bar{\mathbf{u}}) + \epsilon \delta I_j(\bar{\mathbf{y}}, \bar{\mathbf{u}}; \delta \mathbf{y}, \delta \mathbf{u}); \quad j = k+1, k+2, \ldots, q$$

Also

$$I_j(\bar{\mathbf{y}}, \bar{\mathbf{u}}) < 0; \quad j = k+1, k+2, \ldots, q$$

Combining the last two relations for a small enough ϵ in the interval $(0, \delta_1)$ [see Equations (8.34)], we obtain Inequalities (8.35).

8.C   Fourier Transform

Given a function $f(t)$, its **Fourier transform**, or the k-th *Fourier coefficient* is defined as

$$f_{\langle k \rangle} = \frac{1}{\tau} \int_{-\tau/2}^{\tau/2} f(t) e^{-ik\omega t} \, dt$$

where $\omega = 2\pi/\tau$. The function in terms of the Fourier coefficients is then given by the **inverse Fourier transform**,

$$f(t) = \sum_{k=-\infty}^{\infty} f_{\langle k \rangle} e^{ik\omega t}$$

which is the Fourier series.

The Fourier transform of the derivative of f with respect to time is

$$\dot{f}_{\langle k \rangle} = ik\omega f_{\langle k \rangle}$$

The last two equations can be easily verified by using the definition of the Fourier transform.

8.D   Derivation of Equation (8.25)

We need to show that each right-hand side term of Equation (8.24) on p. 254 is equivalent to the corresponding term in Equation (8.25). We will do that for the first term, i. e., derive

$$\frac{1}{2\tau} \int_0^\tau \delta\mathbf{y}^\top \bar{L}_{\mathbf{yy}} \delta\mathbf{y} \, dt = \mathbf{U}^\top \overline{\mathbf{G}}^\top \bar{L}_{\mathbf{yy}} \mathbf{G}\mathbf{U}^\top$$

The rest of the equivalences may be similarly obtained by the reader.

In the left-hand side of the above equation, we substitute $\delta\mathbf{y}$ from Equation (8.21) to obtain

$$\frac{1}{2\tau} \int_0^\tau \left\{ [(\overline{\mathbf{G}}\mathbf{U})^\top e^{-i\omega t} + (\mathbf{G}\mathbf{U})^\top e^{i\omega t}] \, \bar{L}_{\mathbf{yy}} \, [\overline{\mathbf{G}}\mathbf{U}e^{-i\omega t} + \mathbf{G}\mathbf{U}e^{i\omega t}] \right\} dt$$

$$= \frac{1}{2\tau} \int_0^\tau \left\{ (\overline{\mathbf{G}}\mathbf{U})^\top \, \bar{L}_{\mathbf{yy}} \, \mathbf{G}\mathbf{U} + (\mathbf{G}\mathbf{U})^\top \, \bar{L}_{\mathbf{yy}} \, \overline{\mathbf{G}}\mathbf{U} \right\} dt$$

because the integrals of terms involving $e^{\pm 2i\omega t}$ are zero [see Equation (8.23), p. 254]. Since $\bar{L}_{\mathbf{yy}}$ is a symmetric matrix, the two terms of the last integrand are equal so that the last integral simplifies to

$$\frac{1}{2\tau} \int_0^\tau 2(\overline{\mathbf{G}}\mathbf{U})^\top \, \bar{L}_{\mathbf{yy}} \, \mathbf{G}\mathbf{U} \, dt \; = \; (\overline{\mathbf{G}}\mathbf{U})^\top \, \bar{L}_{\mathbf{yy}} \, \mathbf{G}\mathbf{U} \; = \; \mathbf{U}^\top \overline{\mathbf{G}}^\top \, \bar{L}_{\mathbf{yy}} \, \mathbf{G}\mathbf{U}$$

which is the first right-hand side term of Equation (8.25).

Bibliography

S. Bittanti, G. Fronza, and G. Guardabassi. Periodic control: A frequency domain approach. *IEEE Trans. Auto. Cont.*, AC-18(1):33–38, 1973.

W.L. Chan and S.K. Ng. Normality and proper performance improvement in periodic control. *J. Optimiz. Theory App.*, 29(2):215–229, 1979.

F. Colonius. *Optimal Periodic Control*, Chapter VII, pages 115–119. Springer-Verlag, Berlin, Germany, 1988.

M.R. Hestenes. *Calculus of Variations and Optimal Control Theory*, Chapter 1, pages 48–50. John Wiley & Sons, New York, 1966.

F.J.M. Horn and R.C. Lin. Periodic processes: A variational approach. *Ind. Eng. Chem. Des. Dev.*, 6(1):21–30, 1967.

D. Sinčić and J.E. Bailey. Analytical optimization and sensitivity analysis of forced periodic chemical processes. *Chem. Eng. Sci.*, 35:1153–1161, 1980.

Exercises

8.1 Find the necessary conditions for the minimum of

$$I = \frac{1}{\tau} \int_0^\tau F[\mathbf{y}(t), \mathbf{u}(t)] \, \mathrm{d}t$$

subject to

$$\dot{\mathbf{y}} = \mathbf{g}[\mathbf{y}(t), \mathbf{u}(t)], \quad \mathbf{y}(0) = \mathbf{y}(\tau)$$

where τ is the time period that is free to change.

8.2 Derive the stationarity condition given by [Equation (8.3), p. 242].

8.3 Develop a computational algorithm based on the penalty function method to solve Example 8.4 (p. 254) subject to the following constraints:

$$\frac{1}{\tau} \int_0^\tau u_1 \, \mathrm{d}\tau = \bar{u}_1 \quad \text{and} \quad u_2 \leq u_{2,\max}$$

Solve the problem for

$$\bar{u}_1 = 6 \quad \text{and} \quad u_{2,\max} = 10$$

8.4 Apply the pi criterion to the problem in Example 8.3 (p. 241), and derive the condition for the vector of optimal steady state controls to be proper.

8.5 Find the controls that satisfy the pi criterion in Exercise 8.4 and utilize them as initial controls to solve the optimal control problem.

Chapter 9

Mathematical Review

9.1 Limit of a Function

Consider a function $f(x)$ defined in the vicinity of $x = x_0$. Then the limit of $f(x)$ at $x = x_0$ is a real number L approached by $f(x)$ as x approaches x_0. Symbolically,

$$\lim_{x \to x_0} f(x) = L$$

During the approach process, x and $f(x)$ may be greater or less than the respective targets x_0 and L. The limit L may not be equal to $f(x_0)$. Also, the function may be not defined at x_0.

9.2 Continuity of a Function

Consider the plot of $f(x)$ versus x. The function is said to be continuous when

1. the function values are bounded (i.e., they do not shoot to positive or negative infinity) and

2. the plot, which is a curve, is not broken.

In other words, if a function is continuous at $x = x_0$, then we can have a function value $f(x)$ as close to $f(x_0)$ as we wish by moving x near x_0. Using the limit notation

$$\lim_{x \to x_0} f(x) = f(x_0)$$

where x and $f(x)$ may be greater or less than the respective targets, x_0 and $f(x_0)$. Note that the function is defined at x_0, and the limit $L = f(x_0)$.

With the help of absolute differences, $|f(x) - f(x_0)|$ and $|x - x_0|$, the above concepts are expressed more precisely as follows.

A function $f(x)$ is defined to be continuous at $x = x_0$ when given any $\epsilon > 0$ for which $|f(x) - f(x_0)| < \epsilon$ there exists a $\delta > 0$ such that $|x - x_0| < \delta$.

9.2.1 Lower and Upper Semi-Continuity

In the definition of continuity, the inequality $|f(x) - f(x_0)| < \epsilon$ expands to

$$f(x_0) - \epsilon < f(x) \quad \text{and} \quad f(x) < f(x_0) + \epsilon$$

A function is **lower semi-continuous** at x_0 when the left-hand inequality

$$f(x_0) - \epsilon < f(x)$$

is satisfied for $|x - x_0| < \delta$.

Similarly, a function is **upper semi-continuous** at x_0 when the right-hand inequality

$$f(x) < f(x_0) + \epsilon$$

is satisfied for $|x - x_0| < \delta$.

Thus, a function $f(x)$ is continuous when it is both lower and upper semi-continuous.

9.3 Intervals and Neighborhoods

The set of all values of x satisfying

$$a < x < b, \quad a < b$$

is called an **open interval** and is denoted by (a, b). Note that the end points are not included in an open interval.

The set of x satisfying

$$a \leq x \leq b$$

is called an **closed interval** and is denoted by $[a, b]$. The end points are included in a closed interval. Sets $(a, b]$ and $[a, b)$ are half- or semi-open (or closed).

Open and closed intervals generalize to open and closed sets, which respectively exclude and include the boundary elements.

A **neighborhood** of an element x_0 is an open set containing x_0.

9.4 Bounds

A **lower bound** of a set is a number with the following property:

A member of a set is either less than or equal to the lower bound. Thus, 1 and 2 are lower bounds of $[2, 3]$ as well as $(2, 3)$. There may be multiple lower bounds. Moreover, a lower bound may not be a member of the set.

The *greatest of the lower bounds* of a set is called the **infimum**. Thus, 2 is the infimum of $[2, 3]$ as well $(2, 3)$. Note that the first set contains the infimum, while the second does not.

Similarly, a set member is either greater than or equal to the **upper bound**. The *least of the upper bounds* of a set is called the **supremum**.

9.5 Order of Magnitude

The **order of magnitude** of a real number is how many tens the number has. Thus, the order of magnitude of x is the integer part of the $\log_{10} x$.

An n orders of magnitude difference between two real numbers is a difference by a factor of 10^n.

9.5.1 Big-O Notation

Consider a function $f(x)$ as x tends to some value x_0. If the absolute value of the function is bounded by some positive constant c multiplied by the absolute value of x when x is sufficiently close to x_0, then we can write

$$\lim_{x \to x_0} |f(x)| \le c|x|, \quad |x - x_0| < \delta > 0$$

The shorthand for the above expression is

$$\lim_{x \to x_0} f(x) = O(x)$$

where $O(x)$ represents the bound $c|x|$ as x tends to x_0.

9.6   Taylor Series and Remainder

The estimate of a function $f(x)$ at a distance h from the reference point $x = x_0$ is given by the n-th order Taylor series or expansion

$$f(x_0 + h) = f(x_0) + h\left[\frac{df}{dx}\right]_{x_0} + \frac{h^2}{2!}\left[\frac{d^2 f}{dx^2}\right]_{x_0} + \frac{h^3}{3!}\left[\frac{d^3 f}{dx^3}\right]_{x_0} + \cdots + \frac{h^n}{n!}\left[\frac{d^n f}{dx^n}\right]_{x_0}$$

$$+ \underbrace{\frac{h^{n+1}}{(n+1)!}\left[\frac{d^{n+1} f}{dx^{n+1}}\right]_{\zeta}}_{\text{remainder}} \qquad\qquad (9.1)$$

where ζ lies in the interval $(x_0, x_0 + h)$. In the above equation, the last term is the remainder, which represents the sum of the remaining infinite number of terms

$$\frac{h^{n+1}}{(n+1)!}\left[\frac{d^{n+1} f}{dx^{n+1}}\right]_{x_0} + \frac{h^{n+2}}{(n+2)!}\left[\frac{d^{n+2} f}{dx^{n+2}}\right]_{x_0} + \cdots$$

For a fractional value of h, the remainder becomes negligible as n is increased. For sufficiently large n or small h, the remainder is therefore discarded. It is of course assumed that the function is differentiable at least n times.

9.7   Autonomous Differential Equations

An **autonomous differential equation** does not carry the independent variable explicitly, e. g.,

$$\frac{dy_0}{dt} = y_0 - u$$

On the other hand, the following is a **non-autonomous differential equation**:

$$\frac{dy_1}{dt} = y_1 t - ut^2$$

9.7.1   Non-Autonomous to Autonomous Transformation

Any non-autonomous differential equation can be transformed into a set of autonomous differential equations by introducing additional state variables. For example, with the introduction of the new state variable

$$y_2 \equiv t$$

the last non-autonomous differential equation is transformed into the following set of autonomous differential equations:

$$\frac{dy_1}{dt} = y_1 y_2 - u y_2^2 \quad \text{and} \quad \frac{dy_2}{dt} = 1$$

9.8 Differential

Let f be a function of the independent variable x. Then the differential of f at $x = x_0$ is defined as the change df in f corresponding to the change h in x from x_0 meeting the following requirements:

1. The differential df is closer to the actual function change

$$f(x_0 + h) - f(x)$$

   than h, which is a real number close to zero.

2. The differential df is a linear and continuous function of h at x_0.

Remarks

The motivation for the above requirements is simplification. With negligible error, we would like to have a simple function df represent the function change corresponding to a variable change.

Alternatively, we would like to compute the new function value $f_1 = f(x_0 + h)$ in a simpler way* using

$$f_1 = f(x_0) + df(x_0; h)$$

for small enough values of h that obviate any errors. The two arguments of df denote the dependency on the change h at a given x_0.

Putting it all together, the differential df of f at x_0 is a linear and continuous function of the variable change and satisfies

$$f(x_0 + h) = f(x_0) + df(x_0; h) + \epsilon(h)$$

where the error ϵ vanishes faster than h. So if we keep on reducing h, after a while when h takes a certain value h_0, the error ϵ would disappear and df would be an accurate representation of the function change in the interval $[x_0, x_0 + h_0]$.

---

* As opposed to evaluating $f_1 = f(x_0 + h)$ all over again from the functional relationship.

9.9 Derivative

The definition of the derivative follows from the differential. We multiply and divide $df(x_0; h)$ by h in the last equation to obtain

$$f(x_0 + h) = f(x_0) + \underbrace{\frac{df(x_0; h)}{h}}_{\text{derivative}} h + \epsilon(h)$$

where the derivative is the coefficient of h and is defined as

$$\lim_{h \to 0} \frac{df(x_0; h)}{h} \equiv \left. \frac{df}{dx} \right|_{x_0} = \lim_{h \to 0} \frac{f(x_0 + h) - f(x_0)}{h}$$

9.9.1 Directional Derivative

The directional derivative of a function $f(\mathbf{y})$ at $\mathbf{y} = \mathbf{y}_0$ along a unit vector $\mathbf{v}$ is defined as

$$\nabla_{\mathbf{v}} f(\mathbf{y}_0) \equiv \lim_{\alpha \to 0} \frac{f(\mathbf{y}_0 + \alpha \mathbf{v}) - f(\mathbf{y}_0)}{\alpha}$$

Thus, $\nabla_{\mathbf{v}} f(\mathbf{y}_0)$ quantifies the rate of change in f from $\mathbf{y} = \mathbf{y}_0$ along $\mathbf{v}$.

If $f(\mathbf{y})$ is differentiable at $\mathbf{y}_0$, then $\nabla_{\mathbf{v}} f(\mathbf{y}_0)$ exists along any unit vector. Moreover, if $\mathbf{y} \equiv \mathbf{y}_0 + \alpha \mathbf{v}$, then $f = f[\mathbf{y}(\alpha)]$, and

$$\nabla_{\mathbf{v}} f(\mathbf{y}_0) = \lim_{\alpha \to 0} \frac{f[\mathbf{y}(\alpha)] - f[\mathbf{y}(0)]}{\alpha} = \left[\frac{df}{d\alpha} \right]_{\alpha=0}$$

$$= \sum_{i=1}^{n} \left[\frac{\partial f}{\partial y_i} \frac{dy_i}{d\alpha} \right]_{\alpha=0} = \sum_{i=1}^{n} \left[\frac{\partial f}{\partial y_i} \right]_{\mathbf{y}_0} v_i = f_{\mathbf{y}_0}^{\top} \mathbf{v}$$

where $f_{\mathbf{y}_0}^{\top}$ denotes the vector $f_{\mathbf{y}}^{\top}$ evaluated at $\mathbf{y}_0$.

Example

The directional derivative of

$$f(\mathbf{y}) = \frac{y_1^2 + y_2^2}{2}$$

at $\mathbf{y}_0 = \begin{bmatrix} \sqrt{3} & -\sqrt{2} \end{bmatrix}^{\top}$ along the unit vector $\mathbf{v} = \begin{bmatrix} \sqrt{3/5} & \sqrt{2/5} \end{bmatrix}^{\top}$ is

$$\nabla_{\mathbf{v}} f(\mathbf{y}_0) = \begin{bmatrix} y_1 & y_2 \end{bmatrix}_{\mathbf{y}_0}^{\top} \begin{bmatrix} v_1 & v_2 \end{bmatrix} = \frac{1}{\sqrt{5}}$$

9.10 Leibniz Integral Rule

Consider the following definite integral

$$I = \int_a^b f \, \mathrm{d}x$$

Then the derivative of I with respect to its upper limit is given by

$$\frac{\mathrm{d}I}{\mathrm{d}b} = \lim_{\Delta b \to 0} \frac{I(b + \Delta b) - I(b)}{\Delta b} = \lim_{\Delta b \to 0} \frac{\displaystyle\int_a^{b+\Delta b} f \, \mathrm{d}x - \int_a^b f \, \mathrm{d}x}{\Delta b}$$

$$= \lim_{\Delta b \to 0} \frac{\displaystyle\int_b^{b+\Delta b} f \, \mathrm{d}x}{\Delta b} = \lim_{\Delta b \to 0} \frac{f(b)\Delta b}{\Delta b} = f(b)$$

Similarly, the derivative of I with respect to its lower limit is given by

$$\frac{\mathrm{d}I}{\mathrm{d}a} = -f(a)$$

9.11 Newton–Raphson Method

This is a numerical method to find the root of the function, i. e., the value of the independent variable that makes the dependent function zero. Using the first order Taylor expansion, the function value at x_{i+1} can be expressed as

$$f(x_{i+1}) \cong f(x_i) + f'(x_i)(x_{i+1} - x_i)$$

When x_{i+1} is sufficiently close to the root,

$$0 \cong f(x_i) + f'(x_i)(x_{i+1} - x_i) \tag{9.2}$$

Algorithm

The above equation provides the following algorithm:

$$x_{i+1} = x_i - \frac{f(x_i)}{f'(x_i)}; \quad i = 0, 1, \dots$$

which is repeatedly applied to improve an initial guess x_0 for the root.

Quadratic Convergence

Sufficiently close to the root x_r, the second order Taylor expansion gives

$$f(x_r) = 0 = f(x_i) + f'(x_i)(x_r - x_i) + \frac{(x_r - x_i)^2}{2} f''(\zeta)$$

where ζ lies in the interval (x_i, x_r). From the above equation, we subtract Equation (9.2) to get

$$\underbrace{x_r - x_{i+1}}_{\text{new error}} = -\frac{f''(\zeta)}{2f'(x_i)} \underbrace{(x_r - x_i)^2}_{\text{old error}}$$

Thus, the new error is proportional to the square of the old error. If the old error is 0.1, then the new error would be of the order 0.01. The next error would be of the order 10^{-4}, and so on iteratively. The value x_{i+1} is said to converge quadratically to the root x_r.

Extension to Multivariable Functions

The Newton–Raphson algorithm for multivariable functions is

$$\mathbf{x}_{i+1} = \mathbf{x}_i - \mathbf{J}_i^{-1} f(\mathbf{x}_i); \quad i = 0, 1, \dots$$

where $\mathbf{J}_i$ is the Jacobian

$$
\begin{bmatrix}
\left[\dfrac{\partial f_1}{\partial x_1}\right]_{\mathbf{x}_i} & \left[\dfrac{\partial f_1}{\partial x_2}\right]_{\mathbf{x}_i} & \cdots & \left[\dfrac{\partial f_1}{\partial x_n}\right]_{\mathbf{x}_i} \\[2ex]
\left[\dfrac{\partial f_2}{\partial x_1}\right]_{\mathbf{x}_i} & \left[\dfrac{\partial f_2}{\partial x_2}\right]_{\mathbf{x}_i} & \cdots & \left[\dfrac{\partial f_2}{\partial x_n}\right]_{\mathbf{x}_i} \\[2ex]
\vdots & \vdots & \vdots & \vdots \\[2ex]
\left[\dfrac{\partial f_n}{\partial x_1}\right]_{\mathbf{x}_i} & \left[\dfrac{\partial f_n}{\partial x_2}\right]_{\mathbf{x}_i} & \cdots & \left[\dfrac{\partial f_n}{\partial x_n}\right]_{\mathbf{x}_i}
\end{bmatrix}
$$

evaluated at $\mathbf{x} = \mathbf{x}_i$.

9.12 Composite Simpson's 1/3 Rule

The basic Simpson's 1/3 Rule rule provides the value of the integral

$$I = \int_a^b f(x)\,dx = \frac{b-a}{6}\left[f(a) + 4f\left(\frac{a+b}{2}\right) + f(b)\right]$$

by approximating the integrand $f(x)$ with the quadratic function that interpolates the function values $f(a)$, $f[(a+b)/2]$, and $f(b)$.

The value of I is improved by dividing the integration interval into a number of segments of equal length, applying the Simpson's 1/3 Rule to each segment, and summing up the resulting values. This procedure finally yields the following composite Simpson's 1/3 Rule

$$I = \int_{x_0}^{x_N} f(x)\,dx = \frac{h}{3}\left[f(x_0) + 4\sum_{i=1,3,5,\ldots}^{N} f(x_i) + 2\sum_{i=2,4,6,\ldots}^{N} f(x_i) + f(x_N)\right]$$

where h is the distance between any two successive x_is forming a subinterval $[x_i, x_{i+1}]$ where $i = 0, 1, \ldots, (N-1)$.

Note that each segment in the integration interval $[x_0, x_N]$ has two subintervals. Thus, the number of subintervals, N, is even and the number of x_is being $(N+1)$ is odd.

9.13 Fundamental Theorem of Calculus

If a function $f(x)$ is real and continuous in a closed interval $[a, b]$ such that the primitive integral or the antiderivative

$$F(x) = \int_a^x f\,dx$$

is defined, then $F(x)$ is continuous in $[a, b]$, and differentiable in (a, b). The derivative of F with respect to x in (a, b) is given by

$$\frac{dF}{dx} = f$$

Moreover, if in the closed interval $[a, b]$ the function f is real, but not necessarily continuous, and has the antiderivative F, then

$$F(b) - F(a) = \int_a^b f \, dx = \int_a^b \frac{dF}{dx} \, dx$$

9.14   Mean Value Theorem

If $f(x)$ is a continuous function in the x-interval $[a, b]$, then there is a real number c in the interval such that

$$(b - a)f(c) = \int_a^b f \, dx$$

9.14.1   For Derivatives

Let $f(x)$ be continuous in the x-interval $[a, b]$ with the derivative dy/dx defined in the open interval (a, b). Then there is point x_0 in (a, b) such that

$$f(b) - f(a) = (b - a)\left[\frac{df}{dx}\right]_{x_0}$$

This result is known as the law of the mean or the Mean Value Theorem for derivatives.

9.14.2   For Integrals

Let $f(x)$ be continuous in the x-interval $[a, b]$. Then there is point x_0 in $[a, b]$ such that

$$\int_a^b f(x) \, dx = (b - a)f(x_0)$$

This result is known as the Mean Value Theorem for integrals.

9.15   Intermediate Value Theorem

Let $f(x)$ be continuous in the x-interval $[a, b]$ and $f(a) \neq f(b)$. Then according to this theorem, $f(x)$ assumes each value between $f(a)$ and $f(b)$ as x changes from a to b.

9.16   Implicit Function Theorem

Consider the set of functions

$$f_1(x_1, x_2, \ldots, x_n) = c_1$$
$$f_2(x_1, x_2, \ldots, x_n) = c_2$$
$$\vdots$$
$$f_m(x_1, x_2, \ldots, x_n) = c_m$$

where $m < n$. Then according to this theorem, provided that

- the vector of functions

$$\mathbf{f}(\mathbf{x}) \equiv \begin{bmatrix} f_1(\mathbf{x}) & f_2(\mathbf{x}) & \cdots & f_m(\mathbf{x}) \end{bmatrix}^{\top}$$

  is differentiable near $\mathbf{a}$,

- the derivatives of $\mathbf{f}$ with respect to $\mathbf{x}$ are continuous at $\mathbf{a}$, and

- the Jacobian determinant $\mathbf{f_x}$ is not zero at $\mathbf{x} = \mathbf{a}$.

we can obtain the set of solutions

$$x_1 = g_1(x_{m+1}, x_{m+2}, \ldots, x_n)$$
$$x_2 = g_2(x_{m+1}, x_{m+2}, \ldots, x_n)$$
$$\vdots$$
$$x_m = g_m(x_{m+1}, x_{m+2}, \ldots, x_n)$$

near

$$\mathbf{x} = \mathbf{a} = \begin{bmatrix} a_1 & a_2 & \cdots & a_n \end{bmatrix}^{\top}$$

where g_is are differentiable functions of $x_{m+1}, x_{m+2}, \ldots$, and x_n.

9.17   Bolzano–Weierstrass Theorem

According to this theorem, there is at least one point of accumulation (or limit point) in a bounded set having an infinite number of elements. Alternatively, each bounded sequence in the set has a subsequence that converges to a point in the set.

The point of accumulation and subsequence are described as follows.

Point of Accumulation

Not necessarily in a set, its point of accumulation has in its each neighborhood, at least one non-identical point from the set. A closed set contains all of its accumulation points.

Subsequence

Given a sequence

$$x_1, \ x_2, \ x_3, \ x_4, \ x_5, \ \cdots$$

its subsequence is, e. g.,

$$x_1, \ x_3, \ x_4, \ x_6, \ x_8, \ \cdots$$

which is contained in the original sequence. The order of the elements is preserved in the subsequence.

9.18   Weierstrass Theorem

According to this theorem, a function $f(\mathbf{x})$ in a closed and bounded domain must attain minimum and maximum values. This theorem is also known as the Extreme Value Theorem.

9.19   Linear or Vector Space

A linear or vector space is a set of elements known as vectors for which the following two operations:

   1. addition of vectors and

   2. multiplication of a vector by real numbers

are defined. These operations satisfy the following rules:

   1. For any two vectors x and y,

   i. the sum $(x + y)$ is a vector
   ii. $x + y = y + x$
   iii. $(x + y) + z = x + (y + z)$ for any vector z

   2. For any real number α,

   i. the product αx is a vector where x is any vector

ii. $\alpha(x + y) = \alpha x + \alpha y$ for any two vectors x and y

3. For any two real numbers α and β and any vector x
   i. $(\alpha + \beta)x = \alpha x + \beta x$
   ii. $\alpha(\beta x) = (\alpha\beta)x$

4. The space contains
   i. the zero vector, 0, such that $x + 0 = x$
   ii. the unit vector, 1, such that $1x = x$
   iii. the negative of x, denoted by $-x$, such that $x + (-x) = 0$

   for any vector x.

Examples

Examples of vector spaces are

1. The set of real numbers for which the usual addition and multiplication are defined.

2. The set of elements, each made of n components in an order for which vector addition and scalar multiplication are defined. For example, an element **x** is given by

$$\mathbf{x} = \begin{bmatrix} x_1 \\ x_2 \\ \vdots \\ x_n \end{bmatrix}$$

3. The set of real-valued functions defined on a fixed interval of the independent variable. For example, if any two such functions or vectors are $p(x)$ and $q(x)$ for x in $[a, b]$, then the respective operations of addition and multiplication (by real number α) result in vectors

$$r(x) = p(x) + q(x) \quad \text{and} \quad s(x) = \alpha p(x)$$

   which are also vectors.

It is easy to verify that the above examples obey the rules of a vector space.

9.20 Direction of a Vector

The direction of an n-component vector

$$\mathbf{v} = \begin{bmatrix} v_1 & v_2 & \cdots & v_i & \cdots & v_n \end{bmatrix}^{\mathsf{T}}$$

is characterized by the set of n direction ratios

$$\{d_1, d_2, \ldots, d_i, \ldots, d_n\}$$

where

$$d_i = \frac{v_i}{\|v\|}$$

Two vectors along the same direction have an identical set of direction ratios and vice versa.

9.21 Parallelogram Identity

For any two vectors $\mathbf{a}$ and $\mathbf{b}$ and a scalar α

$$\|\mathbf{a} + \alpha\mathbf{b}\|^2 = \sum_{i=1}^{n}(a_i + \alpha b_i)^2$$

$$= \sum_{i=1}^{n}(a_i^2 + \alpha^2 b_i^2 + 2\alpha a_i b_i)$$

$$= \|\mathbf{a}\|^2 + \alpha^2\|\mathbf{b}\|^2 + 2\alpha\mathbf{a}^\mathsf{T}\mathbf{b}$$

9.22 Triangle Inequality for Integrals

Given a function $f(x)$

$$-|f(x)| \leq f(x) \leq |f(x)|$$

Integrating each term in the above relation, we obtain

$$-\int_a^b |f(x)|\,\mathrm{d}x \leq \int_a^b f(x)\,\mathrm{d}x \leq \int_a^b |f(x)|\,\mathrm{d}x$$

which is equivalent to

$$\left| \int_a^b f(x)\,\mathrm{d}x \right| \leq \int_a^b |f(x)|\,\mathrm{d}x$$

9.23 Cauchy–Schwarz Inequality

The inequality is

$$\mathbf{a}^{\top}\mathbf{b} \le \|\mathbf{a}\|\|\mathbf{b}\|$$

where $\mathbf{a}$ and $\mathbf{b}$ are two n-dimensional vectors.

For $n = 2$, let

$$\mathbf{a} \equiv \begin{bmatrix} a_1 \\ a_2 \end{bmatrix} \quad \text{and} \quad \mathbf{b} \equiv \begin{bmatrix} b_1 \\ b_2 \end{bmatrix}$$

Then

$$\underbrace{(a_1^2 + a_2^2)}_{\|\mathbf{a}\|^2} \underbrace{(b_1^2 + b_2^2)}_{\|\mathbf{b}\|^2} - \underbrace{(a_1 b_1 + a_2 b_2)^2}_{(\mathbf{a}^{\top}\mathbf{b})^2} = \underbrace{(a_1 b_2 - a_2 b_1)^2}_{\ge 0}$$

Therefore, $\mathbf{a}^{\top}\mathbf{b} \le \|\mathbf{a}\|\|\mathbf{b}\|$, which can be easily generalized for higher dimensions.

9.24 Operator Inequality

The inequality is

$$\|\mathbf{A}\mathbf{b}\| \le \|\mathbf{A}\|\|\mathbf{b}\|$$

where $\mathbf{A}$ is an $n \times m$ operator (matrix), and $\mathbf{b}$ is an m-dimensional vector.

For $n = m = 2$, let

$$\mathbf{A} \equiv \begin{bmatrix} a_{11} & a_{12} \\ a_{21} & a_{22} \end{bmatrix} \quad \text{and} \quad \mathbf{b} \equiv \begin{bmatrix} b_1 \\ b_2 \end{bmatrix}$$

Then

$$\mathbf{A}\mathbf{b} = \begin{bmatrix} a_{11}b_1 + a_{12}b_2 \\ a_{21}b_1 + a_{22}b_2 \end{bmatrix}$$

and

$$\|\mathbf{A}\mathbf{b}\|^2 = \underbrace{(a_{11}b_1 + a_{12}b_2)^2}_{t_1} + \underbrace{(a_{21}b_1 + a_{22}b_2)^2}_{t_2}$$

Since

$$(a_{11}^2 + a_{12}^2)(b_1^2 + b_2^2) - t_1 = (a_{11}b_2 - a_{12}b_1)^2 \ge 0$$

we get

$$t_1 \le (a_{11}^2 + a_{12}^2)(b_1^2 + b_2^2)$$

Similarly,

$$t_2 \leq (a_{21}^2 + a_{22}^2)(b_1^2 + b_2^2)$$

Adding the last two inequalities, we get

$$\underbrace{t_1 + t_2}_{\|\mathbf{Ab}\|^2} \leq \underbrace{(a_{11}^2 + a_{12}^2 + a_{21}^2 + a_{22}^2)}_{\|\mathbf{A}\|^2} \underbrace{(b_1^2 + b_2^2)}_{\|\mathbf{b}\|^2}$$

The above result in terms of $\mathbf{A}$ and $\mathbf{b}$ is

$$\|\mathbf{Ab}\| \leq \|\mathbf{A}\|\|\mathbf{b}\|$$

which can be easily generalized for higher dimensions.

9.25   Conditional Statement

A conditional statement is a conjunction of a condition and the outcome generated by the condition. Consider the following statement:

If $\underbrace{\text{you drop the glass}}_{\text{condition (A)}}$ $\underset{\longrightarrow}{\text{then}}$ $\underbrace{\text{it will crack.}}_{\text{outcome (B)}}$

which can be symbolically represented as

$$A \longrightarrow B$$

As the arrow indicates, the flow of events is from A to B but not necessarily vice versa. In other words, B $\longrightarrow$ A is not necessarily true. For example, if the glass cracked, then it was not because you dropped it but you poured boiling water into it.

Given a conditional statement A $\longrightarrow$ B, the satisfaction of A causes B. Condition A is known as the **sufficient condition**. Its satisfaction is *sufficient* to cause the outcome B, which in turn is called the **necessary condition**. Outcome B is *necessary* for condition A to have been true. Put differently, if B did not happen, then neither did A. Using $\neg$ for negation,

$$\neg B \longrightarrow \neg A$$

Thus, if the glass did not crack, then you did not drop it for sure.

The conditional $\neg B \longrightarrow \neg A$ is called the **contrapositive** of the original conditional A $\longrightarrow$ B. The two conditionals are equivalent. One is the contrapositive of the other.

9.26   Fundamental Matrix

Consider the homogeneous linear differential equation

$$\frac{d\mathbf{x}}{dt} = \mathbf{A}\mathbf{x}$$

in a time interval where $\mathbf{x} = \mathbf{x}(t)$ is an n-dimensional vector, and $\mathbf{A}$ is an $n \times n$ matrix of constants. Then n-linearly independent solutions of the differential equation exist. Let us denote the solutions by column vectors:

$$\mathbf{x}_1(t), \ \mathbf{x}_2(t), \ \ldots, \ \mathbf{x}_n(t)$$

Then the fundamental matrix is the collection

$$\boldsymbol{\Psi}(t) \equiv \Big[\mathbf{x}_1(t) \quad \mathbf{x}_2(t) \quad \ldots \quad \mathbf{x}_n(t)\Big]$$

The general solution of the differential equation is

$$\mathbf{x} = \sum_{i=1}^{n} c_i \mathbf{x}_i(t) \equiv \boldsymbol{\Psi}(t)\mathbf{c}$$

where $\mathbf{c}$ is a vector of some constants. If the initial condition at $t = t_0$ is $\mathbf{x}(t_0)$, then from the above equation,

$$\mathbf{c} = \boldsymbol{\Psi}^{-1}(t_0)\mathbf{x}(t_0)$$

and the solution in terms of the fundamental matrix is

$$\mathbf{x} = \boldsymbol{\Psi}(t)\boldsymbol{\Psi}^{-1}(t_0)\mathbf{x}(t_0)$$

Bibliography

W.E. Boyce and R.C. DiPrima. *Elementary Differential Equations and Boundary Value Problems*. John Wiley & Sons, Inc., New Jersey, 9th edition, 2009.

S.C. Chapra and R.P. Canale. *Numerical Methods for Engineers*. McGraw Hill, New York, 6th edition, 2010.

A. Guzman. *Derivatives and Integrals of Multivariable Functions*. Birkhäuser, Boston, MA, 2003.

E. Kreyszig. *Introductory Functional Analysis with Applications*. John Wiley & Sons, US, 1978.

W. Rudin. *Principles of Mathematical Analysis*. McGraw-Hill, Inc., New York, 1976.

B.S.W. Schröder. *Fundamentals of Mathematics — An Introduction to Proofs, Logic, Sets, and Numbers*. Wiley, New Jersey, 2010.

A.E. Taylor and W.R. Mann. *Advanced Calculus*. John Wiley & Sons, Inc., US, 3rd edition, 1983.

Index

In this Index, a page number in bold face corresponds to the main information of the index entry. An underlined page number refers to a numerically solved example.